TECHNOLOGICAL DICTIONARY OF PLASTICS MATERIALS

TECHNOLOGICAL DICTIONARY OF PLASTICS MATERIALS

BY

W. V. TITOW

M. Phil., Ph.D., C. Chem., F.R.S.C., F.I.M., C. Text., A.T.I

PERGAMON

U.K. Elsevier Science Ltd, The Boulevard, Langford Lane, Kidlington, Oxford, OX5 1GB, England
U.S.A. Elsevier Science Inc., 660 White Plains Road, Tarrytown, New York, 10591-5153, U.S.A.
JAPAN Elsevier Science Japan, Tsunashima Building Annex, 3-20-12 Yushima, Bunkyo-ku, Tokyo 113, Japan

---

First edition 1998

Library of Congress Cataloging in Publication Data
A Catalogue record for this title is available from the Library of Congress.

British Library Cataloguing in Publication Data
Titow, W. V. (Witold Victor), 1926-
Technological dictionary of plastics materials
1. Plastics - Dictionaries
I. Title
668.4'03

ISBN 0080418910

Printed and bound in Great Britain by
Biddles Ltd, Guildford and King's Lynn

To Margaret, who knows the message, and I hope always will.
Tout passe, l'amitié reste

CONTENTS

ACKNOWLEDGEMENTS

The definitions of technical terms in Part 1 of this dictionary include some - a relatively very small proportion - which are direct quotations from ISO, BSI, and ASTM standards.

Permission has been granted by the International Organisation for Standardization, ISO, to reproduce extracts from ISO standards. The complete standards can be obtained from any ISO member body, including SABS, South African Bureau of Standards, Private Bag X191 Pretoria 0001.

Extracts from British Standards are reproduced with the permission of BSI. Complete copies of the standards can be obtained by post from BSI Publications, Linford Wood, Milton Keynes, MK14 6LE.

It is a pleasure to record my appreciation of these permissions and of the kind responses accorded to my original requests for them. Each definition I have reproduced directly from an ISO standard or a British Standard is identified and acknowledged in the text.

In accordance with specific instructions from the American Society for Testing and Materials (ASTM), a full acknowledgement in the form prescribed by that organisation follows each of the three definitions I have quoted verbatim from ASTM standards.

Mrs. Sandy Marais and Mrs. Joanette Davidtz have been most helpful in producing the typescript of the book and the corresponding computer disks: they have my appreciative thanks for their timely help, cheerfully and skilfully rendered.

I am much indebted to Mr. Alan Wills of Shell South Africa (Pty) Ltd., for items of information on polyaniline and on aliphatic polyketones.

The book was longer in the writing than originally envisaged. I am grateful to the Publisher, and in particular to Dr. David C. Prosser, Publishing Editor, Materials Science, for their patience and indulgence in this connection.

W.V.T.

GENERAL INTRODUCTION

The principal aim of this dictionary is to provide a set of authoritative, explanatory definitions of technical terms relating to the materials aspect of plastics technology. This aim is reflected in the scope and format of Part 1 which constitutes the dictionary's main body. Part 1 is supplemented by two others: of these, Part 2 provides explanations of the abbreviated terms (letter symbols) used for the parent polymers and other constituents of plastics materials, and Part 3 identifies the trade names of many plastics materials and some of their components.

A substantial proportion of the definitions in Part 1, and some of the explanations of trade names in Part 3, are encyclopaedic in nature, in that they include additional relevant technical information.

The information provided in Parts 1and 2 of the dictionary gains an extra dimension from two of its features which, to the best of the author's knowledge, are novel vis-a-vis the technical dictionaries hitherto available in the field of plastics. Thus, firstly, all the definitions of terms in Part 1, and those of abbreviations in Part 2, are in line with the corresponding definitions - where those are available - in the relevant technical nomenclature standards of the three principal groups of standards published in English, viz. the international ISO standards, British (BSI) standards, and American ASTM standards. Secondly, the supplementary technical information in the entries of Part 1 includes numerous references to, or citations of, relevant technical standards of the above three groups. Whilst the first of these two features contributes to the authority of the dictionary's definitions, the second opens up for the reader an extensive source of highly relevant , authoritative technical information and guidance, which the standards jointly constitute. The significance and value of this second feature stem from the cardinally important role of established technical standards in plastics technology today: no practical technologist, or indeed any other person in any way connected with the technical aspects of plastics, can operate effectively without some recourse to the relevant standards, in particular those dealing with the properties of plastics materials, their evaluation and testing.

A relatively new group of international standards, with texts available in English, has been growing in technical importance: these are the standards of the European Union: the EN standards issued by the European Committee for Standardisation (CEN) and the European Committee for Electrotechnical Standardisation (CENELEC) in conjunction with, respectively, ISO and IEC. However , in the field of plastics materials, and within that in matters of technical terminology, the relevant EN standards are not as numerous or extensive in coverage as those of the other three groups (ISO, BSI, and ASTM). Moreover, many British and ISO standards have been adopted directly as EN standards. For these reasons, the EN standards are not featured in this dictionary as a group, although some are mentioned in various connections.

As indicated by the title, and some of the foregoing remarks, the scope of the dictionary is demarcated in two ways: with a few relatively minor exceptions, the focus is on plastics as distinct from other polymeric materials, and - within that ambit - on those aspects of plastics which are directly relevant to their nature, properties and behaviour as materials. Hence the processing aspects (equipment and methods) have not been included, except for a limited number of selected terms in Part 1. However, within the scope thus defined, the aim has been to present in Part 1 technical terms from all the principal areas of the subject, viz. terms relating to:

a) the nature of plastics materials;
b) their composition (including such non-polymeric components as they often contain);
c) their properties (including methods of property evaluation and testing);
d) their applications and service performance;
e) some aspects of their behaviour in processing and effects of processing upon them (cf. e.g. such terms as "plate-out", "melt fracture", "burn mark", "scorching");
f) some of their special features and faults (cf. e.g. "fish eyes", "nibs", "gel count").

Certain terms and abbreviations from the somewhat diffuse border between plastics and rubbers, such as those relating to thermoplastic elastomers, are included in the dictionary, as are some abbreviations and trade names for rubbers (especially those rubbers which are used as polymeric modifiers in plastics materials). Finally, some terms from the field of polymer science (as distinct from plastics technology) also appear in Part 1: those terms have been selected for inclusion because of their relevance in the technological context.

Plastics form a fairly well defined group of polymeric materials. The essential practical distinction - not always clearly appreciated - between polymers and plastics is that, whilst the former are the basic materials of the latter, a plastic may - in addition to its parent polymer - contain also various additives and modifiers introduced to impart particular service and/or processing properties, or such characteristics as colour, surface finish, etc.

Other distinct groups of polymer-based materials are rubbers, fibres, adhesives and surface coatings: their respective technologies are completely outside the ambit of this dictionary. In fact the technical terminology of these groups of materials would provide ample scope for entirely separate, full-scale lexicographic treatment.

I hate false words, and seek with care, difficulty, and moroseness those that fit the thing.

Walter Savage Landor. 1775 - 1864
(Imaginary conversations)

PART 1

TECHNICAL TERMS

PART 1

INTRODUCTION

On a strict classification, the definition of a technical term may be of one of two kinds: *descriptive* or *prescriptive*. A valid descriptive definition is one which accurately reflects correct usage in the field concerned. A prescriptive definition in effect lays down the way in which a particular term should be defined. Prescriptive definitions thus have a raison d'être in cases where descriptive definitions do not exist, or where they can not be readily formulated (e.g. by reason of confusion in usage, or ambiguous terminology in the field concerned).

Save for a relatively very small number, the definitions given in this part of the dictionary are descriptive. As should be the case with definitions of this kind, they are consonant with proper usage in the field of plastics technology; they also accord with the way in which the particular terms are defined or interpreted in the relevant ISO, BSI and ASTM standards (the major groups of standards published in English). However, many of the definitions provided here are more comprehensive than their standard counterparts, and are supplemented by relevant technical information. Where an element of a definition corresponds to a relevant element (or to the entirety) of a standard definition, the correspondence is indicated by an appropriate reference citation. The few definitions quoted directly from terminology standards are clearly identified as such reproductions (see also Acknowledgements).

ABS (ACRYLONITRILE/BUTADIENE/STYRENE) PLASTICS

A wide range of plastics materials based either on terpolymers of acrylonitrile, butadiene, and styrene or on blends containing polymeric and/or copolymeric combinations of these three constituents[1.1,1.5] in various proportions. The terpolymers are typically graft polymers of styrene and acrylonitile on butadiene latex particles. In their usual form, the ABS polymer blends combine a range of percentages of the terpolymers and styrene/acrylonitrile copolymers. Perhaps the most important single advantageous property of ABS plastics, shared by virtually all the numerous grades, is good impact resistance. Otherwise, each grade is designed to meet better than styrene homopolymers (and in some cases also better than toughened polystyrene) some particular requirements, e.g. those for dimensional stability, mechanical strength, resistance to environmental stress cracking (some selected grades), relatively high deflection temperature under load, especial suitability for electroplating, and others. The limitations of ABS plastics include susceptibility to attack by many solvents (aromatic and chlorinated hydrocarbons, esters, ethers, ketones) and by concentrated acids, relatively high flammability (reduced in flame-retarded grades), comparatively limited resistance to weathering (due to the presence of double bonds in the butadiene component), and opacity.

ABS plastics are available in the following commercial ranges (for these and some others see also Section 1 of Part 3 of this book): *Abstat; Arrhadur; Baymod A; Blendex; Cevian; Cycolac; Diamond; Hanalac; Lastilac; Lupos; Lustran; Magnum; Multibase; Plexiglas; Ravikral; Ronfalin; Sniasan; Stylac; Terluran; Ultrastyr.*

ABS polymers are suitable for blending with a number of other polymers, each component contributing something of its characteristic properties to those of the blend. The ABS blends represented among commercial materials are those with polyamides (e.g. *Novaloy A; Triax*), polyesters (e.g. *Cevian B; Cycolin; Latiblend*), polycarbonate (e.g. *Bayblend; Celstran; Cycoloy; Koblend; Lupoy; Radiblend; Terblend B; Toplex; Toylac*), PVC (e.g. *Condux; Cycovin; Dural; Lastiflex; Polyman*), polyurethane (e.g. *Prevail*), polysulfone (e.g. *Mindel A*), and styrene/maleic anhydride copolymer (e.g. *Cadon*).

The following standards are concerned with various aspects of ABS materials.

BS ISO 2580	ABS moulding and extrusion materials (replaces BS 4935, now withdrawn)
Part 1:1990	Designation.
Part 2:1994	Preparation of specimens and determination of properties.
ISO 2507-3:1995	Vicat softening temperature - test conditions for ABS and ASA pipes.
ISO 7245:1984	General specification for extrusion and moulding materials for pipes and fittings (including ABS).
ASTM D 4673-87(1993)	Standard specification for ABS moulding and extrusion materials.

ASTM D 3965-94	Standard specification for rigid ABS materials for pipes and pipe fittings.
ASTM D 1939-94	Test for residual stress in moulded and extruded ABS parts.
ASTM D 4703-93	Compression moulding of thermoplastic (including ABS inter alia) plaques, sheets, etc. for test specimens.

BS standards for ABS pipes and pipe fittings : BS 5255:1989 (waste pipe and fittings, ABS inter alia); BS 5391:Part 1:1976 (pressure pipes); BS 5392:Part 1:1976 (pressure pipe fittings).

ABRASION

The loss of material from a surface through wearing away by frictional forces[1.2,1.3,1.4]. (A very similar definition has been put forward also specifically in connection with textile materials - cf. ASTM D 123-90).

ABRASION RESISTANCE

The extent to which a material is able to resist surface wear under such mechanical action as friction, rubbing, scraping or erosion (cf. ISO 4649-1985 and ASTM D 1242-87). In standard tests for abrasion resistance, the mechanical action is that of a specified abradant applied to test specimens in prescribed conditions. The abradants used in the tests include loose-particle abrasives (ASTM D 673-88; ASTM D 1242-87), abrasives bonded on cloth or paper (ASTM D 1242-87), and standard abrasive wheels (ISO 9352:1989; BS 2782:Part 3:Method 370:1989; ASTM D 1044-90). Abrasion resistance is assessed and expressed in terms of changes in some optical properties of the material tested, such as light transmittance (for transparent materials, cf. e.g. ASTM D 1044-90), gloss and/or haze (see under *mar resistance*), changes in appearance (ISO 9352:1989), and loss of mass or volume after a given number of abrasion cycles (ISO 9352; BS 2782:Part 3:Method 370; ASTM D 1242) or the number of cycles to produce a defined loss (ISO 9352). Test methods for the abrasion resistance of plastics-coated fabrics are given in BS 3424:Part 24:1990, Methods 27A and 27B, and Part 31:1990, Method 34, and test methods for the abrasion resistance of rubber in ISO 4649-1985; BS 903:Part A9:1988; ASTM D 1630-83 and ASTM D 2228-88.

ABRASION RESISTANCE INDEX

An expression of the abrasion resistance of a rubber relative to that of a standard rubber (q.v.). As defined in the relevant test standards (cf. ISO 4649-1985 and BS 903:Part A9:1988), the abrasion resistance index is the volume lost from the surface of a standard rubber in an abrasion test, expressed as a percentage of that lost by a rubber treated under the same specified conditions.

ABRASIVE WEAR

The progressive loss of material from the surface of a specimen, resulting from the action of the abradant in a test (cf. ISO 9352:1989), or analogous loss from the surface of a material or product in service. See also *wear*.

ABSOLUTE COMPLIANCE

The reciprocal of absolute modulus[1.5] (q.v), i.e. the ratio of maximum strain (ε_0) to maximum stress (σ_0) in the relevant determination mode (in tension, in shear, or in bulk or longitudinal compression). Usual symbol: $|C|$. Standard unit: Pa^{-1}.

ABSOLUTE MODULUS

The absolute value of complex modulus (q.v.) determined in the relevant mode (in tension, shear, or in bulk or longitudinal compression). Usual symbol: $|M|$. Standard unit: Pa. Absolute modulus is defined by the relationship[1.5]:

$$|M| = \left(M'^2 + M''^2\right)^{1/2} = \frac{\sigma_0}{\varepsilon_0}$$

where:

σ_0 is the maximum stress.
ε_0 is the maximum strain.
M' is the storage modulus.
M'' is the loss modulus (the imaginary part of the complex modulus).

ABSOLUTE PERMITTIVITY

Synonymous with *permittivity* (q.v.).

ABSORPTION

(1) In fibre optics: that portion of light attenuation (q.v.) which involves conversion of optical power into heat[1.6].

(2) In polymers and plastics: the taking up and retention of a substance (the absorbate or penetrant) by a material (the absorbent)[1.4].

In the plastics context the penetrants are commonly gases, or vapours, or liquids (occasionally semi-solids, e.g. greases), and the absorbents are polymeric solids. When the absorbent is non-porous, the take-up of penetrant proceeds by diffusion (q.v.): if it is in the form of a sheet absorbing the penetrant at one of its two faces, the penetrant may, in time, traverse the thickness of the sheet and emerge at the opposite face (see permeability). If the absorbent is porous, the penetrant will be taken up primarily into the pores, although some may also diffuse into the solid material of the absorbent.

ACCELERATOR; PROMOTER

An additive incorporated in small proportions in a reactive chemical system to increase the rate of reaction[1.1,1.5]. In the field of plastics accelerators are used, typically, in polyurethane systems, and in unsaturated polyester compositions where they co-act with the curing agent to reduce the gel time (q.v.) and/or to increase the rate of hardening. In rubber compounds, accelerators are similarly used in conjunction with curing (vulcanising) agents to increase the rate of vulcanisation; in some cases they also improve the properties of the vulcanisate[1.2,1.3].

Standard designation for accelerators (and hardeners) for epoxy resins are outlined in ISO 4597:Part 1:1983 (E).

ACETAL PLASTIC(S)

Plastic(s) based either on polymers of the polyacetal type - i.e. ones whose molecular chains are made up of repeated acetal type units[1.1,1.5] (i.e. $-\overset{\displaystyle R}{\underset{\displaystyle H}{\overset{|}{\underset{|}{C}}}}-O-$, in which R may be hydrogen, as in polyoxymethylene, or an alkyl group) - or on copolymers in which the acetal units are in predominant proportion (by mass[1.1,1.5]) in the (main[1.7]) chains[1.1,1.5,1.7]. The advantageous properties which their base polymers impart to acetal plastics, and which make these plastics suitable for engineering applications, are high strength, stiffness, hardness, toughness, resistance to creep, fatigue endurance, dimensional stability over a considerable temperature range, good electrical properties, virtual immunity from environmental stress cracking, relatively low water absorption, and fairly good resistance to chemicals and solvents. The natural strenth and stiffness of the base polymers is further improved by fibrous reinforcement. The main disadvantageous characteristics of the base polymers are flammability and comparatively low weathering resistance (which can, however, be improved by incorporation of UV stabilisers, especially carbon black).

Acetal plastics, and their base polymers, are represented among the following commercial material ranges : *Beetle; Celcon; Delrin; Duracon; Ekatal; Hostaform; Iupital; Kepital; Latan; Lucel; Sniatal; Tenac; Texapol; Ultraform* (See also Section 1 of Part 3 of this book).

Standards dealing eith acetal plastics include the following.

ISO 9988	Polyoxymethylene moulding and extrusion materials.
Part 1:1991	Designation.
Part 2:1991	Preparation of test specimens and determination of properties.
ASTM D 3641-93	Injection moulding of themoplastic test specimens (including acetal homopolymers and copolymers).
ASTM D 4181-94a	Specification for acetal (POM) moulding and extrusion materials.

ACID INDEX (AI)

A value used to characterise the acidity of the products of combustion of polymeric materials. Defined as the number of milliequivalents of acid ($pH < 4$) evolved per cubic centimetre of the polymeric material when burned under standard conditions in pure oxygen (Warren 1984). AI values may be used as an aid to estimating respiratory and corrosion hazards presented by burning polymers and plastics. Some examples of AI values are:

polymers: PE approximately O; PPS 12; PVC 22; PVF 28; PVDF 49;
plastics: ABS (with flame retardant) 3; a flame-retarded EVA compound 8; a PVC power-cable insulation 12.

ACOUSTIC EMISSION

A burst of ultrasound emitted by a material under stress, detectable in practice with the aid of a piezo-electric transducer suitably attached to the article or test specimen concerned. Acoustic emissions from some composites, including fibre-reinforced plastics, are sometimes measured to detect the onset of yielding or permanent damage that may occur on repeated stressing by a force which, on a single application, would normally only cause deformation within the elastic limit (q.v.) of the material. See also *Kaiser effect.*

ACRYLIC PLASTICS

Plastics whose base polymers are polymers of acrylic acid[1.1,1.5,1.7] or polymers of monomers structurally derived from acrylic acid[1.1,1.5,1.7], or are copolymers of acrylic acid or its derivatives with other monomers[1.1,1.5], the proportion by mass of such other monomers being lower than that of the acrylic monomers[1.1,1.5]. Probably the best known acrylic plastics are those based on poly(methyl methacrylate) - q.v.

ACRYLIC POLYMER(S)

See under *acrylic plastics.*

ACRYLONITRILE/CHLORINATED POLYETHYLENE/STYRENE (A/CPE/S[2.1,2.3]; ACPES[2.2]; ACS[2.11])

Terpolymers produced by grafting acrylonitrile undergoing copolymerisation with styrene onto an elastomeric grade of chlorinated polyethylene. The terpolymers may be regarded as analogous to ABS, with the unsaturated rubbery butadiene component of the latter replaced by a saturated chlorinated polyethylene elastomer. As in the case of acrylonitrile/styrene/acrylate terpolymers (q.v.), the absence of unsaturation makes for greater stability of ACS to photochemical degradation (and hence better weathering resistance in comparison with ABS), but the material is fairly susceptible to thermal degradation, especially at the high melt-processing temperatures. Otherwise the properties of ACS are in general similar to these of ABS, except for the former's lower flammability associated with its substantial chlorine content. A relevant standard is ISO 6402 - standard specification for impact resistant acrylonitrile/styrene moulding materials (ACS; AES; ASA) : Part 1:1990 - Designation. Part 2:1994 - Preparation of test specimens and determination of properties.

ACRYLONITRILE/ETHYLENE-PROPYLENE-DIENE/STYRENE (A/EPDM/S[2.1,2.3], AEPDS[2.11]; AEPDMS[2.2]); ACRYLONITRILE-ETHYLENE-STYRENE PLASTICS(AES[2.2])

A/EPDM/S is another polymer of the general type most prominently represented in industry by acrylonitrile/butadiene/styrene (ABS) - (q.v.), in that it is a graft terpolymer of acrylonitrile/styrene copolymer on a rubbery component, in this case EPDM (q.v.). As the EPDM has no unsaturation in the main chain, whilst its side-chain double bonds are largely used up in the grafting process, A/EPDM/S is more resistant to weathering than ABS, which it resembles in its other properties. Commercial versions of A/EPDM/S materials have appeared under the trade names *Centrex 401; Hostyren XS; Jet-Flex; and Novodur AES.*

NOTE - As shown in the caption above, two different versions exist of the name and the abbreviated term (letter symbol) for the terpolymer; moreover, the letter symbol more closely reflecting the composition appears in three variants. This is a potential source of confusion. The letter symbol *AES* is widely used in technical and trade literature; it is also featured in ISO 6402-1:1990 (as well as in the new edition of this standard forthcoming in late 1997), and is listed - without explanation - together with the more accurate *AEPDMS* in the relevant ASTM standard for abbreviated terms[2.2]. The corresponding ISO and BS standards[2.1,2.3,2.11] do not list *AES*, and give their own versions of the more extensive abbreviation.

ISO 6402, Part 1: 1990 and Part 2: 1994, provides a standard specification for AES moulding materials (as well as ASA and ACS materials).

ACRYLONITRILE/STYRENE/ACRYLATE (ASA[2.1, 2.2, 2.3, 2.11])

With MBS, MABS, ACS and AES, ASA is yet another graft terpolymer on the general ABS pattern. However, its rubbery constituent (the acrylate) is an acrylic rubber which, unlike the butadiene in ABS, does not introduce main-chain unsaturation into the terpolymer. In consequence, the natural resistance of ASA to photochemical and thermal degradation is significantly greater than that of ABS (it is also further improved by addition of appropriate stabilisers). This makes ASA suitable for various outdoor applications in which its toughness and substantial modulus are also assets. Typical applications include components of garden-watering and sprinkler systems, covers and casings for machinery (including agricultural machines and tractors), road signs, and street-light housings, pipes and fittings.

Commercial ASA materials are available in the *Centrex* (Monsanto), *Geloy* (GE Plastics), and *Luran* S (BASF) ranges.

The following standards are relevant to ASA materials.

ISO 6402	Impact resistant acrylonitrile/styrene moulding materials (ASA; ACS; AES).
Part 1:1990	Designation.
Part 2:1994	Preparation of test specimens and determination of properties.
ISO 7246:1984	General specification for moulding and extrusion materials for ASA pipes and fittings.

ACTIVATOR

A minor additive incorporated in some reactive plastics compositions[1.1,1.5] (e.g. unsaturated polyester resin systems) and in rubber compounds,[1.2,1.3] to enhance the action of the accelerator (q.v.).

ACTUAL CALORIFIC VALUE

The calorific energy per unit mass (J/kg) which is released by the combustion of a material under specified test conditions[1.8].

"A" DIRECTION

In the testing of a laminated sheet (q.v.) the direction of the principal axis of a test specimen in relation to the longitudinal direction of the laminar filler (i.e. the warp direction of a fabric or the machine direction of paper or film)[1.1]. Thus "testing in the A direction" means that the specimens are cut out of the laminate so that their principal axes are parallel to the laminar filler's longitudinal direction. See also "B" direction.

ADHESIVE FAILURE

In the context of materials or components bonded with an adhesive, rupture of the bond that occurs (or appears to occur[1.16]) at a boundary (interface[1.16]) between the adhesive layer and the material(s) being bonded (the adherend(s)[1.16]), and not within either.

AFTER-BAKE

See *postcure*

AFTERFLAME

Persistence of flaming of a material after the ignition source has been removed[1.8]. Some versions of this definition are qualified by inclusion of the phrase "under specified test conditions"[1.5,1.9].

AFTERFLAME TIME (DURATION OF FLAMING)

In the specified conditions[1.5,1.8] of a burning test, the duration (persistence[1.8,1.9]) of afterflame[1.9] after removal of the ignition source[1.5,1.8].

AFTERGLOW

Continuation (persistence[1.5,1.8,1.9]) of glowing of a material after any afterflame has ceased[1.5,1.8,1.9] or, if no flaming took place[1.8,1.9], after removal of the ignition source[1.5,1.8,1.9].

AFTERGLOW TIME (DURATION OF AFTERGLOW)

In the specified conditions[1.8] of a burning test, the duration of a material's continued glowing[1.8] (persistence of afterglow[1.9]) after removal of the ignition source (and termination of any afterflaming)[1.8].

AGEING

In relation to plastics, the usage of this term is analogous to that of "weathering" (the two concepts being also interrelated in some respects), i.e. it can have one of the following two meanings, depending on the context.

1. In the non-transitive sense, ageing is the sum total of the irreversible changes that occur in the properties of a material or product over a period of time[1.4] as a result of chemical and physical processes[1.1,1.5]. With plastics, the nature and rates of operation of the processes are primarily dependent on the environment experienced during the time period concerned[1.7]. Ageing undergone in a normal service environment is termed *natural ageing*[1.2]. Ageing in controlled environments, such as the test-treatment conditions employed in a laboratory (usually to speed up the processes involved) is commonly referred to as *artificial ageing* or *accelerated ageing*.
2. In the transitive sense of the term, ageing is the action of exposing a material or product for a period of time to an environment that will cause it to age in the sense of 1. above[1.2,1.3,1.7].

In both its senses, ageing outdoors is virtually synonymous with natural weathering (q.v.).

The following test standards are concerned with ageing procedures in which elevated temperatures are used with a view to simulating the effects of natural ageing at a faster rate. In some cases additional environmental factors may be employed (e.g. humidity, oxygen, pressure). ISO 2578:1993, ASTM D 3045-92, IEC Publication 216-1987: the procedures of these standards are mutually similar. Under certain limiting assumptions they may be used to assess the likely long-term durability (ageing resistance) of a plastics material at normal service temperatures (in the absence of stress). ISO 1137 (withdrawn in 1990) and ASTM D 1870-91: ageing tests in tubular ovens. ISO 1419:1977:

simulated service tests and ageing tests for fabrics coated with plastics or rubbers. ASTM D 573-88(1994): oven-ageing test for rubber. ASTM D 756-93: test for changes in weight and shape of plastics under accelerated service conditions. ASTM D 1042-93: test for dimensional changes after oven ageing. ASTM D 2126-94: thermal/humid ageing test for rigid cellular plastics. ISO 2440: accelerated heat and/or humidity ageing of flexible cellular polymeric materials. BS 4443:Part 4:1989:Method 11: humidity ageing of flexible cellular materials at elevated temperatures; Method 12: heat ageing of such materials. BS 4618:Section 4.6:1974 gives a procedure for heat ageing of plastics together with suggested tests for the resulting changes in properties.

ALIPHATIC POLYKETONE; POLY(OLEFIN KETONE)

A linear, thermoplastic copolymer of an alkene and carbon monoxide, with the olefin and CO units alternating in a regular 1:1 sequence in the molecular chain. Some terpolymers have also been made (e.g. poly(ethylene ketone) with a minor proportion of propylene). The polymers, produced with the aid of palladium catalysts, form the commercial *Carilon* range. Their development is discussed in a paper by Ash (1994). See also poly(ethylene ketone).

ALKENE

Generic name for acyclic hydrocarbons, branched or unbranched, having one double bond in the molecule.

ALKYD RESINS

The general name for a group of synthetic resins of the polyester type, convertible to thermoset, crosslinked materials with the aid of additives containing double bonds or suitable functional groups[1.7]. The word "alkyd" is a somewhat modified, contracted combination of "acid" and "alcohol", coined in the early days of these resins to indicate their ester nature. This derivation may be the reason why the ASTM terminology standard[1.7] regards the terms "polyester plastics" and "alkyd plastics" as completely synonymous. However, whilst such identification might be considered applicable in the purely etymological context just mentioned, it is not accurate in the modern technical sense (and is deprecated, presumably for that reason, by the ISO and BS terminology standards[1.1,1.5]). This is so, because the alkyd resins are only one group within the wider class of polyesters: thus, although all alkyd resins are polyesters (of a particular kind) not all polyesters are alkyds. For example, a plastic based on a thermoplastic polyester like poly(ethylene terephthalate) or poly(butylene terephthalate) is not correctly classifiable as an alkyd plastic. Alkyd moulding compounds -the commercial raw materials of alkyd plastics - are essentially thermosetting compositions, typically comprising crosslinkable alkyd resins, curing agents (commonly peroxides), and other additives such as fillers, reinforcements, lubricants, and colorants. The compounds are moulded (by compression, transfer, and sometimes injection techniques) into such products as components of electrical equipment (e.g. switches, housings, connector mountings, insulating parts), in which the good electrical properties of the alkyd resins are utilised. Commercial versions of the moulding compounds include some *ARMC* grades (Rogers Corp., Molding Materials Div., USA), and grades in the *Bakelite* and *Durez* ranges.

In surface coatings - a major field on application of alkyd resins (especially of the *glyptal* type, i.e. condensates of glycerol and phthalic anhydride) - the additives incorporated in the resin (and termed "modifiers" in that context) may be some fatty oils ("drying oils", which crosslink by a mechanism involving atmospheric oxygen) or fatty acids; modifying resins may also be included (cf. ASTM D 16-95). Typical commercial alkyd resins are branched polyesters, resulting from polycondensation of glycerol or, less commonly, other polyhydric alcohols, with phthalic or another anhydride (e.g. maleic), or with dicarboxylic acids (e.g. adipic acid) - cf. also ISO 4618:Part 3:1984. Modifying resins include phenolic and amino resins. Styrene is fairly common as an unsaturated reactive additive (as in so-called styrenated alkyd resins).

The general test-method standard for alkyd resins used in paints and varnishes is ISO 6744:1984. Test standards relevant to alkyd moulding compounds are BS 2782:Part 7:Method 720B:1979(1986) - cup flow of phenolic and alkyd moulding materials, and the analogous (but not identical) ASTM D 731-95; and ASTM D 1201-81(1992) - a general specification for thermosetting polyester moulding compounds.

ALLOY (POLYMER ALLOY)

A term often used interchangeably with "blend" (q.v.), but more accurately applicable to an intimate blend of polymers which are not normally compatible, or not fully compatible, and which are therefore interblended by compounding in the presence of a suitable compatibiliser (q.v.). The service properties of an alloy, and its performance in relevant tests, are better than those of a non-compatibilised blend of the components, and also better - by design - than some of the properties of each individual component (see also ASTM D 883-93 and ASTM F 412-94a).

ALLYL PLASTIC

A plastic whose base polymer is an allyl resin[1.1,1.5,1.7] (q.v.).

ALLYL RESIN

The product of polymerisation of an allyl compound or compounds, i.e. chemicals containing one or more allyl groups[1.1,1.5,1.7] ($CH_2{=}CH{-}CH_2{-}$). The compounds most widely used in practice to produce thermoset allyl resins are diallyl orthophthalate, commonly known as diallyl phthalate (DAP), and diallyl isophthalate (DAIP). A particular advantage of the allyl resins and some of the moulding compounds based upon their diallyl precursors is the outstanding dimensional stability of the final products, utilised inter alia in the manufacture of precision parts. Other advantages are good electrical properties, heat resistance, and acceptability for food contact (of particular, approved material grades). Allyl resins are naturally flammable, but flame-retarded compounds are available. Commercial thermosetting DAP and DAIP compounds, including fibre-reinforced grades, for compression, transfer, and injection moulding are represented by the following ranges: *Cosmic*; *Durez*; *Fudowlite*; *Poly-Dap*; and *Neonit*.

The following standards deal with allyl moulding compounds.

ASTM D 1636-81 (1992)	Standard specification for allyl moulding compounds.
ASTM PS 15-94	Provisional standard for thermosetting moulding compounds, including inter alia compounds based on diallyl orthophthalate and diallyl isophthalate.
ASTM D 5224-93	Procedure for compression moulding of test specimens (inter alia from allyl moulding compounds).

ALPHA CELLULOSE

Term traditionally used (since about 1900) for pure cellulose of high molecular weight, represented by that portion of cellulose pulp (or fibres - e.g. cotton linters) which does not dissolve in 17-18% sodium hydroxide solution at 20°C. The term has been criticised as lacking precision (ASTM D 1695-77 (1989)). ASTM favours an alternative version of the concept viz. the definition of an "R10" fraction as that portion of a cellulose pulp which is insoluble in 10% sodium hydroxide solution in the conditions of the standard test of ASTM D 1696-61 (1985). This portion is considered to comprise the long-chain fraction of the cellulose of the pulp, so that the definition relates to essentially the same material as that of alpha cellulose, but is more precise, in that the test conditions are standardised.

AMINO PLASTIC(S)[1.7]; AMINOPLASTICS[1.1,1.5]

Plastic(s) based on amino resin(s)[1.1,1.5,1.7]. Specifications for amino plastics moulding materials are provided in ISO 2112:1990 and BS 1322:1992, and directions for the compression moulding of amino-plastic test specimens are given in ASTM D 5224-93. A method of determination of volatile matter in amino resins is given in ISO 3671:1976 (1988) (identical with Method 461A:1978 (1992) of BS 2782.

AMINO RESIN

The product of polycondensation[1.1,1.5,1.7] of a diamino or polyamino compound (typically urea or melamine[1.1,1.5,1.7]) with formaldehyde or another aldehyde[1.1,1.5,1.7], introduced directly or generated by another compound[1.1,1.5,1.7] (e.g. trioxane, the trimer of formaldehyde, from which the formaldehyde is regenerated on heating). Wood, adhesives based on some amino resins are featured in BS EN 301, BS 1203 and BS 1204.

ANGULAR STRAIN

A synonym (less favoured[1.10]) for *shear strain* (q.v. under *strain*).

ANTIMONY OXIDES

Antimony trioxide (Sb_2O_3) and pentoxide (Sb_2O_5). The former has long been used as a flame-retardant additive for plastics. Especially in the context of PVC it is often referred to simply as "antimony oxide". The pentoxide came into use more recently; some commercial versions are now available in particularly effective very-fine-particle form (cf. e.g. *Nyacel*). In polymers which do not contain chlorine, the presence of an halogenated additive - in practice normally a brominated or chlorinated organic compound - is required to enable the antimony oxide to exert its flame-retardant action. The latter is usually explained in terms of the effect of the corresponding antimony trihalide which is formed in the flame zone, where - being gaseous at the prevailing high temperature - it physically restricts access of oxygen. It can also react with oxygen, to reduce still further the amount available to sustain combustion. The reaction is believed also to regenerate Sb_2O_3 in the form of finely-particulate fume, and to liberate atomic halogen (bromine or chlorine). The Sb_2O_3 particles are thought to suppress to some extent the generation of the highly active hydroxyl free radicals (by reducing the excitation energy for the oxidation process - the so-called "wall effect"); they can also catalyse the recombination of the free radicals in the flame. The halogen is available for further reaction with antimony (probably via the intermediate hydrogen halide). Decabromodiphenyl is an example of an effective halogen donor widely used in polyolefins in conjunction with an antimony oxide. Like other such additives, it is most commonly employed in the amount that provides bromine in the ratio of 3 mol Br to 1 mol of the Sb present in the antimony oxide. A typical joint loading level (the bromine donor and antimony oxide together) is up to about 25% by weight of the total compound. It may be noted however that the Br:Sb molar ratio of 10:1 has been claimed to give the greatest reduction in the heat of combustion of certain radiation-crosslinked polyethylene materials.

ANTIOXIDANT

An additive incorporated in a plastic or rubber to inhibit (retard[1.1,1.2,1.3,1.5]) the degradative effects of oxidation[1.1,1.2,1.3,1.5] of the material's base polymer, which can be thermally actuated - in the presence of atmospheric oxygen - at any stage of processing and service. Where the service conditions of a plastics material involve exposure to light containing a UV component (e.g. sunlight or fluorescent lighting) the stabiliser system should additionally include a suitable light stabiliser, to protect against photochemical degradation. The antioxidants used in plastics may be broadly divided into two general groups:

primary and secondary antioxidants. The former act as free-radical scavengers, thereby disrupting the sequence of degradative oxidation reactions in the polymer, which normally proceed by a free-radical mechanism. Secondary antioxidants contribute to the interruption of the oxidation sequence mainly by their deactivating effect on peroxide groups whose formation is a factor in the process of degradative oxidation of polymers; many can also act synergistically with primary antioxidants. Among the latter, the most important are the compounds containing sterically hindered phenolic hydroxyl groups (the so-called "hindered phenols"), and certain aromatic amines. The amine antioxidants are more widely used in rubber compositions (especially dark-coloured ones) as, with some exceptions, they can cause staining. The secondary-antioxidant group consists mainly of some phosphites and thioethers. Certain antioxidants in both the primary and secondary groups are also effective in preventing such oxidative degradation of some polymers (e.g. polyolefins) as can be promoted, even at room temperatures, by metal ions, notably of copper, iron, cobalt and manganese. A discussion by Scott (1985) provides a useful summary of the salient features of mechanisms of polymer oxidation and the role of the antioxidants.

The commercial antioxidant ranges include *Age Rite; Carstab Cyanox; Cyasorb; Ethanox; Hostanox; Garbefit; Ionol; Ionox; Irganox; Isonox; Keminox; Lowinox; Mark; Naugard; Naugawhite; Nonox; Polygard; Santonox; Santowhite; Suconox; Tinuvin; Topanol; Vulcanox; Wytox.*

Methods of identification or determination of antioxidants in plastics are given in the following standards. BS 2782:Part 4:Methods 434 A to D (antioxidants in polyolefins); ASTM D 1996-92 (phenolic antioxidants and erucamide slip additives in PE-LD); ASTM D 4275-91 (BHT in PE and EVA); ASTM D 5524-94 (phenolic antioxidants in PE-HD).

Thin-layer chromatography methods for identification of antidegradants (antioxidants, antiozonants, and some other stabilisers) in rubber and rubber products are given in ISO 4645:1984.

ANTISTATIC AGENT

An agent (substance[1.1,1.5]) incorporated as a minor additive in a (plastics[1.1]) material[1.1,1.5], or applied to the surface[1.1,1.5] of a plastics material or product, to reduce[1.1] (counteract the build-up of[1.5]) static electricity charge[1.1,1.5] (q.v.).

Many commercial antistatic agents are surface active agents (cf. e.g. *Myvaplex, Myverol, Glycolube, Glycomul, Glycostat*); some higher-molecular-weight agents for blending with polymers are also available (e.g. *Stat-Rite C2300*, B.F. Goodrich); and certain conductive fillers, effective at low levels of addition in dissipating static charges, are within the ambit of the standard definition of antistatic agent (for some trade names see Section 3 of Par 3 of this book). Surface active agents used as antistatic additives for plastics typically work by attracting atmospheric moisture to the surface; the resulting microscopic moisture layer, assisted by the polar groups of the surface active agent, conducts away incipient electrostatic charges so that charge build-up is prevented. Atmospheric humidity is thus a factor in this mechanism.

APHA COLOUR SCALE

See under *platinum-cobalt colour scale.*

APPARENT DENSITY

Virtually synonymous with bulk density (q.v.), i.e. the mass per unit volume of a void-containing material[1.5] (Ref. 1.5 refers to the voids as "both permeable and impermeable"). In the context of practical determinations of apparent density, the material is usually particulate and loose (e.g. a powder or pellets) and the voids are the inter-particle spaces (cf. e.g. ISO 60 and ISO 61). Whilst recognising the synonymity of apparent density and bulk density, the ASTM main terminology standard[1.7] also defines the former term separately as the weight in air of a unit volume of a material (with no reference to any voids). For methods of determination of apparent density see ISO 60:1977 and ISO 61:1976 (respectively identical with method 621A:1978 and Method 621B:1978 of BS 2782:Part 6); and ASTM D 1895-89(1990).

APPARENT THERMAL CONDUCTIVITY

This term is sometimes used for thermal conductivity values quoted for a material in which heat transmission may not be independent of thickness, and/or may proceed not (or not exclusively) through conduction but through (or partly through) other modes of heat transfer (e.g. radiation or convection). The symbols λ_a or k_a are normally used for apparent thermal conductivity (cf. e.g. ASTM C 168-90).

ARAMID

Generic name of a group of aromatic polyamides and polyamide/imides, best known as the materials of synthetic fibres of high strength and modulus, and high thermal, chemical and flame resistance. The main relevant standards (ISO 2076:1989(E) and ASTM D 123-88b) both define an aramid fibre by reference to its constituent material. The essential elements of the more comprehensive ISO definition are that the molecular chain of an aramid consists of aromatic groups joined either exclusively by amide linkages or by some amide and some imide linkages, that at least 85% of the linkages constitute direct connections between two aromatic rings, and that the number of any imide linkages present does not exceed that of the amide linkages. The otherwise virtually identical ASTM definition is more restricted in that it makes no mention of imide linkages. Examples of commercial aramid fibres include *Kevlar* (q.v.), *Nomex* (q.v.), *Twaron* (q.v.) *Conex* (q.v.), *Kermel* (q.v.), and *Apyeil* (q.v.). Aramid fibres find application as reinforcement in high-performance composites, protective clothing of various kinds (including body armour and heat-resistant suits), friction materials, gaskets, filter fabrics, and thermal and electrical insulation.

ARC RESISTANCE

The extent to which a material is able to resist the effects of an electric arc applied to its surface in specified test conditions[1.8,1.9]. In a relevant test, arc resistance is typically measured in terms of time elapsing between the commencement of the test and the onset of failure (as defined by reference to stated effects of the arc upon the test specimen) - cf. ASTM D 495.

ARTIFICIAL WEATHERING (LABORATORY WEATHERING)

The action[1.1,1.5,1.7] or effects (depending on the context in which the term is used) of subjecting a plastics material to laboratory conditions in which the factors principally instrumental in the natural weathering - such as radiant energy, temperature, relative humidity[1.1,1.5,1.7], sporadic contact with liquid water (spray[1.1,1.5]), and any other relevant elements[1.7] - are suitably controlled (and, typically, varied in a cyclic manner[1.1,1.5,1.7]) with a view to affecting the observable properties of the material in ways similar to those in which they would be affected by prolonged, continuos exposure out of doors[1.1,1.5] at a relevant geographical location. The definitions in all three principal plastics-terminology standards[1.1,1.5,1.7] make the point that in artificial weathering the exposure conditions are usually intensified in comparison with their counterparts in natural weathering, in an attempt to accelerate their effects (see also *weathering*). Two of the standards[1.1,1.5] specifically exclude from their definitions (which are mutually identical) such special conditions as the presence in the test environment of gaseous industrial atmospheric pollutants, salt mist (spray), or ozone. It may be noted however, that salt mist is a factor in natural marine and coast environments, and that some versions of modern laboratory-weathering equipment include provision for introducing into the test atmosphere such gases as sulfur dioxide, nitrogen oxides, extra oxygen, or ozone.

The following standards are concerned with artificial weathering of plastics or with some aspects of the subject which are relevant to plastics inter alia. (In most cases the brief contents descriptions given here are not the actual titles).

Laboratory weathering tests

ISO 4892:1994. Methods of exposure to laboratory light sources (Four parts: 1. General guidance. 2. Xenon-arc light sources. 3. Fluorescent UV lamps. 4. Open-flame carbon arc lamps).
BS 2782:Part 5:Method 540 B:1982(1987). Identical with an earlier (1981) superseded version of ISO 4892. Under revision.
ASTM D 1499-92. Exposure of plastics to light and water in carbon-arc type apparatus.
ASTM D 2565-92. As D 1499 but for xenon-arc-type apparatus.
ASTM G 23-93. Operating light-exposure (carbon-arc type) apparatus for non-metallic materials.
ASTM G 26-93. Similar to G 23 but for xenon-arc type apparatus.
ASTM G 53-93. Similar to G 23 but for fluorescent UV/condensation type apparatus.

ASTM D 4674-89. Accelerated tests for colour stability of plastics exposed to indoor fluorescent lighting and window-filtered daylight.

Assessment and evaluation of weathering effects

ISO 4582-1980 (1986). Determination of changes in colour and other properties of plastics after exposure to daylight under glass, natural weathering or artificial light.
BS 4618:Section 4.3:1974. Discussion of resistance of plastics to colour changes produced by exposure to light.
ASTM G 45-75 (Discontinued in 1981). Recommended practice for specifying limits of fading and discoloration of non-metallic materials.

Effects of exposure to marine and coastal-type environments, and relevant tests

BS 4618:Section 4.4:1973. Discussion of effects on plastics of marine exposure.
ISO 4611-1987. Exposure to damp heat, water spray and salt mist (identical with BS 2782:Method 551A:1988 (1992)).
ISO 9227:1990. Corrosion tests in artificial atmospheres; salt spray test.
BS 3900:Part F4:1968 (1985). Resistance (of paints) to salt spray-tests.
BS 7479:1991. Salt-spray corrosion tests (for metallic materials):neutral salt-spray test (NSS); acetic acid salt-spray test (AASS); copper-accelerated salt-spray test (CASS).
ASTM B 117-94. Operation of test apparatus in salt spray (salt-fog) testing.
ASTM G 85-94. Practice for modified salt-spray (salt-fog) tests: acetic acid/salt spray (single and cyclic tests); acidified sea-water test (SWAAT); sulfur dioxide/salt-spray cyclic test; dilute electrolyte cyclic fog dry test.

ASH CONTENT

The residue remaining after controlled combustion (calcination) of a plastic in a test to determine its inorganic material content (e.g. mineral filler; glass-fibre reinforcement). The ash content is expressed as percentage by weight of the original test specimen. The temperatures at which plastics specimens are heated in ash content determinations by standard methods vary within the general range of about 600%C to 950%C (cf. ISO 3451/1:1981). In some tests the specimen is treated with sulphuric acid before or after calcination (cf. e.g. ISO 3451/5:1989). The following standards deal with ash-content determination methods. BS 2782:Part 4:Methods 470 A to 470 E (identical with, respectively, ISO 3451/1 to ISO 3451/5), individually covering general determination methods for ash of polyalkylene terephthalates; ash of unplasticised cellulose acetate; ash of polyamides; ash of PVC. ASTM D 5630-94 (ash content in thermoplastics).

ASHING

A term, not standard but often used, for the controlled combustion (calcination) procedures employed to determine the inorganic-material content of a plastic (in terms of the non-combustible residue). See also *ash content.*

A-STAGE

An early stage in the preparation of some thermosetting resins, at which the material is still soluble in certain solvents and is either liquid at room temperature, or fusible on heating[1.1,1.5,1.7]. In the context of phenolic resins (q.v.) the term "resol" is sometimes used for the A-stage resin. See also *B-stage* and *C-stage*.

ATACTIC POLYMER

A regular polymer the molecules of which have equal numbers of the possible configurational base units in a random sequence distribution.

Reproduced, with permission, from ISO 472:1988.

ATTENUATION (Fibre optics)

The diminution of optical power in an optical waveguide, occurring as a result of radiation losses (mainly by absorption and scattering)[1.6].

AUTO-IGNITION TEMPERATURE

Term sometimes used as a synonym for *spontaneous ignition temperature*[1.8] (q.v.).

AUTOMOTIVE FOGGING

See *Fogging*.

BACKBONE

See *main chain*.

BACTERIOSTAT; BACTERICIDE

See under *Biostat.*

"B" DIRECTION

In the testing of a laminated sheet (q.v.), the direction of the principal axis of a test specimen in relation to the transverse direction of the laminar filler (i.e. the weft direction of a fabric, or the direction at right angles to the machine direction of paper or film)[1.1]. Thus "testing in the B direction" means that the specimens are cut out of the laminate so that their principal axes are parallel to the laminar filler's transverse direction. See also "*A*" *direction.*

BIAXIAL ORIENTATION

In common general usage, this term denotes the stretch imparted to a plastics product (such as a film, or the walls of a blow-moulded bottle) by forcible extension in two mutually perpendicular directions. In this sense, "biaxial stretch" could be considered more accurate, as in commercial biaxially stratched products the actual orientation of the polymer molecules is not solely or fully biaxial. See also *biaxially oriented film.*

BIAXIALLY ORIENTED FILM

Film which in the course of its production has been deliberately stretched in both the machine direction and the transverse direction[1.1]. Cf. also OPP-B and OPP-X in ASTM D 2673-94 (a standard specification for oriented polypropylene film), and ASTM D 3664-95 (standard specification for biaxially oriented poly(ethylene terephthalate) film for electrical applications).

BIODEGRADABLE PLASTIC

A degradable plastic whose degradation takes the form of breakdown of the polymer molecular chains into fragments of lower molecular weight[1.5], this change being caused by the action of micro-organisms (bacteria and/or fungi and/or algae) present naturally in the environment[1.5,1.7]. Standards relevant to biodegradability of plastics include the following:

ISO 11266:1994 (provides guidance on laboratory tests for aerobic biodegradation of organic materials in soil).
ISO 846:1978, revision started in 1995 (determination of effects on plastics of the action of bacteria and fungi).
BS 4618:Section 4.5:1974 (discusses the effects on plastics of soil burial and microbiological attack).

In the following references to ASTM test method standards for biodegradability of plastics materials, only the general nature of the biodegradation and the test medium are indicated (in brackets)

ASTM D 5209-92 (aerobic biodegradation in sewer sludge).
ASTM D 5210-92 (as D 5209, but anaerobic degradation).
ASTM D 5247-92 (aerobic biodegradability by specified micro-organisms).
ASTM D 5271-93 (aerobic biodegradation in activated-sludge-waste-water treatment system).
ASTM D 5338-92 (aerobic biodegradation under controlled composting conditions).
ASTM D 5511-94 (biodegradation under high-solids anaerobic digestion conditions).
ASTM D 5525-94 (biodegradation in simulated active landfill environment).

ASTM D 5526-94 (aerobic biodegradation under accelerated landfill conditions).

Other relevant ASTM standards include the following.

ASTM G 21-90: Standard practice for determining the resistance of plastics to fungi.
ASTM G 22-76 (1990): As G 21, but deals with resistance to bacteria.
ASTM D 5152-91: Procedure for water-extraction of residual solids from degraded plastics, for toxicity determination.

The two most common ways of achieving biodegradability in a plastics material are the use of an inherently biodegradable polymer, e.g. poly (hydroxybutyrate/valerate) (*Biopol*) or one of the aliphatic polyesters suitable in this sense (e.g. *Bionolle*), or filling a conventional polymer with starch in a proportion sufficiently high for the material to disintegrate physically when the starch has been decomposed by bacteria (cf. e.g. *Ecostar* masterbatches, St. Lawrence Starch, Canada).

BIOSTAT; BIOCIDE

In the context of plastics materials, an additive incorporated to counteract microbiological attack (by bacteria and/or fungi). In the strict sense of the terms, a biostat is an agent which deters the growth of micro-organisms and their depredations in any of the possible ways, whilst a biocide is actually lethal to them. The terms "fungistat" and "fungicide" and "bacteriostat" and "bactericide" have analogous meanings specifically in relation to microfungi and bacteria. The only relevant definition to be found in a major terminology standard is that of a fungicide, given in ASTM D 609-81(1991) as "an agent that inhibits the germination of fungus spores or the development of mycelium". Anti-agal agents are also sometimes included in the general category of biostats. Some examples of commercial biostats and biocides are: *BioMet, Cunilate, Estabex, Intercide, Irgasan DP, Metatin, Micro-Chek, Preventol, Vancide,* and *Vinyzene*. The following standards are relevant to evaluation of the resistance of plastics to micro-organisms : ISO 846:1976(1988); BS 4618:Section 4.5:1974; ASTM G 21-90 and ASTM G 22-76 (1990).

BLEND

In the plastics context this term is often used interchangeably with "alloy" (q.v.). However, it has a wider scope, in that an alloy is, strictly speaking, a particular kind of blend (produced with the aid of a *compatibiliser* - q.v.), whereas a compound of mutually compatible polymers (such as two thermoplastics melt-compounded together) is also correctly termed a blend. Furthermore, the term is applicable to yet another kind of polymer-based mixture in such expressions as "powder blend" (q.v.) or "dry blend".

BLOCKING

Unintended surface adhesion (sticking) between plastics sheet materials (including thin films) or between those and other materials[1.1,1.5,1.7] (e.g. leather, lacquered surfaces). The plastics sheets may be either of the regular, unsupported kind, or coating layers on substrates. Blocking occurs most readily where the two surfaces concerned are smooth and flat, are in intimate contact under pressure, and one at least is of soft tacky plastic. The tackiness may be due e.g. to exudation of plasticiser (as with some PVC products), or incomplete removal in manufacture of solvents originally present (e.g. as in surface coatings or solution-cast films), or to undercuring (in thermoset materials). Blocking may be quantitatively assessed by determining the force required to separate the surfaces involved. The relevant standard test methods are concerned mainly with blocking of plastics films - cf. ISO 11502:1995 (identical with BS 2782:Part 8:Method 825A:1996); ASTM D 3354-89; ASTM D 1893 (discontinued in 1991); and the old Method 310A of BS 2782:1970 - and blocking resistance of rubber- and plastics-coated fabrics : cf. BS EN 25978:1993 (equivalent to BS 3424:Part 11:1983). In many cases blocking may be counteracted by incorporation in the plastics material of antiblocking agents (e.g. some fine-particle silicas and silicates, or certain paraffin and amide waxes). Most flatting (matting) additives (q.v.) also have some antiblocking action. Where applicable, suitable modifications of processing conditions, or he application of an appropriate surface coating, can also serve as countermeasures.

BLOOM (BLOOMING)

(1) In the general context of plastics, an exudate or efflorescence (e.g. of stabiliser, lubricant, plasticiser, etc) visible on the surface of a plastic or rubber[1.1,1.3,1.5,1.7]. Exudation of this kind may be caused by weathering. The term is also sometimes used for the scattering of light (strictly speaking, scattering in directions near the specular angle of reflection) by exudates on a surface (cf. e.g. ASTM E 284-89).

(2) In relation to pultruded products, the term is also applied to two kinds of fault. In *fibre blooming*, fibres are visible at the surface of the product, typically imparting the appearance of surface whitening and glistening in the area of the fault. The fibre bloom area may stand slightly pround of the surface and feel rough to the touch: the prominence and roughness can usually be eliminated by buffing. *Undercure blooming* is a dull, often whitish appearance of the surface of a pultruded product, resulting from inadequate curing of the resin at the surface (cf. e.g. ASTM D 3918-80(1990).

BLOWING AGENT

In the broadest sense of the term, any substance whereby, in the course of processing, a polymeric material, thermoplastic or thermosetting, is expanded[1.1,1.5,] to form a hollow or cellular product[1.1,1.2,1.3,1.5,1.7]. The expansion is always affected by a gas, which may itself be the blowing agent if introduced directly, under pressure[1.1,1.5], into the material to be expanded.

Otherwise, the blowing agent is an additive (in most cases a compounding ingredient[1.1,1.2,1.3,1.5,1.7]) incorporated in the material to generate the blowing gas at an appropriate stage in processing. Depending on the nature of the additive, the gas may be generated in one of a number of ways, viz. by evaporation of a volatile liquid additive[1.1,1.5], by chemical reaction undergone by the additive[1.1,1.2,1.3,1.5,1.7] (alone, or in interaction with some component of the material to be expanded), or by decomposition of the additive (thermally actuated or initiated as a rule[1.2,1.3,1.7]). Direct-blowing gases and volatile-liquid additives are sometimes termed *physical blowing agents.* Additives which evolve gases by own chemical reaction or decomposition are usually referred to as *chemical blowing agents* or, in common parlance, simply "blowing agents". Although not strictly conforming to the standard definitions of the widest scope[1.1,1.5], the latter usage accords with the definitions of a blowing agent given in the major rubber-terminology standards[1.2,1.3] and in one plastics-terminology standard[1.7], in that those definitions do not extend to physical blowing agents. With chemical blowing agents, a distinction is often made, especially by the commercial suppliers, between exothermic and endothermic types: the former evolve a gas or gases (most often nitrogen, in some cases with carbon monoxide and/or carbon dioxide and some water vapour) while undergoing exothermic decomposition at elevated temperature, and the latter are commonly mixtures of a carbonate or bicarbonate with an acid which, on heating, react endothermically to generate carbon dioxide and some water vapour. An example of a widely used exothermic chemical blowing agent is azodicarbonamide (cf. e.g. *Celogen AZ; Porofor ADC*). One of the original endothermic blowing agents is a mixture of sodium bicarbonate and citric acid (cf. eg. *Hydrocerol BIH*). Some chemical blowing agents also act as nucleating agents in certain polymeric materials. Apart from azodicarbonamide (ADC) itself, other exothermic chemical blowing agents in frequent use in plastics are modified ADC (the modifiers used are various additives); azobis(isobutyronitrile) (AZDN); dimethyldinitroso tetraphthalamide (DMDNT); diphenylsufone-3,3'-disulfonylhydrazide (DPSDSH); isatoic- anhydride (IA); 4,4'-oxybis(benzenesulfone hydrazide) (OBBSH, OBBS, OBSH); 5-phenyltetrazole (PTAZ); trihydrazinotriazine (THTA); toluene sulfonylhydrazide (TSH); toluene sulfonylsemicarbazide (TSSC). The commercial ranges of chemical blowing agents for plastics include *Activex; Azobis; Azobul; Celogen; Chemocel; Epicell; Epicor; Expandex; Ficel; Foam-Kon; Hostatron; Hydrocerol; Kempore; Kemtec; Microfine; Onifine; Polybound; Polycor; Porofor; Safoam; Tego AB-M; Unicell; Vinstab;* and *Vinyfor.*

BLOWN FILM OR SHEETING

See *lay-flat film.*

BRANCHED LINEAR POLYOLEFINS

Polyolefins produced with the aid of metallocene "single-site" polymerisation catalysts; characterised inter alia by closely controlled, narrow molecular weight distribution. Early development was carried out by Exxon Chemical and Dow ("constrained geometry" olefin copolymers). Cf. e.g. *Achieve, Affinity, Dowlex, Exact, Luflexen, Ultra Tuflin.*

BRITTLENESS TEMPERATURE

The temperature, estimated statistically on the basis of a specified test, at which there is *either* a specified calculated percentage probability of brittle failure of a test specimen[1.5], *or* a calculated probability that a specified percentage of test specimens would suffer brittle failure (see ASTM D 746-95). The ISO terminology standard[1.5], which links its version of this definition to the test method of ISO 974, specifies 50% as the relevant probability of a test specimen failing in brittle fashion, whereas ASTM D 746-95 refers to 50% of test specimens probably suffering brittle failure. ASTM D 1790-94 requires the probability that 80% of the specimens would pass the test 95% of the time. The determinations of brittle temperature dealt with in the above standards are applicable respectively to plastics generally (ISO 974:1980), plastics and elastomers (ASTM D 746-95), and to plastics sheeting (in particular plasticised PVC sheeting) of thickness up to 1 mm (ASTM D 1790-94). In all these cases the tests concerned are impact tests.

BS SOFTNESS

See under *Hardness.*

B-STAGE

An intermediate stage in the formation of certain thermoset resins (e.g. phenolic or epoxy resins), in which the material swells when in contact with some liquids, and softens when heated, but may not entirely dissolve or fuse[1.1,1.5,1.7]. In the context of phenolic resins (q.v.), the term "resitol" is sometimes used to denote a resin in the B-stage. See also *A-stage* and *C-stage.*

BT EPOXY SYSTEMS

Epoxy resin systems incorporating bismaleimide-triazine.

BULK COMPRESSION (VOLUME COMPRESSION)

The reduction in volume ($-\Delta V$) of a material under hydrostatic (i.e. isotropic, all-round) pressure, expressed as a fraction of the original volume (V),[1.5] i.e. $\chi = -\Delta V/V$, where χ is the bulk compression (dimensionless, being a ratio).

BULK DENSITY

The density (q.v.) of a particulate material (such as moulding power[1.5,1.7], pellets, granules, etc.[1.5]), determined under specified test conditions[1.2]. The bulk density value thus includes the effect of voids, permeable and impermeable (mainly the inter-particle spaces; also any intra-particle porosity) present ("inherent"[1.7]) in the material during the determination[1.5,1.7]. See also *apparent density* and *powder density*. For a method of determination of bulk density of compacted PVC resins see ISO 1068:1975.

BULK FACTOR

This is expressed as the ratio V_m/V_M (or its equivalent ρ_M/ρ_m)[1.1,1.5], where

V_m is the volume of a given mass of moulding material;
V_M is the volume of this mass of material after moulding;
ρ_M is the density of the moulding; and
ρ_m is the bulk density of the original moulding material.

For methods of determination see ISO 171:1980 (identical with BS 2782:Part 6:Method 621C:1983), and ASTM D 1895-89(1990). See also *density*, *bulk density*, and *apparent density*.

BULK MOULDING COMPOUND (BMC)

An intimate mixture of resin (typically an unsaturated polyester) and chopped fibre reinforcement, suitable for moulding under heat and pressure. The compound, which is normally supplied in bulk, may contain such additives as catalysts, fillers, and colorants, and will usually incorporate chemical thickeners to impart the requisite high viscosity (cf. ISO 8604:1988).

BURN

(1) As transitive verb : to combust.
(2) As intransitive verb : to undergo combustion[1.8,1.9,1.12] (q.v.).

BURN (BURN MARK)

On a moulding or extrudate, or poltrusion (cf. ASTM D 3918), visible sign of local thermal decomposition, typically manifested as darkening, to the point of blackness in severe cases[1.5]. In such cases a burn may be associated with local distortion or disruption of the surface. In the context of glass-reinforced plastics laminates, a relevant ASTM standard (ASTM D 2563-94) uses the term "burned". On injection mouldings, burn marks may be caused by local overheating in the presence of hot gases or air trapped in the mould. Burns are sometimes referred to as "scorching" or "scorch marks", especially when they appear on PVC mouldings or extrusions, where they may be caused by locally excessive shear or friction, or stagnant spots, in the melt. "Scorching" (q.v.) has also a special meaning in connection with the processing of crosslinkable polymers.

BUTYLENE (BUTENE) PLASTICS

Plastics whose base polymers may be homopolymers of butylene (see polybutylene plastics) or copolymers in which butylene is the major monomeric component, predominant by mass[1.1,1.5,1.7].

NOTE - Preferred IUPAC term for butylene is butene.

The main butylene plastics products are polybutylene pipes, which are suitable for hot-water installations (cf. e.g. BS 7291-2:1990).

CAPILLARY RHEOMETER

A rheometer, usually of the piston type, in which a thermoplastic is converted to a melt by heating in controlled conditions in the body (the cylinder or barrel) of the apparatus, and then extruded through a capillary die. The dimensions of the cylinder, the piston and the die appropriate to a particular application must be closely specified and accurately known, and the temperature, pressure applied to the melt, and the output rate must be suitably controlled and/or measured (cf. e.g. ISO 1133:1991; BS 2782:Part 7:Method 720A:1979; ASTM D 3835-90).

CAST FILM

A film made by depositing a layer of plastic, which is molten, in solution or in a dispersion, on to a surface, solidifying and then removing the film from the surface.
Reproduced, with permission, from ISO 472:1988:Amendment 3:1993.

Casting from a polymer solution in a solvent is, in general, a slower and somewhat more involved method of film production than extrusion or calendering. At the same time, properly solvent-cast films offer much lower minimum thickness (down to single-micrometre figures in some cases) with good thickness uniformity, and possibility of using high-molecular-weight material. Lack of molecular orientation, and hence freedom from mechanical and optical anisotropy, can be another useful feature of such films. A concise outline of the mechanical format of the film-casting process has been published by Ricklin (1983).

CATALYST

A substance which, when present in only a relatively small quantity influences (augments[1.1,1.5]) the rate of a chemical reaction, but itself remains apparently (in theory[1.1,1.5]) chemically unchanged throughout the reaction (and at its end[1.1,1.5]).

CELL

One of the cavities in a cellular plastic (q.v.), demarcated by walls which may be complete or incomplete[1.1,1.5,1.7] (see *closed cell* and *open cell*).

CELLULAR PLASTIC (EXPANDED PLASTIC; FOAMED PLASTIC)

A plastic in which numerous small cavities (cells), relatively uniformly distributed and generally mutually comparable in size, have been deliberately created[1.7] in the course of production. Due to the presence of the cells - which may be open, i.e. interconnecting, or closed[1.1,1.5,1.7] - the density of a cellular plastic is lower than that of the same plastic in the completely solid, cell-free state[1.1,1.5]. See also *syntactic cellular plastic; microcellular plastic, closed-cell cellular plastic,* and *open-cell cellular plastic.*

Main standards relevant to cellular plastics are listed in tables 1 and 2 below. See also urethane foams under *urethane plastics.*

TABLE 1 STANDARD SPECIFICATIONS FOR CELLULAR PLASTICS MATERIALS AND PRODUCTS

MATERIAL OR PRODUCT	STANDARDS
Thermal insulating materials - application categories and basic requirements	ISO/TR 9774:1990
Rigid expanded PVC for thermal insulation and building applications.	BS 3869:1965
Rigid phenolic (PF) foam for thermal insulation (slabs and profiled sections).	BS 3927:1986 (1996)
Rigid cellular materials used in thermal insulation of buildings. Phenol-formaldehyde cellular plastics.	ISO 4898:1984 Addendum 1:1988
Evaluating high-density rigid cellular thermoplastics.	ASTM D 3748-94
Flexible cellular PVC sheeting.	ISO 6453:1985 BS 4023:1975 (1986)
Extruded cellular PVC-U profiles.	BS 7619:1993
Expanded polystyrene boards Boards produced from expandable beads Extruded boards.	BS 3837 Part 1:1986 (1996) Part 2:1990 (1996) (corresponding to ISO 2896 and ISO 7616)
Pre-formed cellular polyethylene insulation for pipework.	BS 7523:1991
Flexible cellular polyolefin plastics materials	ASTM D 4819-88

TABLE **2** STANDARD TEST METHODS FOR DETERMINATION OF PROPERTIES OF CELLULAR PLASTICS

PROPERTIES	CELLULAR MATERIAL R = RIGID F = FLEXIBLE	RELEVANT STANDARDS		
		ISO	BS	ASTM
CELL CHARACTERISTICS Cell count	F	-	BS 4443:Part 1:1988:Method 4	-
Cell size	R	-	-	ASTM D 3576-94
Open/closed cell content (volume %)	R	ISO 4590:1981 (Method 1 - pyknometer; Method 2 - volume expansion)	BS 4370:Part 2:1993:Method 10 (Related to ISO 4590)	ASTM D 2856-94 (Pyknometer)
PHYSICAL PROPERTIES Apparent density	R	ISO 845:1988	BS 4370:Part 1:1988(1996):Method 2 BS EN ISO 845:1995	ASTM D 1622-93 ASTM C 271-94 (Cellular core materials of sandwich panels)
	F		BS 4443:Part 1:1988:Method 2	-
Linear dimensions	R	ISO 1923:1981 (Cellular plastics and rubbers)	BS 4370:Part 1:1988:Method 1 BS EN ISO 1923:1995	-
	F	-	BS 4443:Part 1:1991:Method 1	-
MECHANICAL PROPERTIES Tensile properties	R	ISO 1926:1979 (1986)	BS 4370:Part 2:1993:Method 9	ASTM D 1623-78 (1995) (including tensile adhesion)
	F	ISO 1798:1983	BS 4443:Part 1:1988:Method 3	-

PROPERTIES	CELLULAR MATERIAL R = RIGID F = FLEXIBLE	RELEVANT STANDARDS		
		ISO	BS	ASTM
Compressive properties - general	R	ISO 844:1978	BS 4370:Part 1:1988:Method 3	ASTM D 1621-94
Compressive strength and modulus of cellular cores of sandwich panels	R	-	-	ASTM C 365-94
Compressive creep	R	ISO 7616:1986; ISO 7850:1986	-	ASTM D 2221-68 (1984) (Creep of package cushioning)
	F	ISO 10066	-	-
Stress/strain characteristics in compression Low-density materials High-density materials	F	ISO 3386 Part 1:1986 Part 2:1984	BS 4443:Part 1:1988:Method 5	-
Compression set	F	ISO 1856:1980	BS 4443:Part 1:1988:Method 6	-
Compression set under humid conditions	F	-	BS 4443:Part 7:1992:Method 18	-
Flexural properties	R	ISO 1209:Part 2:1990 (Part 1:1990 - Bending test)	BS 4370:Part 1:1988:Method 4 (Cross-breaking strength)	ASTM C 393-62 (1980)*
	R	-	BS 4370:Part 4:1991 (1996) Method 14 (Not in line with ISO 1209-2)	
Shear properties: strength : fatigue	R R	ISO 1922:1981 -	BS 4370:Part 2:1993:Method 6 -	ASTM C 273-94* ASTM C 394-94*
Friability	R	-	BS 4370:Part 3:1988 (1996):Method 12	ASTM C 421-88 (Tumbling test for thermal insulation)

*Cellular core materials of sandwich panels

PROPERTIES	CELLULAR MATERIAL R = RIGID F = FLEXIBLE	RELEVANT STANDARDS		
		ISO	BS	ASTM
Hardness (indentation resistance)	F	ISO 2439:1980	BS 4443:Part 2:1988:Method 7	-
	R	-	-	ASTM C 569-83 (Discontinued in 1989)
Tear strength	F	ISO 8067:1989	BS 4443:Part 6**:1989 (1996):Method 15 BS EN ISO 8067:1995	-
	F	-	BS 4443:Part 7:1992:Method 17 (Test for material with integral skin)	-
Creep	F	-	BS 4443:Part 3:1988 (1996):Method 8	-
Fatigue (determination by constant-load pounding)	F	ISO 3385:1989	BS 4443:Part 5:1980 (withdrawn) BS EN ISO 3385:1995	-
Dynamic cushioning performance	F	ISO 4651:1988	BS 4443:Part 3:1988 (1996):Method 9	ASTM D 1596-78 (1983) (Package cushioning materials)
Resilience	F	ISO 8307:1990	-	-
Apparent impact strength (thermal insulation materials)	R	-	-	ASTM C 589-68 (1975) (Discontinued in 1983)
Resistance to dropping (thermal insulation materials)	R	-	-	ASTM C 487-64 (1976) (Discontinued in 1983)
THERMAL PROPERTIES Apparent thermal conductivity	R	ISO 2581:1975	BS 4370:Part 2:1993:Method 7	ASTM C 177-85 (1993)

**Part 6 partially replaced by BS EN ISO 8067:1995, but remains current (in 1997)

PROPERTIES	CELLULAR MATERIAL R = RIGID F = FLEXIBLE	RELEVANT STANDARDS		
		ISO	BS	ASTM
Coefficient of linear thermal expansion at sub-ambient temperature	R	ISO 4897:1985	BS 4370:Part 3:1988 (1996):Method 13	-
Dimensional stability	R	ISO 2796:1986 (Cf. also ISO/TR 2799:1978)	BS 4370:Part 1:1988:Method 5	-
PERMEABILITY				
Air flow	F	ISO 4638:1984	BS EN ISO 4638:1995	-
Air flow value (at constant pressure drop)	F	ISO 7231:1984	BS 4443:Part 6:1991 (1996):Method 16	-
Water vapour transmission rate (WVT)	R	ISO 1663:1981 (under revision in 1996)	BS 3837:Part 1:1986 (1996) (Expanded polystyrene boards)	ASTM E 96-94
MISCELLANEOUS PROPERTIES				
Water absorption	R	ISO 2896:1987	-	ASTM D 2842-94 ASTM C 272-91 (1996) (Cellular core materials of sandwich panels)
Preparation of water extract	F	–	BS 4443:Part 6:1991 (1996):Method 14	-
Resistance to accelerated ageing	F and R	ISO 2440:1983 (Under revision in 1997)	BS 4443:Part 4:1989 (1996):Method 12	-
Resistance to thermal/humid ageing	F	-	BS 4443:Part4:1989(1996):Method 11	-
	R	-	-	ASTM D 2126-94

PROPERTIES	CELLULAR MATERIAL R = RIGID F = FLEXIBLE	RELEVANT STANDARDS		
		ISO	BS	ASTM
Burning characteristics (small specimen, small flame) Horizontal burning	F R	ISO 3582:1978 ISO 9772:1994	-	ASTM D 4986-95
Vertical burning	R	-		ASTM D 3014-94 (Thermoset cellular plastics)
Solvent swelling	F	-	BS 4443:Part 4:1989 (1996):Method 10	-
Methods of test for cellular polyethylene	R	ISO 7214:1985	Various relevant methods of BS 4370	-
	F	-	-	ASTM D 3575-93
Methods of test for skinned, high-density cellular materials	R	ISO 9054:1990	Various relevant methods of BS 4370	-

CELLULOSE ACETATE (CA)

In the strict, formal sense, the term applies to any acetic acid ester of cellulose, and this is the substance of the standard definitions[1.1,1.5] (c.f. also ASTM D 1695-77(1989)). As the basic (glucose) unit of the cellulose molecular chain contains three hydroxyl groups, there is scope for the acetylation of one, or two, or all three of those (i.e. degrees of substitution of 1, 2, or 3). Within the entire molecular chain, and over all the chains within a given amount of the original cellulose, this provides, in theory, further scope for various combinations of these degrees of substitution. In practice, only the cellulose acetates with average degrees of substitution of 2 and 3 (originally also known, respectively, as secondary and primary cellulose acetates) are of industrial significance, as the basic polymers of some plastics and textile fibres. In this context, the ester with the average degree of substitution of about 2 is often referred to simply as cellulose acetate (occasionally also as "diacetate"), whilst the fully, or nearly fully, acetylated material is styled cellulose triacetate (CTA) - q.v., or sometimes just "triacetate". Of the two, the diacetate has seen wider use as the basis of CA plastics, because of the comparatively greater ease of melt processing (when suitably plasticised), whilst CTA textile fibres offer certain advantages over CA fibres. However, both these grades of CA have lost much of their former importance in both fields with the advent of newer, entirely synthetic polymers. In CA technology, the average degree of acetylation may be expressed in one of the following two ways.

(a) As the weight percentage of combined acetic acid, styled "acetic acid yield" and determined as the weight in grams of acetic acid liberated by complete hydrolysis of 100g of dry CA (cf. ISO 1597:1994).
(b) As the weight percentage of the acetyl group, calculated on the basis of the amount of acetic acid liberated by hydrolysis

The two values typical for cellulose diacetate are: acetic acid yield 55%; acetyl content 39%.

The following standards deal with specifications or tests for cellulose acetate. Specifications for moulding and extrusion of CA materials: BS 1524:1993 and ASTM D 706-82(1989). Test methods for CA flake: BS 2880:1991(1996). Determination of free acidity: ISO 1061:1990 (identical with BS 2782:Method 459A:1991). Determination of moisture content: ISO 585:1990 (identical with BS 2782:Method 431C:1991). Determination of in-soluble particles: ISO 1598:1990 (identical with BS 2782:Method 480A:1991). Determination of light absorption: ISO 1600:1990 (identical with BS 2782:Method 553A:1991). Determination of viscosity number and viscosity ratio of CA in solution: ISO 1157:1990. Determination of acetic acid yield: ISO 1597:1994. Determination of viscosity loss on moulding: ISO 1599:1990. Determination of ash content: ISO 3451-3:1984. Determination of plasticiser content: ISO 1875:1982. An early monograph by Yarsley and Flavell (1956) is a useful source of information on cellulose acetates, as well as some other esters and ethers of cellulose. Commercial cellulose acetate materials include *Bioceta*, *Cellidor CA*, *Dexel*, *Setilithe*, and *Tenite CA*.

CELLULOSE ACETATE BUTYRATE (CAB)

A mixed ester of cellulose with acetic and butyric acids[1.1,1.5] (cf. also ASTM D 1695-77 (1989)). Methods of testing of CAB are given (together with those for cellulose acetate propionate) in ASTM D 817-91. Specifications for CAB moulding and extrusion compounds, and CAB film and sheeting for fabricating or forming, are contained respectively in ASTM D 707-84(1990) and ASTM D 2411-89.

Commercial versions of CAB include *Cellidor* CAB and *Tenite CAB*.

CELLULOSE ACETATE PROPIONATE (CAP)

A mixed ester of cellulose with acetic and propionic acids[1.1,1.5] (cf. also ASTM D 1695-77(1989)). Methods of testing of CAP are given in ASTM D 817-91. Commercial versions of CAP are *Cellidor CAP* and *Tenite CAP*.

CELLULOSE NITRATE (CN)

An ester of cellulose and nitric acid[1.1,1.5]. The term "nitrocellulose" often used for such esters is not strictly correct (cf. ASTM D 1695-77(1989)). CN (with degrees of substitution of about 2 - cf. cellulose acetate) was the base polymer of the earliest thermoplastics (which typically incorporated camphor as a plasticiser). However, because of their high flammability, their applications have dwindled, despite the useful performance properties they offered in such products as billiard and table-tennis balls.

CELLULOSE PROPIONATE (CP)

An ester of cellulose and propionic acid[1.1,1.5] (cf. also ASTM D 1695-77 (1989)). A standard specification for CP-based extrusion and moulding compounds is given in ASTM D 1562-82(1989). CP is available within the commercial *Cellidor* range.

CELLULOSE TRIACETATE (CTA)

A cellulose acetate (q.v.) in which all, or nearly all, available hydroxyl groups of the parent cellulose have been acetylated[1.1] (i.e. the degree of substitution is, or is close to, 3 - see under *cellulose acetate*), so that the acetic acid yield approaches the theoretical value of 62.5%[1.1] (in commercial materials it may range from 61.5% upwards). Unlike cellulose diacetate, the triacetate is not soluble in acetone (the acetone-solubility range corresponds to that of degrees of acetylation represented by acetic acid yields between 50% and 60%). See also ASTM D 1695-77 (1989). CTA-based plastics materials require plasticisation[1.1] for melt processing.

CELLULOSIC PLASTICS

Plastics whose base polymers are derivatives of cellulose[1.1,1.5], such as cellulose esters or ethers[1.7].

CHALKING

A powdery, chalk-like appearance of the surface of a plastics or rubber material or product[1.1,1.3,1.5,1.7]: this may be associated with the actual presence of a fine pulverulent layer[1.7], some of which may be removable by rubbing. Chalking can be a manifestation of surface degradation[1.2], or it may be caused by migration to the surface of an additive present in a plastics or rubber material : occasionally both causes may operate simultaneously. Chalking is often among the early symptoms of the weathering of badly formulated PVC products (e.g. rainwater goods, window frames).

CHAR

(1) A residue, carbon-rich or essentially consisting of carbon, formed as a result of pyrolysis (q.v.) or incomplete combustion[1.8,1.9,1.12].

(2) To form char or/and undergo charring.

CHIRALITY

The property of non-identity of a molecule with its mirror image.

NOTE - A molecule in a given configuration or conformation is termed chiral when it is not identical with its mirror image. All asymmetric molecules are chiral; however, not all chiral molecules are asymmetric, since some molecules having axes of rotation are chiral. Chiral and prochiral atoms are sites or potential sites, respectively, of stereoisomerism.

Reproduced, with permission, from ISO 472:1988.

CHLORINATED POLYETHYLENE (PE-C[1.5]; CPE[2.2])

Polyethylene, in which a proportion of the hydrogen atoms of the molecular chains has been substituted by chlorine in a chlorination process[1.5]. High-density polyethylene (PE-HD) is widely used for the chlorination, which can be run with the polymer in solvent solution or in dispersion. Solid-phase chlorination, with chlorine gas, is also possible. Chlorination in solution gives the greatest uniformity of distribution of chlorine atoms along the PE-HD molecular chains. The mechanical properties of PE-C vary considerably depending on the chlorine content, and on the molecular weight and degree of crystallinity of the original polyethylene. To the extent to which the generalisation if applicable, chlorination may be said to impart elastomeric character to the polyethylene so modified. This is utilised, inter alia, in an important application of PE-C, viz. as an impact modifier in rigid PVC products such as profiles (including window frames), cladding, and rain-water goods. In all these applications the good UV and general weathering ressistance of PE-C is an additional advantage; other outlets include films and sheeting. Suitable elastomeric grades of PE-C, which can be crosslinked (typically with peroxides), are used as rubbers offering good resistance to ageing, weathering, ozone and oils.

Trade names under which some PE-C materials are available include *Elaslen; Hostapren; Kelrinal; Lotigren; Tyrin.*

ASTM D 5021-89 provides a specification for a thermoplastic elastomer based on chlorinated polyethylene, for moulding, extrusion and thermoforming.

CHLORINATED POLYVINYL CHLORIDE (PVC-C[1.5]; CPVC[2.2])

Modified polymer resulting from chlorination of poly(vinyl chloride)[1.5]. The chlorination is carried out either with the vinyl chloride homopolymer dissolved, at an elevated temperature, in a suitable solvent (typically a chlorinated hydrocarbon), or - at a lower temperature - in aqueous dispersion, in the presence of swelling agents and actinic light. The original commercial PVC-C polymers produced by the chronologically earlier solution process were *Igelit PC* (I.G. Farbenindustrie) and *Rhenoflex* (Dynamit Nobel). *Trovidur HT* (Dynamit Nobel) and *Geon HT* (B.F. Goodrich) were early commercial products of the dispersion provess.

In both processes chlorination takes place mainly at the $-CH_2-$ groups of the poly(vinyl chloride) molecular chains (i.e. the 1,2-chlorinated configuration, $-CHC\ell-CHC\ell-$, is preferentially formed). Thus complete chlorination would give a polymer very similar to symmetrical poly-(dichloroethylene):

$$\sim\!\!\sim CHC\ell—CHC\ell—CHC\ell—CHC\ell \sim\!\!\sim$$

However, the degree of chlorination of commercial PVC-C polymers is considerably less than that. A fairly typical chlorine content of such polymers, in terms of percentage by weight, would be about 65.7%, as compared with 56.8% for vinyl chloride homopolymer, and 73.2% for fully chlorinated PVC (or the theoretical polymer of 1,2-dichloroethylene).

In comparison with poly(vinyl chloride), PVC-C has a higher T_g, so that its compounds have a higher Vicat softening temperature, deflection temperature under load, and maximum service temperature than PVC-U compounds. The higher maximum service temperature is the property instrumental in the main application of PVC-C compounds, viz. as the materials of pipes and fittings for the conveyance of hot water (in some cases under pressure). The T_g and density of PVC-C polymers both increase with increasing chlorine content. The chemical resistance of PVC-C materials is generally similar to that of PVC-U, except for their somewhat greater susceptibility to a number of organic solvents (chlorinated hydrocarbons, ketones, esters, higher alcohols, aromatic hydrocarbons). The higher chlorine content of PVC-C makes its compounds less readily flammable than PVC-U compounds.

Commercial PVC-C materials are available in the *Hostapren*, *Lucalor*, and *TempRite* ranges (see also Section 1 of Part 3 of this book).

The following standards are relevant to PVC-C materials and products.

PVC-C *materials*

ASTM D 1784-95	Standard specification for rigid PVC and CPVC compounds.
ASTM D 5260-92	Classification of chemical resistance of compounds of vinyl chloride polymers and copolymers, and CPVC.
ASTM D 3915-92	Standard specification for PVC and related compounds (including CPVC) for pressure pipes and fittings.

Pipes and fittings

Specifications and tests : ISO 3514:1976 (specification and determination of density); *ISO 2507-2:1995* (Vicat softening temperature - test conditions for PVC-U and PVC-C pipes and fittings); *ISO 7675:1991* (pipes and fittings for soil and discharge).

Standards for fittings : ISO 727:1985 (fittings of PVC-C, PVC-U, and ABS); *ISO 2045:1988* (single sockets for PVC-C and PVC-U pressure pipes); ISO 8283-1:1991 (sockets and spigots of PVC-C, PVC-U, and ABS).

CHLORINATED POLY(VINYL CHLORIDE) PLASTIC

A plastic whose base polymer, or principal polymeric component, is chlorinated poly(vinyl chloride) (PVC-C) - q.v.

NOTE - The definition in the ASTM terminology standard[1.7] (the ISO and BS standards[1.1,1.5] do not define the term) appears to be more limited in scope, stating as it does that in a PVC-C plastic the PVC-C polymer "is in the greatest amount by weight". This would appear to exclude from the definition any plastic in which PVC-C is the sole polymeric component, but in which the content of additives (e.g. fillers) exceeds 50%.

CHLOROFLUOROCARBON PLASTIC

A plastic based on a polymer or polymers whose monomer units contain only carbon, chlorine and flourine[1.1,1.5,1.7]. An example is polychlorotrifluoro-ethylene (PCTFE).

CHLOROFLUOROHYDROCARBON PLASTIC

A plastic based on a polymer or polymers whose monomer units contain only carbon, chlorine, fluorine and hydrogen[1.1,1.5,1.7]. An example is a copolymer of ethylene with chlorotrifluoroethylene (ECTFE).

CHLOROSULFONATED POLYETHYLENE; CHLOROSULFONYL-POLYETHYLENE[2.5] (CSM[2.4,2.5,2.6])

A chlorinated polyethylene containing also some sulfonyl chloride groups in the polymer chains; produced by chlorination of polyethylene in solution, in the presence of sulfur dioxide and under activation by UV light or ionising (high-energy) radiation. The material is produced in a number of grades - all non-crystalline and rubbery - containing 25% to 43% chlorine, and 1% to 1,5% sulfur in the sulfonyl chloride groups. Because of its rubbery nature and suitability for vulcanisation (via the sulfonyl chloride groups), chloro-sulfonated polyethylene is classified as a rubber.

NOTE - Hence the standard letter symbol is *CSM*, in accordance with the relevant abbreviated nomenclature for rubbers[2.4,2.5,2.6].

CSM is characterised by low flammability and good mechanical properties. Vulcanisation imparts particularly good resistance to heat, weathering, ozone, oils, and many chemicals, but the properties in the unvulcanised state are also suitable for many applications, such as e.g. sheeting used as geomembranes, reservoir linings and the like. Other main uses of CSM include cable coverings and fabric coatings (cf. e.g. BS AERO F131 : CSM-coated nylon fabric for wind socks, and BS AERO F136 : CSM-coated nylon fabrics for aircraft covers). Chlorosulfonated polyethylene is available commercially in the Du Pont *Hypalon* range.

CIELAB COLOUR DIFFERENCE

Numerical expression of colour difference calculated with the aid of CIE 1976 *L*a*b** opponent-colour scales, based on application of the appropriate cube-root transformation to CIE 1931 tristimulus values (X, Y, Z) or to CIE 1964 tristimulus values (X_{10}, Y_{10}, Z_{10})[1.14]. See also ASTM D 2244-89 (standard method for calculating colour differences from instrumental measurements of colour coordinates); ASTM D 1729-89 (standard practice for visual evaluation of colour differences of opaque materials); and ASTM E 308-95 (computation of colours of objects with the aid of the CIE system). A useful discussion of colour-order systems (including the CIE system), the nature and measurement of colour, and other aspects of colour technology, has been published by Billmeyer and Saltzman (1966).

CLARIFYING AGENT

A reagent used in the production of high-clarity grades of polypropylene (cf. e.g. *Millard*)

CLOSED CELL

In a cellular plastic (q.v.), a cell fully enclosed[1.1,1.5] by complete, solid walls (which may consist of relatively thin membranes) so that it does not interconnect with neighbouring cells[1.1,1.5], and - if located in the surface layer of the cellular plastic - is not open to the exterior[1.1]. See also *open cell, closed-cell cellular plastic,* and *open-cell cellular plastic.*

CLOSED-CELL CELLULAR PLASTIC

A cellular plastic (q.v.) containing relatively few, if any, interconnecting cells[1.1,1.5,1.7]. See also *closed cell.*

CLOSING TIME

See under *moulding index.*

COEFFICIENT OF CUBICAL THERMAL EXPANSION

The reversible change in volume of a material (or, specifically, of a test specimen in a relevant test) per unit original (reference) length per one degree change in temperature (cf. the analogous definition of the *coefficient of linear thermal expansion*). A standard method of determination of the coefficient was given in ASTM D 864 (standard discontinued in 1989).

COEFFICIENT OF FRICTION

In general terms, the coefficient of friction between two surfaces in contact is defined[1.1,1.5] as the ratio of the frictional force (F) to the force (L) acting normally (i.e. perpendicular[1.1,1.5]) to the surfaces to maintain their contact (the latter force is gravitational in most cases[1.1,1.5]). I.e.

$$\mu = F/L$$

where μ is the IUPAP symbol for the coefficient of friction (alternative symbol f).

In the context of plastics materials, two coefficients of friction are distinguished, viz. the static coefficient (μ_S) and the dynamic (kinetic) coefficient (μ_D), defined respectively by the expressions $\mu_S = F_S/L$ and $\mu_D = F_D/L$, where F_S is the minimum force required to initiate sliding, and F_D is the minimum force to maintain sliding (at a particular speed). The coefficients of friction are discussed in BS 4618:Section 5.6:1975. Methods of their determination for some plastics materials are given in the following standards. *Coefficients of friction of film and/or sheeting* : ISO 8295:1995; BS 2782:Part 8:Method 824A:1984(1990); ASTM D 1894-95. *Kinetic coeficient of friction of plastics solids* : ASTM D 3028-93. See also IEC Publication 648-1979.

COEFFICIENT OF LINEAR THERMAL EXPANSION

The reversible fractional change in length of a material[1.1,1.5] (or, specifically, of a test specimen in a relevant test) on heating, defined by the expression

$$\alpha = \Delta L/(L_O.\Delta T)$$

where α is the usual (IUPAP) symbol for the coefficient, ΔL is the change in length, L_O is the original (reference) length, and ΔT is the temperature range over which ΔL is measured. The value of the coefficient may not be the same when measured over different temperature ranges[1.1,1.5]. As can be seen from the above expression, the unit of α is the reciprocal of temperature, i.e. K^{-1}, $°C^{-1}$, or $°F^{-1}$, but the values of the coefficient are also often quoted as length per unit length per degree of temperature, e.g. $mm.mm^{-1}.°C^{-1}$, $in.in^{-1}.°F^{-1}$, etc.

The coefficient of linear thermal expansion of plastics materials is discussed in BS 4618:Section 3.1:1970; methods of its determination can be found in the following standards.

ASTM D 696-91 (dilatometer method, for plastics materials with $\alpha > 1\times10^{-6}$ per °C; determinations in the temperature range –30°C to 30°C).
ASTM E 228-85 (1989) (dilatometer method; determinations at temperatures outside the range –30°C to 30°C).
ASTM E 831-93 (determination by thermomechanical analysis).

COHESION

Ability of the elementary constituents of a substance (such as particles[1.5], molecules) to hold together so as to maintain the substance's integrity. In polymers, as in other chemical materials, cohesion is the result of intermolecular forces[1.5]. In the case of materials or components bonded with adhesives, cohesive failure of the bond (as distinct from adhesive failure - q.v.) is formally defined[1.16] as rupture of the bond occurring, or appearing to occur, within the adhesive layer.

NOTE - If the internal cohesion of the material(s) being bonded (the adherend(s)[1.16]) is lower than that of the adhesive, a bond may also fail by rupture occurring within the adherend(s), i.e. through cohesive failure of the latter.

COHESIVE FAILURE

see under *cohesion.*

COLOPHONY

The resinous, non-volatile residue remaining after distillation of turpentine from oleoresin (the fresh secretion of pine and fir trees). The residue is a mixture of resin acids and resin acid esters, containing a high proportion of abietic acid (cf. Latin *abies* - fir tree), itself a product of transformation of natural primary resin acids. Colophony may be modified in various ways to increase its stability. The principal methods of modification are: partial neutralisation of the acid groups ("hardening"); esterification with polyhydric alcohols (producing "ester gum"); hydrogenation to reduce unsaturation; and partial polymerisation. Some standards treat "colophony" as a synonym for "rosin", and indeed prefer the latter term (cf. ASTM D 804-92 and D 907-93). At the same time, such usage is deprecated by IEC because of possible confusion of "rosin" within "resin" (cf. IEC 68-5-2:1990).

COMBUSTION

The exothermic reaction of a substance with an oxidising agent (i.e. a chemical process of oxidation[1.12,1.13]) typically proceeding at a rate fast enough[1.12,1.13] to manifest some or all of the following: flame (flaming combustion), glowing (glowing combustion), and/or evolution of smoke[1.5,1.8,1.9]. Two ASTM publications provide extensive lists of combustion test methods, viz. ASTM D 3814-93 (listing over 270 standard test methods of ASTM, CSA, NFPA, SAE, UL, MIL, IEC, and some official technical publications); and ASTM book on fire testing - the 4th (1993) edition (1 125 pages) lists 149 relevant current ASTM standards.

COMPATIBILISER

Mixing-aid additive used in the production of a polymer alloy (q.v.). Functionally akin to a wetting agent in aqueous systems, in that its presence, normally in a relatively small proportion, promotes interfacial adhesion and hence intimate intermixing in the blending of polymers which are not otherwise compatible, or not fully compatible. The identity of compatibilisers used in the production of commercial polymer alloys is not often disclosed, but some are known to be block copolymers incorporating groups compatible respectively with each of the components of the alloy. Examples of commercial compatibilisers (from the BASF range) used in two-component mixtures of reclaimed polymers are: *RCM 1* (for PE+PS alloys), *RCM 2* (for PP+PS alloys), and *RCM 3* (for PP+ABS alloys). Some grades of chlorinated polyethylene were found to have compatibilising action in ternary blends PVC+EPDM+PE-C (Lee and Chen 1987). See also *Fusabond; Polybond; Unite; Vector.*

COMPLEX COMPLIANCE

C^* (Pa^{-1}) : The reciprocal of complex modulus:

$$C^* = C' - iC''$$

where $i = \sqrt{-1}$.

NOTE - It may be measured in tension: $D^* = \frac{1}{E^*}$; shear: $J^* = \frac{1}{G^*}$; bulk compression: $B^* = \frac{1}{K^*}$; longitudinal compression: $O^* = \frac{1}{L^*}$.

The compliances D, J, B and O of linear-viscoelastic material subjected to a non-periodic strain are time-dependent owing to the delayed equilibrium behaviour of these materials.

Reproduced, with permission, from ISO 472:1988.

COMPLEX MODULUS

M^* (Pa); **dynamic modulus (deprecated)** : The stress-strain ratio of a viscoelastic material that is subjected to sinusoidal loading :

$$M^* = M' + iM''$$

where $i = \sqrt{-1}$.

NOTES

1. The definition of M^* takes into account the phase shift between stress and strain.
2. The complex modulus may be measured in tension (E^*), shear (G^*), bulk compression (K^*) or longitudinal compression (L^*):

$$E^* = E' + iE''$$

$$G^* = G' + iG''$$

$$K^* = K' + iK''$$

$$L^* = L' + iL''$$

The storage modulus (E' or G' or K' or L') portion of the complex modulus is the ratio of the steady state stress in phase with the strain to the magnitude of the strain. The loss modulus

(*E*" or *G*" or *K*" or *L*") portion of the complex modulus is similarly defined, but uses the component that leads to the steady state stress at an angle 90% out of phase from the steady state strain.

The storage moduli are measures of the energy stored and regained during a loading cycle, whereas the loss moduli are proportional to the energy lost (dissipated) during this time.

The moduli *E, G, K* and *L* of linear viscoelastic materials subjected to a non-periodic stress are time-dependent owing to the delayed equilibrium behaviour of these materials.

See also *loss modulus* and *storage modulus*..

Reproduced, with permission, from ISO 472:1988.

Complex modulus can be determined by some dynamic test methods - cf. e.g. ISO 6721:1994 and ASTM D 4065-94.

COMPLIANCE

C (Pa^{-1}): The quotient of strain by stress:

$$C = \frac{\varepsilon}{\sigma} = \frac{1}{M}$$

D is the tensile compliance; *J* is the shear compliance; *B* is the bulk compliance; *O* is the longitudinal compression compliance.

Reproduced, with permission, from ISO 472:1988.

COMPOSITION

In the two of its senses relevant to plastics materials, "composition" is the make-up of a material that consists of more than one component, or the material so made up. In the latter sense, the term is sometimes used - especially in reference to PVC materials - interchangeably with "compound" (q.v.), both in general technical parlance and in some standards (cf. e.g. ASTM D 2115-92, *Standard Practice for Oven Stability of Poly(Vinyl Chloride) Compositions*, or ASTM 2151-95, *Standard Practice for Staining of Poly(Vinyl Chloride) Compositions by Rubber Compounding Ingredients)*. Analytical procedures are available for determining the composition (in the sense of make-up) of plastics materials. Sources of relevant information include books by Antelman (1974), Arnold and Willis (1972), Crompton (1977 and 1984), Haslam et al. (1973), Hummel and Scholl (1988), and Lawson (1961).

See also *analysis* under *vinyl chloride plastic.*

COMPOUND

An intimate mixture of one or more polymers (in this context usually termed "the base polymer(s)") with various requisite additives[1.1,1.5,1.7]. Appropriate additives are incorporated in a compound as aids or stabilisers in processing, or to impart to the ultimate product particular properties or features. Relatively cheap fillers are also often included simply to lower material cost. Some examples of additives functional in the processing of thermoplastic compounds are lubricants (q.v.), polymeric melt-processing aids (as used e.g. in PVC compounds), and heat stabilisers (e.g. for polyolefins or PVC). Examples of additives affecting ultimate product properties include plasticisers, reinforcing fillers, impact modifiers, flame retardants, and UV stabilisers. Among common physical forms of compounds used as materials for the manufacture of plastics products are thermoplastic pellets and powders (for injection moulding and extrusion), thermosetting powders (typically for compression and transfer moulding), various prepregs (q.v.), and liquid compositions (including PVC pastes - q.v., and thermosetting compositions based on UP and epoxy resins).

COMPRESSIVE STRESS

See under *normal stress.*

CONFIGURATIONAL BASE UNIT

A constitutional repeating unit, the configuration of which is defined at one or more sites of stereoisomerism in the main chain of a polymer molecule.

NOTE - In a regular polymer, a configurational base unit corresponds to the constitutional repeating unit. In the regular polymer molecule

$$-[CH(CH_3)CH_2]_n- \text{ (polypropylene)}$$

the constitutional repeating unit is $-CH(CH_3)CH_2-$ and the configurational base units are

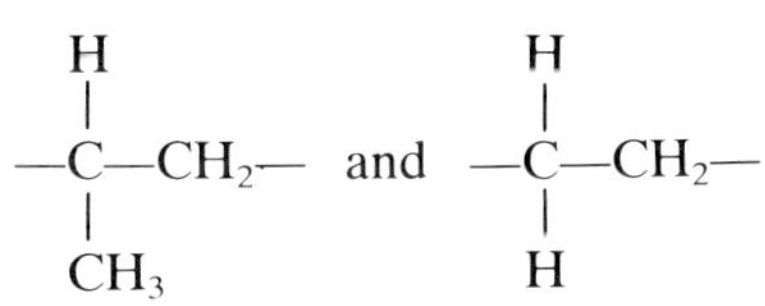

These two configurational base units are enantiomeric to each other.

Reproduced, with permission, from ISO 472:1988.

CONFIGURATIONAL REPEATING UNIT

The smallest set of one, two, or more successive configurational base units that prescribes configurational repetition at one or more sites of stereoisomerism in the main chain of a polymer molecule.

See notes to *isotactic polymer* and *syndiotactic polymer.*

Reproduced, with permission, from ISO 472:1988.

CONFIGURATIONAL UNIT

A constitutional unit having one or more sites of defined stereoisomerism.

Reproduced, with permission, from ISO 472:1988.

CONSTITUTIONAL REPEATING UNIT

The smallest constitutional unit the repetition of which describes a regular polymer.

NOTE - The polymer chain

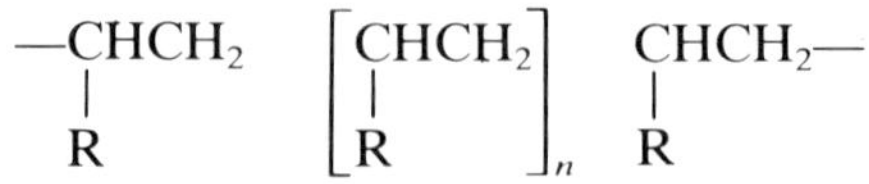

can provide the following constitutional units:

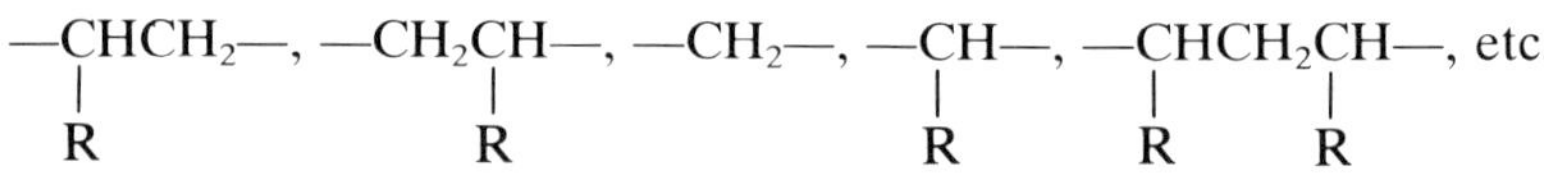

Only the first two constitutional units are the smallest ones completely describing this polymer chain. Either of these two constitutional units is a constitutional repeating unit and the polymer whose molecules can be described by the polymer chain shown above is a regular polymer.

Reproduced, with permission, from ISO 472:1988.

CONSTITUTIONAL UNIT

A species of atom or group of atoms present in a chain of a polymer or oligomer molecule.

See note to *constitutional repeating unit.*

Reproduced, with permission, from ISO 472:1988.

"CONSTRAINED GEOMETRY" POLYOLEFINS

See *branched linear polyolefins.*

CONTACT STAINING

See under *migration*

CORROSIVITY INDEX (OF PLASTICS AND FILLERS)

The mean specific conductance (S ohm^{-1} cm^{-1} $\times 10^{-6}$) of an aqueous extract of a plastic or filler, determined under prescribed test conditions. In the context of the relevant standard (ASTM D 4350-86(1990)) the index may be regarded as reflecting the degree of probability that, in humid conditions, metal surfaces in contact with the material(s) concerned could suffer galvanic corrosion or direct chemical attack. The S values for some epoxy resins, cited in the standard by way of example range between 25.5×10^{-6} ohm^{-1} cm^{1} and

104.0×10^{-6} ohm^{-1} cm^{-1}. The value for a defined grade of distilled water after storage in borosilicate glass bottles is given as less than 2.0×10^{-6} ohm^{-1} cm^{-1}.

CRAZING

The formation of crazes at the surface or in the interior of a polymer or plastic. Whilst to the naked eye crazes may appear as fine cracks[1.1, 1.7], there is in fact an important distinction between the respective natures of a craze and a crack in a plastics material. In this context, even the finest crack is a clear gap in the material (typically wedge-shaped if extending from the surface, or plate-like in the interior), whereas a craze is a gap which may resemble a crack in general shape, but which is bridged by strands of substantially oriented polymer material, so that the region of the craze is microcellular in character and - unlike a clear gap - can transmit stress (albeit less effectively than the surrounding solid polymer) - cf. e.g. Kambour (1963), and Walters (1983). The ISO terminology standard's[1.5] definition of a craze (this standard does not define "crazing"), as well as the mutually identical definitions of crazing in the corresponding BS and ASTM standards[1.1,1.7] (neither of which defines "craze"), do indicate the essential difference between a craze and a crack by referring to crazes as "apparent cracks" and mentioning their bridging by polymeric material[1.5] which is of lower density[1.1,1.5,1.7] than the solid bulk polymer around it. It may be noted that the definition of crazing in a rubber terminology standard[1.3], as well as one - still provisional at the time of writing - proposed in a draft addendum to another such standard[1.2], both describe crazing as a random pattern of cracks on a rubber surface, usually symptomatic of ageing by light. This is reminiscent of the traditional application of the term to a network of fine cracks on the surface of glass or pottery glaze. In the context of GRP laminates, a standard for classification of visual defects (ASTM D 2563-94) defines "crazing" as fine cracks at or under the surface of a laminate. See also *environmental stress cracking and crazing*.

CREEP LATERAL CONTRACTION RATIO

See *Poisson's ratio*.

CROSS-BREAKING STRENGTH

See *flexural strength*.

CROSSLINK (noun)

A bond, chemical[1.2,1.3] or ionic, or bridging atom or group of atoms[1.2], that joins two polymer chains[1.2,1.3] or two parts of the same chain[1.2].

NOTE - The ISO definition (in Addendum No. 7 to Ref. 1.2) relates specifically to crosslinks in rubber, formed by vulcanisation.

CROSSLINK DENSITY

The number of crosslinks present, on the average, in a unit volume or unit mass of a crosslinked polymer.

NOTE - The only major terminology standard to offer a definition of crosslink density (Addendum No. 7 to Ref. 1.2) confines the definition to crosslink density in rubber.

CROSSLINKED POLYETHYLENE

Several properties of polyethylene (e.g. strength, temperature resistance) are improved considerably by crosslinking (q.v.). In practice, suitable grades of polyethylene and some ethylene copolymers (e.g. EVA) may be crosslinked in one of two general ways: by exposure to high-energy ionising radiation, or chemically (by grafting, or by copolymerisation of ethylene with vinyl-trimethoxy silane). Crosslinking by radiation or by peroxides produces covalent inter-chain bonds

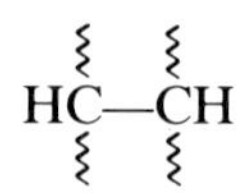

with a bonding energy of 85 kcal per mole. The silane route results in silane bridges

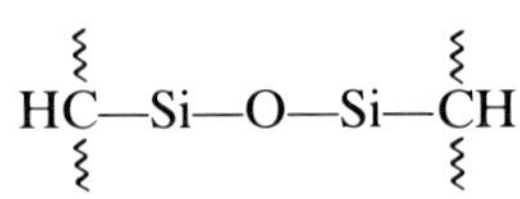

with an effective bonding energy of 106 kcal per mole.

Chronologically, silane crosslinking is more recent than the peroxide process, and does not require the high temperatures and pressures necessary in the latter (typically about 200°C and 15 bar). See also *Monosil* and *Sioplas processes, Si-Link, Aquathene,* and *Visico NCPE 4421.*

CROSSLINKING

The process of forming[1.1, 1.5], or the presence of, a multiplicity of covalent or ionic bonds[1.1,1.5] (or relatively short molecular bridges, each involving a sequence of such bonds) between the molecular chains of a polymer. Typically, the polymer-chain segments between such crosslinks are much longer than the crosslinks themselves. In sufficiently advanced stages of crosslinking, the polymer chains and the crosslinks jointly form a three-dimensional network[1.3,1.7]. Crosslinking on a significant scale modifies some of the polymer's physical[1.7], chemical, and thermal properties (increased tensile strength and modulus, reduced extensibility and creep, reduced solubility, higher softening point and temperature of deflection under load). The degree of crosslinking may be expressed in terms of crosslinking coefficient, crosslink density, or network parameter. It is also reflected in the values of gel content and (for cross-linked polyethylene) the hot set, as measured in the relevant tests (cf. ASTM D 2765; IEC 811/2/1; BS 6469, Section 2.1, Clause 9).

CROSSLINKING AGENT

A reagent that promotes or regulates crosslinking[1.1,1.5] (q.v.), aids it or participates in it. Some examples are: peroxides which promote formation of covalent bonds between polyethylene chains in peroxide crosslinking; vinyltrimethoxy silane which forms inter-chain bridges in chemical crosslinking of polyethylene or EVA; and such aids to radiation crosslinking as e.g. triallyl cyanurate (used with polyethylene), trimethylolpropane trimethacrylate (used with PVC as a reactive plasticiser in crosslinkable compositions). In the field of rubber, crosslinking agents are represented by vulcanising agents and aids. See also *crosslinked polyethylene*.

"CROW-FOOT' MARKS; CROWSFEET[1.3]

Relatively small, V-shaped marks on calendered sheeting, resembling birds' footprints; sometimes also referred to as "pine trees". The marks are usually associated with incomplete dispersion of particulate additives (such as fillers or pigments) in the sheet material.

CRYSTALLINE POLYMER

A non-standard term sometimes used in reference to a polymer containing a crystalline phase. The standard term, favoured by ISO[1.5] is "semi-crystalline polymer" (q.v.).

C-STAGE

The final stage in the reaction of some thermosetting resins, in which the material has become practically insoluble and infusible. The resin of a fully cured thermoset moulding is in this state[1.1,1.5,1.7]. In the context of phenolic resins (q.v.) the term "resite" (q.v.) is sometimes applied to the material in the C-stage. See also *A-stage* and *B-stage*.

CUP FLOW FIGURE

The time in seconds taken by a mould of standard design to close completely under pressure when loaded with a charge of phenolic moulding material. Used to indicate behaviour, in respect of flow under heat and pressure, of such a material in industrial moulding conditions.

Reproduced, with permission, from BS 1755:Part 1:1982.

See also *moulding index*.

CUP FLOW TIME

See under *moulding index*.

CURE (noun)

The process or effect of curing.

CURE (verb) (a polymer)

To convert a prepolymeric or polymeric composition into a more stable, usable condition by polymerisation and/or crosslinking.

NOTE - For example, the cure of a bifunctional urethane system takes place by polyaddition, that of a rubber system by crosslinking, and that of a phenol-formaldehyde system by both polycondensation and crosslinking.

Reproduced, with permission, from ISO 472:1988.

The term is also sometimes applied to the solidification of a liquid system by removal of solvent, or by phase inversion (as e.g. in the fusion of a PVC paste - q.v.). Depending on the nature of the system concerned, curing may be effected by heating[1.7], irradiation[1.7], or the action of reactants[1.7] incorporated in the system.

CURING AGENT ("CURATIVE")

A chemical employed to bring about the curing of a prepolymeric or polymeric composition. The function of a curing agent may include regulation as well as promotion of the curing process[1.1,1.5]. See also *cure*.

CYCLO-OLEFINS

Term applied to thermoplastic olefin copolymers, usually based on ethylene with typically 25% to 50% of a cyclic olefin comonomer. Cyclo-olefins are produced with the aid of metallocene catalysts. Commercial examples include *Zeonex*, *Arton*, *APO* and *Apel*.

DECORATIVE LAMINATE

See under *high-pressure decorative laminate*.

DEFLECTION TEMPERATURE UNDER LOAD (DTL)(T_f)

The temperature at which, in specified test conditions of three-point loading, a standard test specimen is deflected by a specified distance under a prescribed applied load[1.5].

The relevant general test-method standards are ISO 75-1:1993 (identical with BS 2782:Part 1:Method 121:1994) and ASTM D 648-95. The latter standard refers specifically to applied loads required to produce in the test specimen maximum fibre stress of either 455 kPa (66 psi) or 1820 kPa (264 psi). Very closely similar maximum fibre-stress values (respectively 0.45 MPa and 1.80 MPa) are specified - with an additional, highest stress value of 8.00 MPa - by the ISO/BS test standard for the DTL of plastics and ebonite (ISO 75-2:1993, identical with BS 2782:Part 1: Methods 121 A to C:1994). An ISO/BS standard applicable to the determination of DTL of high-strength laminates and long-fibre reinforced plastics (ISO 75-3:1993, identical with BS 2782:Part 1:Method 121D:1994) calls for the applied load to be such as to generate in the test specimen a bending stress equal to one-tenth of the requirement for the flexural strength of the material as laid down in a relevant material standard or as determined in accordance with an appropriate flexural-strength test standard.

Data from DTL tests are intended to be used as aids in material development and control. They are not directly applicable in product design, or as a basis for predicting behaviour in service, except - to a limited extent - in situations where the service conditions are sufficiently closely similar to those of the test.

DEGRADABLE PLASTIC

The standard definitions[1.5,1.7] effectively exclude plastics naturally susceptible to degradation, in that they are confined to a plastic *so designed* that particular environmental conditions can bring about an appreciable change in its chemical structure, with consequent change (loss[1.5,1.7]) in some properties. The two relevant terminology standards[1.5,1.7] note further that, as determined by applicable standard test methods, the change in (loss of) the properties may vary, and that the period of time concerned is significant in this connection. See also *biodegradable plastic, photodegradable plastic, hydrolytically degradable plastic* and *oxidatively degradable plastic*.

DEGRADATION

Of a polymer or plastic: an irreversible detrimental alteration of the chemical structure[1.5,1.7] (e.g. polymer chain scission; oxidative breakdown), typically associated with permanent changes in physical properties[1.5,1.7] (e.g. loss of strength and extensibility; embittlement) and, in many cases, with changes in appearance[1.7] (e.g. discolouration; darkening; cracking; blistering).

DEGREE OF POLYMERISATION

(1) The (average) number of base units per molecule if the molecules are composed of regularly repeating units.

(2) The (average) number of mers (real or hypothetical) per molecule if the molecules have been (or could hypothetically be) produced by polymerisation from identical monomers.

NOTE - The two definitions are not necessarily equivalent; for example, for polyethylene the base unit is CH_2; the mer is C_2H_4.

Reproduced, with permission, from ISO 472:1988.

DENDRIMER

A polymer whose molecules are not linear but highly branched in a radial fractal pattern, so that their overall shape is substantially spherical. Other, relatively simple molecules may be physically accommodated in the spaces among the branches: this can provide a basis for the use of a dendrimer as a controlled-release agent, e.g. for biomolecules in diagnostic devices. Early examples of dendrimers include a hydrocarbon one ($C_{1134}H_{1146}$) based on phenylacetylene as the monomer, and the first dendrimer to be suitable for bulk manufacture (produced by the Dutch firm DSM). In the production of the latter, the amino groups of diaminobutane are reacted with acrylonitrile, and the terminal nitrile groups of the resulting four branches are reduced to amino groups which provide eight new reactive sites. The reaction with acrylonitrile and the reduction step are carried out repeatedly to multiply and "grow" the branches of the dendrimer molecule.

DENSITY

In the general sense, this is the mass (weight[1.7]) of a unit volume of a material. In the context of polymers and plastics the term itself, as well as any numerical density value quoted, normally relate to density determined under particular, defined conditions (such as those of a standard test). The usual symbol for density is ρ, and the SI unit is $kg.m^{-3}$; other units in frequent use are $g.cm^{-3}$, $g.ml^{-3}$, $lb.in^{-3}$, and $lb.ft^{-3}$. For practical purposes, density may be regarded as numerically equal to *relative density* (sometimes also called *specific gravity*), which is the ratio of the density of a material to that of water at the same temperature. See also *apparent density*, *bulk density* and *powder density*. Methods of determination of density of plastics materials are given in the following standards. ISO 1183:1987 (identical with BS 2782:Part 6:Methods 620A to 620D) - the methods given in these standards are for density determinations of non-cellular plastics by immersion, in a pyknometer, by titration, and in a density gradient column. BS 2782:Part 6:Method 620E:1980 is for pyknometric determination of the density of liquid resins. ASTM D 792-91 (liquid-displacement methods). ASTM D 1505-85(1990) (density-gradient column). The density of plastics materials is discussed in some depth in BS 4618:Section 5.1:1970.

DIBUTYLTIN DILAURATE (DBTDL)

An organotin compound, $[CH_3(CH_2)_{10}COO]_2.Sn.(C_4H_9)_2$, used as a catalyst, or part of a catalyst system, in formulations for polyolefin (in particular polyethylene) materials crosslinked with vinyltrimethoxy silane. The materials are made into such products as cable coverings and heat-shrink tubing. Other applications of DBTDL, are in formulations for certain kinds of polyurethane foams (typically flexible foams based on polyether polyols), and as a heat stabiliser for PVC.

DIE SWELL

The passage of a polymer melt through an extrusion die normally involves constriction of the flow, because the cross-sectional area of the die is less than that of the melt-filled space behind it. In an extruder, that space is the channel formed by the screw flights and externally bounded by the inner surface of the barrel. The flow constriction imposes compression and shear upon the melt. As a result, and because of its viscoelastic nature, the melt undergoes some elastic extension in the die. If this longitudinal deformation is not (or not fully) relieved by relaxation inside the die, then elastic recovery takes place along the flow direction when the melt emerges from the die: the net result of this "elastic memory" effect is a degree of lateral expansion of the emerging extrudate, known as *die swell.*

DIELECTRIC

A substance or medium (including a vacuum) in which an electric field can be maintained with little supply of energy from an external source. All insulating plastics materials are dielectrics.

DIELECTRIC CONSTANT

See *relative permittivity.*

DIELECTRIC STRENGTH (ELECTRIC STRENGTH)

In the general sense, the ability of a dielectric (in the plastics context, a plastics insulating material) to resist breakdown by disruptive discharge[1.1,1.5]. Dielectric strength is measured and expressed in terms of the intensity of the electric field[1.1,1.5] (i.e. the potential gradient, the ratio of applied voltage to material thickness) required to produce breakdown[1.1,1.5] under specified test conditions (cf. also ASTM D 1711-94). The units of dielectric strength are volts (or volt multiples) per unit thickness, e.g. $V.m^{-1}$; $kV.m^{-1}$; $V.mm^{-1}$; $V.mil^{-1}$. The following standards are relevant to the concept and determination of dielectric strength. IEC Publication 50 (International Electrotechnical Vocabulary - Group 05 : Fundamental Definitions). BS 2782:Part 2:Methods 220 and 221:1983; ASTM D 149-94.

DIFFERENTIAL SCANNING CALORIMETRY (DSC)

A procedure (technique[1.1,1.5,1.15]) whereby the difference in energy inputs into a specimen of a test substance and one of a reference material is measured[1.5,1.15] (the energy input necessary to establish zero temperature difference between the two specimens is recorded[1.1]) as a function of temperature[1.1,1.5,1.15] or time[1.1], while the specimens are being subjected to identical temperature regimes in an environment heated or cooled at a controlled rate[1.1] governed by an appropriate temperature programme[1.5,1.15]. The test results are plotted as the DSC curve, in which the ordinate values represent the amounts of heat applied per unit time, with the corresponding temperature (T) or time (t) values as the abscissae[1.1]. The ISO terminology standard[1.5], and a relevant ASTM standard[1.15], point out that a DSC determination can be run in one of two modes, viz. *power compensation DSC* and *heat-flux DSC*.

Differential scanning calorimetry is an alternative to differential thermal analysis (DTA) for determinations of transition temperatures (T_g and T_m) of polymers (cf. e.g. ASTM D 3418-82(1988), especially where a determination has a quantitative aspect, for example determination of both the T_m and the heat of fusion (and/or crystallisation) - cf. ASTM D 3417-83(1988). Another example is the determination of transition temperature(s) associated with curing of thermosetting resins in composites (cf. ASTM D 5028-90). ASTM D 472-86(1991) provides directions for reporting thermoanalytical data.

DIFERENTIAL THERMAL ANALYSIS (DTA)

A procedure (technique[1.1,1.5,1.15]) for measuring[1.5,1.15] (recording[1.1]) - as a function of temperature[1.1,1.5,1.15] or time[1.1] - the difference in temperature between a specimen of the test substance and one of a reference material, while the two specimens are being subjected to identical temperature regimes in an environment heated or cooled at a controlled rate[1.1] governed by an appropriate temperature programme[1.5,1.15]. The test results are recorded in the form of a plot - the differential thermal analysis (DTA) curve, on which the temperature difference (ΔT) should be plotted on the ordinate (with endothermic reactions downwards), and the temperature (T) or time (t) on the abscissa, increasing from left to right[1.1,1.5].

The BS and ISO terminology standards[1.1,1.5], and a relevant ASTM standard[1.15], additionally recognise the term *quantitative differential thermal analysis (quantitative DTA)* as applicable to determinations with versions of DTA apparatus which are designed to provide quantitative results (in terms of energy or another relevant physical quantity). The BS standard[1.1] points out that, according to the International Confederation for thermal Analysis (ICTA), the term "differential thermal analysis" or its abbreviated version "DTA" are not applicable to differential calorimetry. The ASTM terminology standard[1.7] provides no definition of DTA. Differential thermal analysis is widely used for determining transition temperatures (T_g and T_m) of polymers (cf. e.g. ASTM D 3418-82(1988)). ASTM D 472-86(1991) provides directions for reporting thermoanalytical data.

DIFFUSION

In the most general terms, diffusion is the transport of components of a system (e.g. molecules, ions) from one part of the system to another, in consequence of the random molecular motion actuated by kinetic (thermal) agitation (cf. e.g. Fujita 1961; Meares 1964). In the plastics context, the system most often consists of a polymer (which may contain various additives) and a diffusant (penetrant) of much smaller molecular size, typically a gas, vapour or liquid (when in bulk; the molecules are transported individually once they have entered the polymer). In some cases the transported species may be ions, as e.g. in desalination of water with the aid of polymeric membranes. In polymer/penetrant systems, the rate of transfer of the penetrant across a unit area of a section taken through the system should - in the simplest case - be proportional to the penetrant concentration gradient measured normal to the section; i.e. for flow in one dimension:

$$F = -D(\partial C/\partial x) \tag{1}$$

where x is distance in the flow direction, F is the rate of transfer per unit area, C is the concentration, and D is the appropriate diffusion coefficient (sometimes also called *diffusivity*). This relationship is known as Fick's law of diffusion. If both F and C are expressed in the same unit of quantity, D is independent of the unit and its dimensions are $length^2\ time^{-1}$. For the same case of one-dimensional diffusive flow of a penetrant in a material (which is assumed to be homogeneous and isotropic), continuity considerations give:

$$\partial C/\partial t = \partial/\partial x[D(\partial C/\partial x)] \tag{2}$$

where t is the time: and hence, for a constant D,

$$\partial C/\partial t = D(\partial^2 C/\partial x^2) \tag{3}$$

Equation (3) is often called Fick's second law of diffusion.

In simple terms, the higher the diffusion coefficient, the more rapid the diffusion. In polymer/penetrant systems in which the penetrant is a swelling agent for the polymer D is rarely a constant, but normally varies with concentration. For such systems, or for non-homogeneous ones in which D may vary from point to point, the applicable partial differential diffusion equations are of the second order and second degree (i.e. non-linear), and as such are not soluble implicitly even when simple specific functions are assumed for the concentration-dependent D. However, mathematical treatments applicable to polymer/penetrant systems are available: of necessity they are phenomenological, but in general relatable to the concepts embodied in Fick's laws (Crank 1967; Crank and Park 1968; Fujita 1961; Meares 1964).

In any polymer/penetrant system, the diffusion coefficient is characteristic of the system as a whole, in the sense that it is influenced by both the penetrant and the polymer. It is also affected by temperature, a factor external to the system, because temperature influences both the movement of penetrant molecules and the random motions of the polymer-chain segments, which are jointly responsible for the diffusion process. The general way in which various factors can affect the nature of the diffusion coefficient is to make it non-constant (and variable in particular ways) and/or to alter its value. Apart from the temperature of the system, the significant factors are penetrant concentration, chemical structures of the polymer and penetrant (with special reference to the functional groups and polarity), molecular shape and configuration of the penetrant and its molecular weight and size, and morphology of the polymer (for example, in a semi-crystalline polymer only the non-crystalline regions are normally accessible to the penetrant).

DILATOMETRY

See *thermodilatometry.*

DIMER

An oligomer (q.v.) molecule consisting of two monomeric units of the same species[1.1,1.5], or the substance composed of such molecules[1.1]. The ISO standard definition[1.5] additionally points out that a dimer may result either from appropriate oligomerisation (i.e. dimerisation), or from scission of a longer molecular chain.

NOTE - The definition of a dimer molecule in the BS terminology standard[1.1] appears to extend to the case where the two monomer units are of different species.

DISPERSION

In the most general sense, a two-part system comprising a material in a state of fine division (the disperse phase) distributed, typically in a uniform manner, through the continuous matrix of another material (dispersion phase; dispersion medium). The term "polymer dispersion" is most commonly applied to dispersions in which the disperse phase is a polymeric material, and the dispersion medium is water - cf. e.g. ISO/DIS 12000 (a proposed nomenclature and test method standard still under discussion in 1996). Stable colloidal aqueous dispersions of rubbers[1.2,1.3] (natural or synthetic) and of certain other polymers (such as e.g. polystyrene; some styrene copolymers; some polymers and copolymers of vinyl chloride) are commonly known as *latices* (singular "latex"; alternative plural "latexes"). Standards dealing with various aspects of polymer dispersions and latices include specifications for aqueous dispersions of PTFE (ISO/DIS 12086 - still under discussion in 1996; BS 7485:1991; ASTM D 4441-91); test standards for various properties of aqueous dispersions of polymers and copolymers (ISO 1147, 1148, 1625, 2115, 3219, 3499); standards relating to the properties of natural and synthetic rubber latices, such as BS 6057 (a multi-part standard for rubber latices), ASTM D 1076 (natural rubber latex), ASTM D 1417 (standard test methods for synthetic-rubber latices); and rubber and latex letter-symbol standards[2.4,2.5].

DOUBLE-SKIN SHEET

This is the term favoured by ISO (cf. e.g. ISO 12017). The product is also sometimes known as double-walled sheet, twin-walled sheet, or double-web panel (German "Stegdoppelplatte"). It is extruded plastics sheet consisting of two mutually parallel layers ("skins") connected by internal ribs which define the inter-layer spacing and the size and configuration of the internal channels formed by the ribs between the skins. Double-skin sheets are produced from a number of thermoplastics, viz. polycarbonate, poly(methyl methacrylate), ABS, blends of ABS with polycarbonate, cellulose acetate butyrate, and polypropylene.

DOUGH MOULDING COMPOUND (DMC)

A bulk moulding compound (q.v.) in which the requisite high viscosity is imparted not by chemical thickeners but by the appropriately high content of a suitable filler (cf. ISO 8604:1988).

DRAW RATIO

An expression of the amount of stretching undergone by a plastics material in a stretching or drawing process[1.1,1.5] (e.g. uniaxial or biaxial stretching of film, or vacuum forming of sheet):

$$DR = A/a$$

where DR is the draw ratio, and A and a are, respectively the cross-sectional areas of the material before and after stretching.

DRAW-DOWN RATIO

The ratio of the width of the extruder die opening to the corresponding ultimate thickness dimension of the extrudate[1.1,1.5].

DRY BLEND

See under *powder blend*.

DYNAMIC MODULUS

A synonym for complex modulus (q.v.), deprecated by the ISO terminology standard[1.5] (and not featured in the other principal English-language plastics terminology standards).

DYNAMIC STRAIN

A strain (with elastomers normally in the tensile mode), sinusoidal in character, varying with time at a defined frequency or rate of repetition.

DYNAMIC THERMOMECHANICAL[1.5] (THERMAL MECHANICAL[1.15]) MEASUREMENT (DTMM)

A technique in which a material is subjected, in a suitable mode, to an oscillatory load under an appropriately controlled temperature programme, and the dynamic modulus and/or damping is determined as a function of temperature[1.5,1.15]. Standards relevant to DTMM include the following: ASTM E 1142-93b (terminology relating to thermophysical properties): ASTM D 4092-90 (terminology relating to dynamic mechanical measurements on plastics); ASTM D 5023-94, D 5024-94 and D 5026-94 (measurement of dynamic mechanical properties of plastics in, respectively, three-point bending, compression, and tension); ASTM D 4065-94 (determination and reporting of dynamic mechanical properties of plastics).

DYNAMIC THERMOMECHANOMETRY

Term deprecated by the relevant ASTM terminology standard in favour of "dynamic thermal mechanical measurement" (q.v.).

DYNAMIC VULCANISATE (DYNAMICALLY VULCANISED ALLOY)

An essentially thermoplastic elastomer produced by blending a thermoplastic polymer and a vulcanisable elastomer in such conditions that the elastomer is vulcanised in the course of the processing, and the ultimately resulting alloy is a microdispersion of the vulcanisate in a continuous matrix of the thermoplastic polymer. Cf. polypropylene/EPDM alloys under *thermoplastic elastomers*.

EFFECTIVITY QUANTITY (PLASTICISERS)

See under *efficiency factor*.

EFFICIENCY CONCENTRATION

See under *efficiency factor*.

EFFICIENCY FACTOR (PLASTICISERS)

In the context of formulating PVC compositions, the efficiency factor is the ratio of the *efficiency concentration* of a given plasticiser to that of dioctyl phthalate (DOP). The efficiency concentration is the amount of plasticiser (expressed in phr, i.e. parts by weight per hundred parts of PVC resin) that in a PVC formulation will give a particular level of a chosen property: the properties most often used in determining efficiency concentrations are the "hundred percent modulus" (q.v.), hardness (q.v.) or BS softness (q.v.), or tensile strength and elongation. The more effective the plasticiser, the lower the efficiency concentration and hence also its efficiency factor.

As a general example, if a certain value of the 100% modulus (say 6.9 MPa) may be imparted to a PVC composition by plasticising *either* with 63.5 phr DOP *or* with 58.8 phr DOS (dioctyl sebacate), then these effective efficiency concentrations give 58.8/63.5 = 0.926 as the efficiency factor for DOS (in relation to the 100% modulus). The relevant efficiency factors are used to calculate the correct amount of one plasticiser to replace another when modifying a PVC composition for particular properties. Thus, for example, it may be desirable to replace 20 phr of the DNP (dinonyl phthalate) in a DNP-plasticised PVC composition with a plasticiser imparting good low-temperature properties (such as DOS), whilst keeping the composition's 100%

modulus at the same value. Then, given the efficiency factors for DOS and DNP as, respectively, 0.926 and 1.17, the amount of DOS required will be: 20x(0.926/1.17) = 15.8 phr.

ELASTIC LIMIT

The greatest stress that a material is capable of sustaining without any permanent strain remaining upon complete release of stress[1.1,1.3,1.5,1.10]. In practical determinations of the elastic limit, measurements of strain taken with a small load rather than zero load are normally used as the initial and final reference[1.1,1.5,1.10]. Elastic limit should not be confused with *proportional limit* (q.v.).

ELASTOMER

A macromolecular material that, at room temperature, returns rapidly to its initial dimensions and shape after substantial deformation by a weak stress and release of the stress[1.1,1.5,1.7]. The quality manifested by such behaviour is sometimes referred to as "nerve" (q.v.) and is a characteristic of many rubbers.

ELECTROMAGNETIC (ELECTRIC AND/OR MAGNETIC) SHIELDING (EMS)

Presence (or the action) of a conductive, protective encasement around an item of electrical or electronic equipment or some of its components (or around a specimen in a test), installed to exclude or reduce electric and/or magnetic flux within or beyond the equipment (or the test specimen), or one or more of their elements. EMS is ofter provided by conductive plastics casings. (Cf. also ASTM D 1711-91)

ELECTROMAGNETIC INTERFERENCE (EMI)

Interference, generated by electronic and electrical equipment. Conductive plastics are used as materials of housings required to shield equipment that can produce EMI or is affected by it (e.g. computers, television sets, office machines, etc.). Whilst metal housings provide excellent shielding, conductive plastics offer lower part weight and greater facility of production. Very high conductivities (volume resistivity less than one ohm.cm) have become achievable at low loading levels of some conductive metal fillers (e.g. about 1% of stainless steel fibres in the *Faradex* range of conductive polymers produced by DSM) - cf. volume resistivities of about 100 ohm.cm and 10^3-10^6 ohm.cm, fairly typical, respectively for some carbon-fibre-filled compounds and conventional compounds with carbon black fillers. See also *EMS*.

ENANTIOMERIC CONFIGURATIONAL UNIT

Either one of two configurational units that are mirror images at the plane containing the main-chain bonds.

Reproduced, with permission, from ISO 472:1988.

ENGINEERING STRESS

See under *stress*.

ENVIRONMENTAL STRESS CRACKING AND CRAZING (ESC)

Stress cracking (q.v.) or crazing (q.v.) of a polymer or plastic under the joint action of stress and a substance in this context termed "active environment" or "stress-cracking agent". The agent, which in practical situations will typically be a fluid (i.e. a gas, vapour or liquid) or semi-liquid (e.g. a grease), plays a crucial role, in that in its presence the susceptible polymeric material will crack or craze under stress much lower (in some cases by a factor of about 100) than the stress that would normally be safely withstood at the same temperature in air. Whether or not a particular substance will act as a stress cracking agent for a particular polymer depends primarily on their respective chemical natures. For example, many normally glassy polymers (e.g. polystyrene, polycarbonate, poly(methyl methacrylate)) are susceptible in this sense to various organic liquids and their vapours, whilst environmental stress-cracking agents for polyethylene include solutions of surface active agents. Environmental stress cracking and crazing are significant and not infrequent modes of failure of plastics products. Useful reviews of various aspects of ESC of plastics have been published by Kambour (1973), Haward (1973), and Deanin and Hauser (1981).

Several standard ESC tests are available, directed to one or the other of two main objectives, viz. assessment of susceptibility of a polymer or plastic to ESC, or detection of significant levels of internal stress in moulded and extruded plastics products. The following standards are noteworthy.

Standard tests for ESC effects in plastics:

ISO 4599:1986 (determination of resistance to ESC - bent strip method); ISO 4600:1992 (determination of resistance to ESC - ball or pin impression method); ISO 6252:1992 (determination of resistance to ESC under tensile stress). These three ISO test standards are identical, respectively, with the following test methods of BS 2782:Part 8 - Method 832A:1991; Methods 831A and B:1993; and Methods 833A to C:1993. ASTM D 1693-95 (test for ESC of ethylene plastics); ASTM F 484-83 (test for ESC of acrylic plastics); ASTM F 791-82(1988) (determination of critical crazing stress of transparent plastics); MIL-C-87936A:1985 (test for resistance of acrylic plastics to cleaning compounds).

Standard ESC tests for the presence of internal stresses in plastics products:

BS 2782:Part 2:Method 1109A:1989 (resistance to ESC of polyethylene pipes and fittings); BS 5255, Appendix B (ESC test for assembled fittings made from ABS, PVC-U, PE or PP); BS 7029:Part 2:1993, Annex A (ESC test for polycarbonate mouldings); ASTM D 1939-94 (ESC test for extruded or moulded ABS parts); ASTM D 2561-91 (ESC test for blow-moulded polyethylene containers); ASTM D 5419-93 (ESC test for threaded plastics closures); ASTM F 1248-89 (ESC test for polyethylene pipes).

EPDM

Initials for the words "ethylene, propylene, diene monomer", designating a rubbery terpolymer of ethylene, propylene and a diene. If the diene comonomer is linear, the double bond of its residues in the polymer molecule is located in the brach chain, not the main-chain backbone[2.4,2.5,2.6] (cf. (I) below). If the diene is cyclic or cyclo-aliphatic, the unsaturation is in, or on, the part of the ring not linked into the main chain (cf.(II) and (III) below).

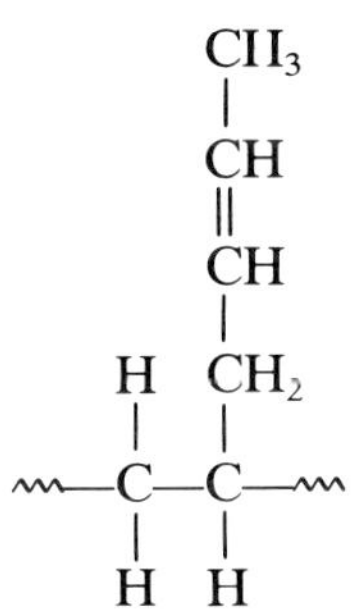

(I)

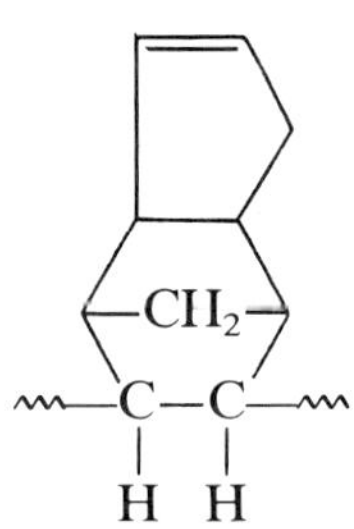

(II)

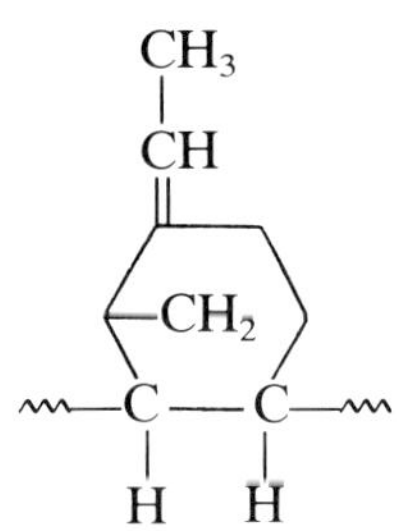

(III)

Absence of unsaturation from the main molecular chain is largely responsible for EPDM'S good resistance to chemicals, ozone, heat and weathering, whilst the side-chain unsaturation provides sites for vulcanisation with sulfur systems.

The commercial versions of EPDM rubbers vary in the nature and/or content of the diene comonomer residues, the content, which is generally low, ranging between about 2% and 10%. The dienes most widely used are hexa-1,4-diene (giving rise to residue (I) - see above), dicyclopentadiene (residue (II), and 4-ethylidenenorbornene-2 (residue (III). Apart from the identity and proportion of the diene residues, the other principal parameters governing the material and processing properties of EPDM rubbers are the ethylene/propylene ratio and the molecular weight.

The major standards relevant to EPDM rubbers are ISO 4097 (evaluation procedures); ASTM D 3568 (test methods); and ASTM D 3900 (determination of ethylene units in EPDM).

EPOXY PLASTICS

See under *epoxy resins.*

EPOXY (EP) RESINS

A class of resins, also known as *epoxide resins*, crosslinkable[1.1,1.5] by virtue of the presence in their molecules of epoxy groups[1.1,1.5] (—CH—CH—, with O bridging the two CH). In their initial state - before crosslinking or molecular-chain build-up - the resins are essentially short-chain, liquid prepolymers, with the epoxy groups (more than one per molecule) typically located at the chain ends, although in some cases they may also be present elsewhere in the chain. The simplest representative of common commercial liquid epoxy resins is the diglycidyl ether of bisphenol-A:

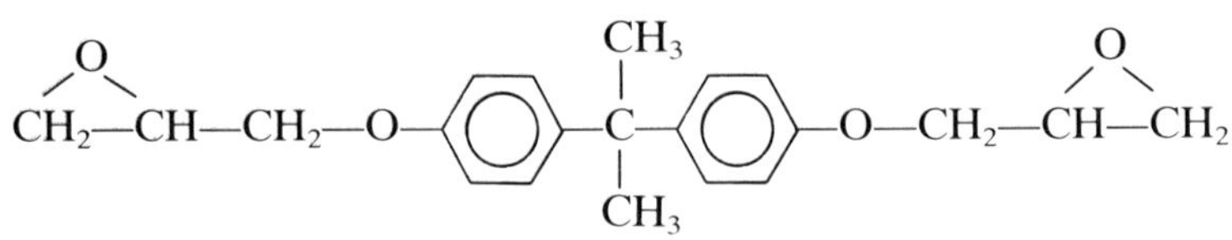

Longer-chain glycidyl ether epoxy resins result from chain extension of this compound via reactions of the terminal epoxy groups, in which, inter alia, hydroxyl groups are formed along the extended chain. These groups can later participate in crosslinking reactions. Linear-chain epoxy resins of sufficiently high molecular weight are thermoplastic solids with measurable melting temperatures; some are available in liquid form as solutions in solvents.

Crosslinking of epoxy resins via their reactive epoxy groups (and any hydroxyl groups that may be involved) is effected with the aid of curing agents (hardeners) which may be catalytic, or reactive, or both in some cases. The resulting thermoset products are also often referred to as epoxy resins, albeit the original epoxy groups will have been used up in crosslinking. *Epoxy plastics* is another term sometimes used either for the thermosets, or generally for plastics based on epoxy resins[1.1,1.5].

NOTE - The formal definition of epoxy plastics in the ASTM terminology standard[1.7] would appear to be at some variance with this usage, in that it refers to these materials as "thermoplastic or thermosetting plastics". On a strict interpretation, this description cannot be regarded as including the *thermoset* forms (cf. the definitions in this book of *thermosetting plastic* and *thermoset plastic*).

Thermosetting epoxy resins are particularly important as the basic materials of epoxy plastics widely used in electrical and electronic applications (e.g. for printed-circuit boards, as potting compounds, and in cable jointing), in special-purpose composites (e.g. carbon-fibre/epoxy composites have been used in the satellites of the NAVASTAR navigational guidance system, and in components - including cargo-bay doors - of the space shuttle), and in thermoset mouldings (especially ones in which low shrinkage is an important requirement). Other large application areas are adhesives and surface coatings.

The commercial ranges of epoxy resins and resin compounds include the following (see also Section 1 of Part 3 of this book; Section 3 of Part 3 for trade names of some catalysts and curing agents for epoxy resins, and Section 4 of Part 2 for some abbreviated terms for such hardeners).

Araldite; Atlac; Basoset; Beckopox; Cosmic; Devcon; Dobeckot; Epic; Epocast; Eponite; Eposand; Farboset; Fibredux; Grilonit; Hysol; Lecutherm; Lytex (SMC); Paraplex; Polyset; Premi-Glas (SMC); Quatrex; Ravepox; Scotchcast; Stycast; Trolitax.

The following standards deal with epoxy materials, or tests for their various properties.

Materials specifications

ISO 3673-1:1980	Epoxide resins - Designation.
ASTM D 1763-94	Standard specification for epoxy resins.
ASTM D 3013-94	Standard specification for epoxy moulding compounds.
ASTM D 15-94	Provisional standard specification for thermosetting moulding compounds (including epoxy compounds inter alia).

Properties and tests

ISO 4895:1987	Liquid epoxide resins - Determination of tendency to crystallise.
ISO 3001:1978	Determination of epoxide equivalent.
ISO 4573:1978	Epoxide resins - Determination of organic chlorine.
ISO 4583:1978	Epoxide resins - Determination of easily saponifiable chlorine.
ISO 4615:1979	Unsaturated polyester and epoxide resins - Determination of total chlorine content.
ISO 3521:1976	Unsaturated polyester and epoxide resins - Determination of total volume shrinkage.
ISO 7142:1984	Epoxy resins as binders for paints and varnishes - Methods of test.
BS 2782:Part 8:Method 835C:1980 (1986)	Determination of gelation time of unsaturated polyester and epoxide resins with a gel timer.

Curing agents

ISO 4597-1:1983	Hardeners and accelerators for epoxy resins.
ISO 7327:1994	Determination of free acid in acid-anhydride hardeners and accelerators.

EUROPEAN ORGANIZATION FOR TESTING AND CERTIFICATION (EOTC)

Organisation set up by the European Community to serve as the focal point for all issues relating to conformity assessment of products and services. EOTC operates through Sectoral Committees, responsible for all aspects of management and finance, and through Agreement Groups formed by testing laboratories and certification bodies on the basis of mutual recognition.

ETHYL CELLULOSE (EC)

An ethyl ether of cellulose[1.1,1.5] (cf. also ASTM D 1695-77(1989)). ASTM D 787-81(1988) provides a specification for EC-based moulding and extrusion compounds.

ETHYLENE/ACRYLIC ACID (EAA)

Copolymer of ethylene and acrylic acid $\left(\begin{array}{l} CH_2{=}CH \\ \quad\quad | \\ \quad O{=}C{-}OH \end{array}\right)$.

A range of these copolymers is produced, containing various proportions (up to about 18%) of acrylic acid, for applications utilising the combination they offer of toughness and strength with excellent adhesion to other polymers (including polyethylene) and to metals. The applications include heat-sealing layers in film laminates, bonding layers in polyethylene/aluminium-foil laminates for sachets, pouches, and the bags of bag-in-box beverage packages, and skin packaging, inter alia for medical instruments and accessories. Commercial materials are available in the *Escorene Ultra* (Exxon Chemical) and *Primacor* (Dow) ranges.

Acrylic acid can also be used as a comonomer with ethylene in the production of ionomers (q.v.).

ETHYLENE/BUTYL ACRYLATE (E/BA)

A copolymer of ethylene and n-butyl acrylate $\left(\begin{array}{l} CH_2{=}CH \\ \quad\quad | \\ \quad O{=}C{-}O{-}CH_2.CH_2.CH_2.CH_3 \end{array}\right)$

A number of grades is produced with various comonomer contents. The copolymers are similar to ethylene/ethyl acrylate (q.v.), especially in their melt-adhesion properties. They are used as the material of tie layers in coextrusion, and in heat-sealable coatings. Moulding grades are also available. Commercial E/BA materials include *Ebac*, *Enathene*, and the *NCPE 6400* series of Neste Polymers.

ETHYLENE-CHLOROTRIFLUOROETHYLENE (E/CTFE) COPOLYMER

As in ETFE, the comonomer units in the molecular chain of E/CTFE alternate in almost completely regular 1:1 arrangement. The commercial version (*Halar 900* series) is semi- crystalline (T_m, by DTA, 245°C), with density between 1.68 g cm^{-3} and 1.70 g cm^{-3}. It is processable as a thermoplastic (at melt temperatures between 260°C and 300°C), mostly into such products as cable insulation, chemical-resistant linings and containers, films, and fibres used in flame-resistant fabrics (the material is non-flammable in terms of small-scale burning tests). The service temperature range of E/CTFE products is usually given as –80°C to 140°C (up to about 170°C on short-term exposure without stress). The electrical properties are generally good, but inferior to those of PTFE (the volume and surface resistivities in particular are considerably lower). The all-round chemical resistance of E/CTFE is good, but the copolymer is susceptible to attack by some amines at elevated temperatures, and to swelling by chlorinated hydrocarbons. E/CTFE is among the fluropolymers covered by the proposed ISO standard for these materials (ISO/DIS 12086, still under discussion in 1996). Moulding, extrusion, and coating E/CTFE compounds are the subject of ASTM D 3275-91. Such finished products as sheets and braided expandable sleeving are featured in the combined British/European standard BS EN 60684-3-343 to 345. Along with many other fluoropolymers, E/CTFE is featured in the ASTM standard (D 4591-93) which deals with the determination of their heats and temperatures of transitions.

ETHYLENE/ETHYL ACRYLATE (E/EA[2.1,2.3]; EEA[2.2]; EEAK[2.11])

A copolymer of ethylene and ethyl acrylate 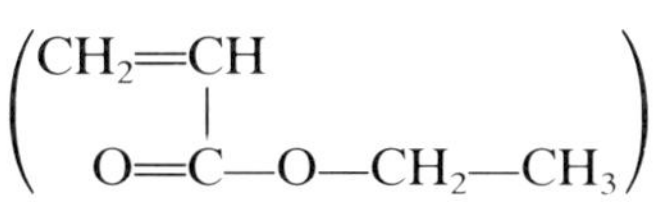

Copolymers are produced with various ethyl acrylate (EA) contents. They resemble ethylene/vinyl acetate (EVA) copolymers and, like the latter, range - with increasing proportion of EA - from materials similar to low-density polyethylene (but somewhat tougher and more resilient) to rubbery polymers. On the whole, they have the edge over EVA in resistance to heat, environmental stress cracking, and wear, but - grade for grade - are less transparent and slightly inferior in mechanical properties. They are fairly readily flammable, require stabilisation against UV radiation, and are susceptible to a number of organic solvents (including esters, ketones, chlorinated hydrocarbons and aromatic hydrocarbons), whilst being essentially unaffected by flats, oils, dilute acids and alkalis.

Some applications of E/EA copolymers are also similar to those of EVA (including the use of some grades in or as melt adhesives). Other applications include antistatic components and films. Commercial E/EA materials are available in the *DPD* and *DPDA* ranges marketed by Union Carbide. They were originally also produced by Dow under the trade name *Zetafin*. ASTM D 3594-93 provides a test method for the determination of the EA component in E/EA copolymers (this is relevant inter alia in connection with the FDA requirement that E/EA materials, such as films, for food-contact applications should contain not more than 8% EA).

ETHYLENE/METHYL ACRYLATE ELASTOMERS

These are terpolymers of ethylene, methyl acrylate

$$\left(\begin{array}{l} CH_2{=}CH \\ \qquad | \\ \quad O{=}C{-}O{-}CH_3 \end{array}\right)$$

and a relatively small proportion of a comonomer containing carboxyl groups which act as cure sites. The materials are classified as acrylic rubbers of the standard ACM group[2.4,2.5,2.6]. A commercial example is *Vamac* (Du Pont).

ETHYLENE/METHACRYLIC ACID (E/MA[2.1,2.3]; EMA[2.2,2.11])

Copolymer of ethylene and methacrylic acid

$$\left(\begin{array}{l} CH_2{=}C{-}CH_3 \\ \qquad | \\ \quad O{=}C{-}OH \end{array}\right).$$

A range of copolymers is produced (with up to about 20% of the methacrylic acid comonomer) These materials are characterised by particularly good heat stability, resistance to environmental stress cracking, heat-sealability, and adhesion properties only slightly lower than those of ethylene/acrylic acid copolymers. Applications are in extrusion coating, as tie layers in coextruded multi-layer films, films and sheeting for packaging, injection and blow mouldings (inter alia squeeze bottles and toys), extruded tubing, and as impact modifier in polyethylene. Commercial grades for all these applications are available in the *Nucrel* (Du Pont) and *Escorene Ultra* (Exxon Chemical) ranges.

ETHYLENE PLASTICS

Plastics whose base polymers are homopolymers of ethylene, or copolymers in which ethylene is the major monomeric component, predominant by mass[1.1,1.5,1.7].

NOTE - The ISO and BS terminology standards[1.1,1.5] include "ethene" as the alternative term for "ethylene" (in line with the IUPAC recommended terminology).

The following standards deal with ethylene plastics as such. See also under *polyethylene, polyethylene plastics, polyolefin, polyolefin plastics*, and individual names of relevant ethylene copolymers.

ISO 1872-2:1989	Polyethylene and ethylene copolymer thermoplastics - Preparation of test specimens and determination of properties.
ASTM D 1601-86(1991)	Determination of dilute-solution viscosity of ethylene polymers (refers to the standard test practice given in ASTM D 2857).
ASTM D 2647-94	Specification for crosslinkable ethylene plastics.

ASTM D 2765-90	Gel content and swell ratio of crosslinked ethylene plastics.
ASTM D 2952-81(1987)	Specification for ethylene plastics.

ETHYLENE-TETRAFLUROETHYLENE (ETFE) COPLYMER

Coplymer of ethylene and tetrafluoroethylene, with almost entirely 1;1 alternating structure and the resulting tetrafluoroethylene content of about 76% by weight. Commercial versions are semi-crystalline (T_m about 270°C) and of density 1.7 g cm^{-3}.

The ethylene content makes ETFE processable as a thermoplastic (at melt temperatures between 300°C and 350°C). The products (mouldings, weather-resistant sheeting, electrical insulation) have a service-temperature range of –80°C to 160°C (up to about 190°C on short term exposure without stress). The electrical properties of ETFE are very good (albeit slightly inferior to those of PTFE), and its flame resistance is comparable with that of PTFE. The copolymer has a higher modulus and impact resistance than PTFE, but somewhat lower chemical resistance, in that it is not immune from attack by concentrated oxidising acids, and can be swollen by some oils and greases. Its weathering resistance is good, as are its wear characteristics coupled with low friction (neither impaired by glass-fibre reinforcement). ETFE is included in the forthcoming ISO standard for fluoropolymer materials (ISO/DIS 12086 - still under discussion in 1996), and - inter alia - in ASTM D 4591-93, a test - method standard for determination (by DSC) of the temperatures and heats of transition of fluoropolymers. ASTM D 3159-91 provides a standard specification for ETFE moulding and extrusion materials; such materials are available under the trade names *Halon ET, Hostaflon ET, Neaflon EP,* and *Tefzel* (the earliest commercial version of ETFE was brought out by Du Pont in the early 1970's under this name).

ETHYLENE/VINYL ACETATE (E/VAC[2.1,2.3], EVA[2.2]) COPOLYMER

The commercial versions of EVA are random copolymers, the various grades available representing a very wide range of vinyl acetate (VA) contents (from a few percent to about 95 percent, by weight). As would be expected, with increasing VA content the properties of the copolymer become increasingly different from those of polyethylene. In comparison with PE-LD, there is a progressive rise in density, toughness (both at room temperature and below), transparency, gloss (in films), fatigue resistance, flexibility, softness, solubility in solvents (above about 35% vinyl acetate), resistance to environmental stress cracking and weathering, surface energy (with consequent improvement in adhesion properties), and heat-sealability. The main properties which are reduced as the VA content increases are modulus, softening temperature, hardness, chemical resistance, and resistance to creep. The properties associated with different VA contents determine the applications of various grades of EVA copolymers. At up to about 10% VA, the properties are essentially those of PE-LD with improved flexibility, transparency and heat-sealability. The consequent main applications of the material are in packaging films (including shrink-wrap, stretch-wrap, and cling films), sacks, and the like. Copolymers with about 15% to 30% VA find outlets in flexible mouldings (e.g. gasketing, beading) and cable coverings. At about 30% to 40% VA contents the copolymers are used principally in adhesives (hot-melt, and solvent-based) and coatings. Above about 40% VA the copolymers are rubbery, curable with peroxides. Applications include many of those in which rubbers are used (including footwear soling), and as impact modifiers for PVC (EVA copolymers with VA contents between about 70% and 95% are used chiefly in emulsion paints and adhesives, e.g. for tiles). Various grades of EVA copolymers are available commercially in the following product ranges: *Alcudia, DQDA, Elvaloy, Elvax, Escorene, Evatane, Levapren, Lupolen, NCPE, Petrothene, Ultrathene, Vinnapas.*

The following standards are relevant to EVA polymers and plastics. ISO 4613:1993 (specification, in two parts, for EVA moulding and extrusion materials), ISO 8985:1989 (under revision in 1996) and ASTM D 5594-94 - determination of VA content in EVA copolymers; ASTM D 4275-91 (determination of butylated hydroxytoluene antioxidant content in EVA and PE); ASTM D 5227-92 (hexane-extractable content of polyolefins, including EVA); ASTM E 1090-91 (standard test method for dicumyl peroxide and its decomposition products in resins - applicable, inter alia, to PE and EVA crosslinkable or crosslinked by peroxides).

ETHYLENE/VINYL ALCOHOL (EVOH[2.2,2.11];E/VAL[2.1,2.3]) COPOLYMER

Copolymer of ethylene and vinyl alcohol. As vinyl alcohol cannot be processed directly (see poly(vinyl alcohol)), the copolymer is produced by copolymerisation of ethylene with vinyl acetatate followed by conversion (by alcoholysis - see poly(vinyl alcohol)) of the acetyl groups to hydroxyl groups. In commercial EVOH materials the vinyl alcohol content typically ranges between about 52 and 73 mol%. (For such materials it is the ethylene content that is more usually quoted: on this basis the range is 27 to 48 mol%). The melting temperature increases with decreasing ethylene content; some values quoted for one commercial make (*Eval*, Eval Corporation of America (EVALCA); Kuraray Eval Europe GmbH, Germany) are

EVOH grade :	*Eval 105*	*Eval H101*	*Eval F101*	*Eval XEP-400*
Mol% ethylene :	44	38	32	32
T_m (by DSC), °C :	165	175	183	183

Other commercial EVOH materials include *Clarene, Levasint, Selar,* and *Soarnol.*

EVOH has very good barrier properties against gasses, many solvents, and aromas. In general, these effects increase with increasing vinyl alcohol content; however, as the hygroscopicity - associated with the hydroxyl groups - also increases, there is a reduction in the barrier functions against water vapour, and increase in moisture absorption. The internal expansion (increase of free volume) associated with the latter also reduces the barrier effect against gasses. To minimise moisture absorption and its effects in the main application of EVOH, viz. as the barrier layer in multi-layer coextruded packaging films and sheeting (used inter alia for production of sachets, pouches, bags for bag-in the-box assemblies, and thermoformed containers) the EVOH layer is flanked by protective layers of other polymers, e.g. PA, PE, PP, and PET; these also help preserve the integrity of the EVOH layer, which is normally quite thin.

An ISO standard for EVOH moulding and extrusion materials (ISO/DIS 14663, Parts 1 and 2) is in preparation (in 1997).

EXTRA-THICK MOULDING COMPOUND (XMC)

A sheet moulding compound (q.v.) substantially thicker than a regular thick moulding compound (q.v.).

EXTRUSION PLASTOMETER

A version of capillary rheometer (q.v.) of the dead-load, piston type, thermostatically controlled and of closely specified, appropriate dimensions of cylinder, piston and die (cf. ISO 1133:1991; BS 2782:Part 7:Method 720A:1979; ASTM D 1238-90b), used for the determination of melt flow rates (q.v.) of thermoplastics.

FELICITY RATIO

See under *Kaiser effect.*

FIBREGLASS

A term sometimes used colloquially for plastics composites consisting of thermosetting resins reinforced with glass fibres. Although not very accurate - and hence potentially misleading in some contexts - because it appears to relate solely to glass fibres, the term is promoted (in its American-spelling version : "fiberglass") by a number of technical ASTM standards relating to GRP pipes (cf. e.g. ASTM D 2924, D 2925, D 2992, D 2996, D 2997, D 3262, D 3681, D 3754, and D 3840). In the titles of these standards the term appears in inverted commas, and is followed by the explanatory phrase ("glass-fiber-reinforced thermosetting-resin"). The standards also define *fiberglass pipe* as a tubular product containing glass-fibre reinforcement embedded in or surrounded by cured thermosetting resin (with or without such additives as aggregate (q.v.), fillers, and colorants).

FIBRE OPTICS

See *polymer-fibre optics.*

FIBRE-REINFORCED PLASTIC(S) (FRP)

The abbreviation FRP is often restricted to those plastic/fibre composites in which the plastics component (resinous matrix) is a thermosetting resin. This usage is recognised by some product standards - cf e.g. ANSI C136.20-1990. Glass fibres are most widely used as the fibrous reinforcement in FRP (see also *GRP*), but other fibres - e.g. carbon, aramid - are also employed, in particular where the composite has to meet special strength and/or modulus requirements. See also *reinforced plastic*.

FILLER

A solid material, relatively inert[1.1,1.5,1.7], commonly in a state of fine division[1.3] (and in particulate[1.2] or fibrous form), which may be incorporated in a plastic in comparatively large proportions[1.2,1.3], to modify the properties or to reduce material cost[1.1,1.2,1.3,1.5,1.7]. Cost reduction is often the primary reason for the use of a filler, and because of this the term is occasionally treated, incorrectly, as if it were synonymous with 'cheapening extender'. In fact all fillers - when present in significant quantities - affect in some measure the material and/or processing properties of the plastic, and some - which may be termed 'functional fillers' - are indeed used, often at increased cost, expressly as property modifiers, e.g. glass fibres as reinforcing filler in thermoplastics, antimony trioxide or alumina trihydrate as flame retardants. See also *reinforcing filler* and *reinforced plastic*. ASTM D 4350-86(1990) provides a method for determining the "corrosivity index" of plastics and fillers (as a means of assessing their potential for corrosive effects in humid atmospheres).

FILM-STACKED COMPOSITE

A structurally designed composite with a thermoplastic matrix, produced from fibre prepregs (q.v.) interspaced with thermoplastic films. Composites of this kind represent advanced technology used inter alia in the production of space satellites (cf. Hollaway et al., 1992).

FIRE RETARDANT

A chemical incorporated in a material as an additive[1.8,1.9], or applied by way of a treatment[1.8,1.9], to retard (i.e. to suppress[1.8,1.9], significantly [1.8] reduce[1.8,1.9], or delay[1.8,1.9]) the material's combustion[1.8,1.9]. Use of this term as a noun is deprecated by one of the relevant standards[1.12].

FIRST-ORDER TRANSITION

In a substance, a transition - typified by a change of state - associated with sharp changes in specific volume, heat content and entropy. Some examples of first-order transitions are : evaporation, vapour-to-liquid condensation, crystallisation, and crystal melting (in the polymer context especially the latter two)[1.2,1.3,1.5,1.7]. See also *melting temperature; second order transition; glass transition*.

FISH-EYE

A small globule of polymer which has not merged completely with the surrounding material[1.1,1.5,1.7] in processing (because it consists of polymer of high molecular weight, or partially oxidised or partially crosslinked, or not completely solvated in the case of e.g. solvent-cast films). Fish-eyes are particularly noticeable in transparent and translucent products[1.1,1.5]. In the context of such products (and especially of transparent films) the term is virtually synonymous with "gel" (q.v.) - c.f. e.g. ASTM D 3596 - but it is more closely descriptive of the characteristic appearance of this kind of inhomogeneity. Fish-eyes can occur not only in thermoplastics but also in some thermosetting materials (e.g. GRP - cf. ASTM D 2563).

FLAME RETARDANT

A chemical incorporated in a material as an additive[1.5,1.8,1.9], or applied by way of a treatment[1.8,1.9], to retard markedly[1.5] (significantly reduce[1.8]), suppress[1.8,1.9] or delay[1.8,1.9] the appearance[1.9] and/or the propagation[1.5] rate[1.8,1.9] of flame. Use of the term as an adjective is not recognised by some of the relevant standards[1.5,1.8,1.9], and is deprecated by others[1.12,1.13] (except in compound terms like "flame-retardant treatment"). The ISO terminology standard[1.5] additionally recognises internal flame retardants, i.e. chemical groups with flame-retardant action introduced into a polymer in the course of its production.

FLAME RETARDED (FLAME-RETARDANT TREATED[1.13])

Treated with[1.8,1.9], or containing, a flame retardant (q.v.).

FLAMMABILITY

Ability of a material or product to burn with a flame under defined conditions (such as those specified for a test)[1.5,1.8,1.9]. Flammability is associated with those characteristics of a material which influence also its relative ease of ignition and relative capacity for sustaining combustion (q.v.)[1.5,1.8,1.9].

FLASH-IGNITION TEMPERATURE

In the context of the hot-air ignition furnace test of ASTM D1929-91a, this term denotes the lowest initial temperature of air passing around the test specimen in the furnace, at which a sufficient amount of combustible gaseous material is evolved to be ignited by a small external pilot flame. The term is thus virtually synonymous with "flash point" (q.v.).

FLASH POINT

The minimum temperature to which a material must be heated under specified test conditions, for the vapour(s) emitted to ignite momentarily on application of an ignition source[1.8,1.9] (such as an external pilot flame). See also *flash-ignition temperature*, and *flash temperature*.

FLASH TEMPERATURE

Alternative term[1.8] for flash point (q.v.).

FLATTING (MATTING) AGENT

An additive incorporated in a plastics material to reduce surface gloss and/or produce a matt "silk" finish.

NOTE - This term is not defined in any of the three main plastics terminology standards[1.1,1.5,1.7]. A definition in another standard source[1.4] relates to flatting agents for paints, varnishes and the like (cf. also ASTM D 16-95).

Most flatting agents have also some antiblocking action (cf. *blocking*). Examples of flatting (matting) agents are precipitated silicas (e.g. the Degussa *Flatting Agent OP 278)* and certain acrylic resins, such as *Paraloid KF 708 or 710* (Rohm and Haas), or *Maltex* (Mercury Plastics Ltd., UK).

FLEXIBILITY

The ability of a material readily to undergo, and to withstand[1.5], repeated bending and flexing without breaking or suffering visible damage[1.5].

FLEXURAL STRENGTH (CROSS-BREAKING STRENGTH; TRANSVERSE STRENGTH)

(1) In the most general sense, the ability of a material to withstand the stresses arising in bending[1.1].

(2) MAXIMUM SURFACE STRESS IN BENDING (see also *modulus of rupture in bending*). The maximum stress experienced by a beam at failure (rupture) in bending[1.1] (or, in the specific context of tests on plastics, by the test specimen in a flexural test to rupture[1.5]). Although such stress at failure may, in general terms, be styled "flexural"[1.5], it is in fact a tensile or compressive stress arising - under the action of the bending force(s) - in the outermost layer (fibre) of the beam, that it responsible for the rupture[1.10].

For an ideal elastic solid, the stress at failure in bending is given by the expression

$$\sigma_{max} = \frac{M_{max}}{W} \tag{1}$$

where M_{max} is the flexural moment at the maximum load, and W is the section modulus (q.v.). In practice, eq. (1) is widely used for calculating the flexural strength of plastics materials from relevant data obtained in standard tests.

Thus e.g. for the uniform, rigid plastics test specimens used in the determination of flexural properties by the method of ISO 178-1975, the flexural moment is calculated from the relationship

$$M_{max} = \frac{F_{max} L}{4} \tag{2}$$

where L is the length of the specimen between the supports (span length) in the three-point loading arrangement used in the test, and F_{max} is the maximum force (load) at failure. The value of W is assumed to be given, with sufficient accuracy, by the expression

$$W = \frac{bh^2}{6} \tag{3}$$

where b and h are, respectively, the width and thickness of the specimen.
Substitution from eqs. (2) and (3) into eq. (1) then gives

$$\sigma_{max} = \frac{3F_{max} L}{2bh^2}$$

FLOW DRILLING

A method (originated in metal processing) of producing in a thermoplastic sheet a drilled hole reinforced by material built-up round the periphery. This thickening of the sheet around the hole results from outward flow of material (softened by work-heat) under the action of a special drill used in appropriately controlled drilling conditions (in particular the drilling rate and drill speed). Flow-drilled holes provide additional support for inserts, and are particularly advantageous - especially in thin sheets - where subsequent threading and insertion of threaded bolts or screws is required. Crawford et al. (1991) provide practical information on drill designs and drilling conditions suitable for PVC, PA, PP, ABS, and PMMA sheets.

FLOW MOULDING

(1) A method of injection moulding which enables large mouldings to be produced on relatively small injection machines, by utilising pressure developed by the rotation of the screw (instead of that produced by the injection stroke) as mould-filling pressure. The method may be regarded as a combination of extrusion (as the principal means of supplying melt to the mould) and injection moulding (in that a conventional - though modified - injection-moulding machine is used). The modifications include adjustment of the hydraulics to keep the screw in the forward position (feeding the melt through a gap between the screw tip and the rear face of the nozzle); provision for adjustment of the size of this gap (where the main working of the material to produce plastication is effected through the shear imparted) to regulate the amout of shear; special provisions for screw and barrel cooling; and overall control of temperatures and pressures. The process is claimed to be particularly suited to heat-sensitive materials such as PVC.

(2) The term is also occasionally used for the operation of impressing a surface pattern (e.g. leather-like grain) on a heated plastics material (which may be a coating on a substrate) by pressure-contact with a suitably patterned mould.

FLOW VALUE

See under *ultra-high-molecular-weight polyethylene.*

FLUORINATED ETHYLENE/VINYL ETHER POLYMERS (FEVE)

Thermosetting fluoropolymers, copolymers of fluoroethylene and vinyl ethers. Some of the alkyl groups of the vinyl ether component can contain hydroxyl or carboxyl groups which confer solubility in solvents and adhesion to substrates. An FEVE is typically used as the resin component of a two-pack surface-coating system, with an isocyanate or melamine curing agent as the second component.

FLUOROCARBON PLASTIC

A plastic based on a polymer or polymers that consist of perfluoromonomer[1.7] units, i.e. whose monomer units contain only carbon and fluorine[1.1,1.5]. One of the three major terminology standards[1.7] deplores the use of the abbreviation "TFE" to designate polymers (properly abbreviated to PTFE) whose monomer is essentially tetrafluoroethylene, but sanctions the use of "FEP" to designate copolymers of tetrafluoroethylene and hexafluoropropylene.

FLUOROELASTOMERS

These are fluorine-containing elastomers, most of which fall into one or another of the following type groups: copolymers of vinylidene fluoride with other fluorinated or chlorofluorinated monomers, certain perfluoro(alkoxy alkane) copolymers, and fluoroslicone elastomers. Those fluoroelastomers which are classifiable as fluororubbers - as the majority of them are - do not form a single group in the standard rubber-nomenclature system, but are individually included in some of its classification groups (termed "classes" in the relevant ASTM standard[2.4]) in accordance with their chemical types.

NOTE - As mentioned briefly in Section 1 of Part 2 of this book, in an abbreviated term (letter symbol) according to the standard rubber nomenclature[2.4,2.5,2.6], the final letter denotes the main classification group (e.g. "M" for rubbers having a saturated chain of the polymethylene type), with the preceding letters indicating more closely the nature of the particular member of the group. However, in contrast with the standard letter symbols for the base polymers of plastics, the letters in the symbols for rubbers are to some extent arbitrary, in the sense that in many cases thcy are not recognisable initials of the chemical components of the material. For example, whilst the letter symbol for polychloro-trifluoroethylene according to the standard abbreviated terminology for plastics is *PCTFE*[2.1,2.2,2.3,2.11], the corresponding designation in the BS and ASTM rubber-nomenclature systems[2.4,2.6] is *CSM* (see also table below).

FLUOROELASTOMER TYPE	FLUOROELASTOMER MATERIAL	STANDARD LETTER SYMBOLS IN RELEVANT PLASTICS TERMINOLOGY*	STANDARD LETTER SYMBOLS GROUP DESIGNATION IN RUBBER NOMENCLATURE STANDARDS	EXAMPLES OF TRADE NAMES
Copolymer of vinylidene fluoride	Vinylidene fluoride/ hexafluoropropylene**	VDF/HFP	FKM[2.4,2.5] (FFKM[2.4,2.5])	*DAI-EL G701* - Daikin Koygo Co. Ltd. *Viton A* (and other grades) - Du Pont *Fluorel* (various grades) - 3M *Technoflon* FOR - Montecatini (Montell) *Kynar Flex* - Elf Atochem
	Vinylidene fluoride/ chlorotrifluoroethylene	VDF/CTFE	CFM[2.4,2.6]	*Voltalef* - Elf Atochem *Kel-F* - 3M *Solef 3000* - Solvay
Terpolymer of vinylidene fluoride	Vinylidene fluoride/ tetrafluoroethylene/ hexafluoropropylene**	VDF/TFE/HFP	FKM[2.4,2.5] (FFKM[2.4,2.5])	*DAI-EL G501* - Daikin Kogyo Co. Ltd. *Hostaflon TFB* - Hoechst *Viton B* (and other grades) - Du Pont *THV* - 3M, and Dyneon (Germany)
Copolymer of tetrafluorethylene	Tetrafluoroethylene/ propylene**	TFE/P	FEPM[2.4,2.5]	*Aflas* - 3M
Fluorosilicone elastomer copolymers	methyl-1,1,1-trifluoro-propylsiloxane/ dimethylsiloxane	———	FMQ[2.4,2.6]	
	as above with a small propotion of cure-site monomer (methyl-vinyl siloxane)	———	FVMQ[2.4,2.5,2.6]	*Sylon* - 3M
Perfluoro(alkoxy alkane) copolymer	tetrafluoroethylene/ perfluoro (methyl vinyl ether)**	TFE/PFA	FFKM[2.4,2.5] FPM[2.6]	*Kalrez* - Du Pont *Hyflon MFA* - Ausimont
Phosphonitrilic fluoro-elastomer (fluorophosphazene rubber)	Hydrofluoroalkoxy derivative of poly(phosphonitrilic chloride)	———	FZ[2.4,2.5,2.6]	*Eypel* - Albermarle Corp.

* The most directly relevant standard, ISO 12086-1, not yet finalised (i.e. still a discussion document - ISO/DIS) in 1997.
** Fluororubber version contains a small proportion of cure-site comonomer.

In the case of some fuoroelastomeric copolymers (e.g. those of vinylidene fluoride and chlorotrifluoroethylene), the ratio of the comonomers largely determines whether the material is an elastomer or a tough thermoplastic. Fluoroelastmers classifiable as rubbers normally contain crosslinking (cure) sites provided by the incorporation of a relatively small proportion of a suitable comonomer (cf. definitions of *elastomer* and *rubber*). In commercial terms, certain copolymers and terpolymers of vinylidene fluoride, notably VDF/HFP and VDF/TFE/HFP, are the most important types of rubbery fluoroelastomers. Two other fluoroelastomer types significant in this sense are crosslinkable copolymers of perfluoro(methyl vinyl ether), and fluorosilicone rubbers. The best known fluoroelastomers are listed in the table.

A few other kinds of fluoroelastomer have been produced , but have not became commercially established. These include some fluoroacrylate polymers (with fluorine atoms in the esterifying groups), poly(2-fluorobutadiene-1,3), originally marketed (by Du Pont) as *Fluoroprene*, copolymers of vinylidene fluoride with hydropentafluoropropylene and similar terpolymers additionally containing tetrafluorethylene as the third monomeric component (respectively *Technoflon S* and *Technoflon T*, originally produced by Motecatini). The established fluoroelastomers combine good mechanical properties (especially in the cured state) with stability to heat and resistance to a wide range of chemicals, which includes virtual immunity from any effects of fuels, lubricants and hydraulic fluids; these characteristics are utilised in their applications as the materials of seals, gaskets, o-rings, and other components of vehicles, aircraft, and industrial equipment.

ASTM D 4591-93 a provides a standard test method for the determination (by DSC) of the temperatures and heats of transitions of several fluoropolymers, including some copolymers and terpolymers.

FLUOROHYDROCARBON PLASTIC

A plastic based on a polymer or polymers whose monomer units contain only carbon, fluorine and hydrogen[1.1,1.5,1.7] - cf. e.g. polyvinyl fluoride (PVF), polyvinylidene fluoride (PVDF), and copolymers of ethylene with tetrafluoroethylene (ETFE).

FLUOROPLASTIC

Generic term applicable to any plastic based on a polymer or polymers whose monomer units each contain at least one fluorine atom, or/and on copolymers incorporating such fluorine - containing monomer units, and in which these units account for a major proportion of the polymer mass[1.1,1.5,1.7]. Examples of the main kinds of fluoroplastics are[1.7] *fluorocarbon plastics*, *fluorohydrocarbon plastics*, and *chlorofluorohydrocarbon plastics*.

FLUOROPOLYMER

A term not included in any of the three major terminology standards[1.1,1.5,1.7], but which is commonly applied to the fluorine-containing polymers or copolymers used as base polymers of fluoroplastics (q.v.), and which is encountered in standard specifications, e.g. ASTM D4591-93a and ISO/DIS 12086. The former, a test-method standard for the determination of heats of transition of fluoropolymers by differential scanning calorimetry (DSC), lists the fluoropolymers falling within its ambit as ECTFE, EFEP, ETFE, FEP, PCTFE, PFA, PTFE, PVDF, PVF, VDF/CTFE, VDF/HFP, VDF/TFE and VDF/TFE/HFP (for the meanings of these abbreviations see Part 2 of this book). The same fluoropolymers are listed in ISO/DIS 12086-1, with the addition of TFE/PDD. This proposed two-part ISO standard (still under discussion in 1996) covers designation, specification and determination of properties of fluoropolymer dispersions, and moulding and extrusion materials. Two of the informative annexes in Part 1 respectively provide a useful discussion of the fluoropolymer family, and a list of ASTM standards for fluoropolymers (mainly PTFE). According to some non-standard definitions, a fluoropolymer should contain at least 40% by weight of fluorine to be classified as such. *Lumiflon* (q.v.) is an example of a fluoropolymer coating applied from solution.

FOGGING (AUTOMOTIVE FOGGING)

Condensation of oily deposits on surfaces inside the passenger compartments of motor vehicles; particularly pronounced on the relatively cooler windscreen and other glass surfaces, where the deposit is also more readily noticeable as a light-scattering film. Fogging was originally attributed solely to volatile plasticiser from PVC crash-pads and leathercloth, but later analyses of fogging deposits indicated also other origins, including plasticisers and volatile plasticiser impurities (e.g. alcohol residues) from adhesives as well as PVC, some pigment-dispersing media, catalyst residues and other materials present in polyurethane foam, volatile emissions from certain stabilisers and lubricants (including such components of some polypropylene interior trim materials). Fogging is promoted by elevated temperature, and relatively high temperatures (typically up to about 100°C) are employed in fogging tests for materials and components for use in motor-vehicle interiors. Apart from the vehicle manufacturers' own tests, available standard tests include those of ISO 6452*, ASTM D 5393, and the German DIN 75 201. The fogging propensity of a material as shown up in a test may be expressed in terms of a "fogging number" (cf. ASTM D 5393-93). This is the average gloss value of a glass plate fogged in the test, expressed as a percentage of the original gloss value. A similar expression, termed "fogging value", is used in DIN 75 201. Aspects of the origin and investigation of fogging are discussed in recent papers by Howick (1994, 1995).

*Initiated in 1985 and still under discussion (ISO/DIS 6452) in 1996.

FORMING

A shaping operation performed on a plastics product (typically a thermoplastic tube, sheet, or rod), whereby a desired configuration is imparted to the product[1.1,1.5,1.7]. The term does not apply to normal moulding, casting, or extrusion[1.7].

FRACTURE TOUGHNESS

In fracture mechanics, a material's resistance to the propagation of a crack, expressed in terms of the *critical stress intensity factor* (cf. e.g. ASTM E 399-90).

FREE VOLUME

That part of the total volume of a solid or liquid substance at a given temperature, which is not occupied by the molecules making up the substance. This general definition encompasses the following, more specific interpretations of free volume, each depending on the particular concept of the volume occupied by a molecule.

1. *Empty volume* (V_E), defined by the expression

$$V_E = V_T - V_W$$

where V_T is the observed molar volume at temperature T, and V_W is the volume per mole of the substance calculated from the van der Waals molecular dimensions.

2. *Expansion volume* (V_F), given by the expression

$$V_F = V_T - V_O$$

where V_T is the observed molar volume, and V_O is the volume occupied by the molecules at 0 kelvin in a close-packed crystalline array. Visualised in this way, the expansion volume represents extra free space generated by thermal motion of the molecules.

3. *Fluctuation volume* (V_f), defined as

$$V_f = N_A V_Q$$

where N_A is Avogadro's number, and V_Q is the volume swept out by the centre of gravity of a molecule as a result of thermal vibration.

The free-volume concept is helpful in the understanding of the mechanisms of such phenomena of practical importance and theoretical interest in the field of polymers, as plasticisation, glass transition and concentration - dependent diffusion of penetrants. The free volume considered in these contexts is of the same nature as V_F above, but it is usually defined in terms of the WLF equation (q.v.), and hence is sometimes termed "WLF free volume".

"FUNCTIONALISED" POLYMER

A polymer grafted with a monomer containing functional groups which confer compatibility with polymers that would otherwise be incompatible. Many compatibilisers (q.v.) are graft copolymers of this kind. The following are two examples of functionalised polymers. EPDM grafted with maleic anhydride, maleic acid, or acrylic acid, for use as an impact modifier in polyamides.

Polypropylene grafted with maleic anhydride (cf. *UNITE MP*), used as compatibiliser in blends of polyolefins with polyamides, polyesters, polyvinylidene chloride, and ethylene-vinyl alcohol copolymers, as coupling agent for fillers and glass-fibre reinforcements in polyolefins, and as a tie layer in co-extrusions.

FUNGISTAT; FUNGICIDE

See under *biostat.*

FUSE (verb)

(1) To liquefy by heating; to melt.

(2) To combine by melting together.

See also *fusion point*, and fusion of PVC pastes under *paste (1)*.

FUSION POINT

Along with *solubility temperature* (q.v.), one of a number of terms for essentially the same concept, viz. the temperature at which, in specified test conditions, particles of PVC resin are first transparentised, gelled, or dissolved by a plasticiser. The other relevant terms are "clear point", "solid-gel transition temperature", and "apparent wetting temperature". Like the solubility temperature, the fusion point provides a numerical indication of the compatibility of a plasticiser with, and its solvent power for, a PVC resin: both these effects are the higher the lower the fusion point. Determinations are typically carried out under the microscope, but procedures vary to some extent, mainly in the respective proportions of plasticiser and resin used: for example, 5% suspensions of resin in plasticiser may be employed (cf. DIN 53 408-1967), as well as individual resin particles immersed in plasticiser (cf. e.g. Anagnostopoulos et al. (1960), and Bigg (1975).

GARDNER COLOUR SCALE

A colour scale based on glass standards numbered 1 to 18 (the "Gardner colour standard numbers"), each standard (and its number) representing a particular combination of chromaticity coordinates, luminous transmittance and transmittance tolerance. The scale serves as means of description and determination (by comparison under specified conditions) of the colours of transparent liquid resins and resin solutions, drying oils, varnishes, and the like (see ISO 4630 : 1997 and ASTM D 1544-88).

GARDNER IMPACT TEST

So called after the makers of one version of the test apparatus. The test is intended primarily as a means of relative ranking of plastics materials, in terms of the energy required to fracture flat, rigid specimens impacted by a striker actuated by a falling weight (see ASTM D 5420-93).

GAS TRANSMISSION RATE (GTR)

See under *permeability*.

GEL (noun)

(1) In the most general sense, a solid/liquid system, typically of jelly-like consistency, in which the liquid is dispersed in a network of the solid.

(2) In the context of liquid-derived resins, the initial jelly like solid phase that develops during the formation of a resin from a liquid[1.1,1.5,1.7].

(3) A soft, weak semi-solid which constitutes an early transitional phase in the heat fusion of a PVC paste[1.7], sometimes also formed by the paste on prolonged storage[1.7], especially in warm surroundings.

(4) In transparent or translucent plastics products: a nodule of polymer incompletely dispersed in the course of processing; particularly evident when present in transparent films or thin mouldings (in opaque materials gels may show up as slight protrusions on the surface, sometimes referred to as "nibs"). In common plastics terminology there is no basic distinction between a gel and a fish-eye (q.v.), especially when the gel appears as an imperfection in a plastics film. A standard method for the determination of the number of gels per unit area of film (gel count) is given in ASTM D 3351-93; a gel count procedure applicable to blow-moulding compounds has also been described (Latham and Mendham 1973). ASTM D 3596-92 gives a standard method for the determination of gels (sizes, and numbers per unit area of prescribed specimen) in general-purpose PVC resins. The gel nodules can originate at the polymer-production stage: in thermoplastic films they may consist of polymer of molecular weight significantly higher than that of the surrounding material. In some cases they may contain polymer partly oxidised and/or cross-linked as a result of degradation by heat in melt-processing.

(5) In a crosslinked polymer: that part of the material which is sufficiently highly crosslinked to become insoluble in a solvent capable of dissolving the rest of the polymer. In gel content (q.v.) determinations by solvent extraction, the insoluble fraction is normally penetrated and swollen by the solvent, so that it becomes a gel in the sense of (1) above.

GELATION

A material's conversion to the state of a gel[1.5] (q.v.). For gelation of PVC paste see under *paste (1)*.

GEL BLOCK

See under *temperature-gradient bar*.

GEL CONTENT (GEL FRACTION)

In a solvent-extraction test on a polymer crosslinked chemically or by radiation, that part by weight of the polymer which is insoluble in the solvent in the conditions of the test. The soluble part is sometimes referred to as the *sol fraction*. In gel content determinations on crosslinked PVC the solvent is usually tetrahydrofuran. Xylene or decahydronaphthalene (decalin) are the solvents used in standard extraction tests on crosslinked polyethylene (cf. ASTM D 2765-90). See also *gel (5)*.

When a polymer is crosslinked at relatively low radiation doses, or low concentration of a chemical crosslinking agent, the effect of the comparatively low number of crosslinks formed is to increase the average size of polymer chains and their degree of branching, and hence also the number of interchain entanglements: this is reflected in improvements in some mechanical properties of the polymer, as well as in an increase of its melt viscosity. However, at this stage the crosslinks are too few to give rise to any consolidated three-dimensional networks, and the material remains soluble, albeit less readily, in appropriate solvents. With increasing radiation dose level, or crosslinking agent concentration, further crosslinks are formed between the enlarged chain molecules, resulting in the formation of more or less extensive regions in which the molecular chains are firmly linked into a continuous three-dimensional structure, no longer soluble in the previously effective solvent(s). It is this part of the polymer that constitutes the gel fraction in a solvent-extraction test. Thus the gel content provides an indication of the degree of crosslinking. Another value indicative of the degree of crosslinking - the so-called "swell ratio" (q.v.) - may also be obtained in a solvent extraction test.

GEL COUNT

Determination of the number of gels per unit area of a plastics film or a processed polymer specimen - see *gel* (4). Cf. ASTM D 3596-92.

GEL POINT

(1) In the most general sense, the onset of pseudoelastic behaviour in a liquid[1.1,1.5,1.7] (e.g. as a result of chemical reaction). The onset is typically marked by an inflection point on a plot of viscosity versus time. In the practical context of plastics materials, the liquid will normally be a thermosetting resin system (e.g. a phenolic or unsaturated polyester resin composition). When the cure progress of such a system is followed by dynamic mechanical measurement and the data obtained are appropriately plotted, the point of intersection of the plots of the storage modulus (G') and loss modulus (G'') versus time identifies the gel point. Since $G''/G' = \tan\delta$, the point is one for which the value of $\tan\delta$, (the loss factor - q.v.) on a corresponding plot of $\tan\delta$ is unity (c.f. e.g. ASTM D 4473-90).

(2) Specifically in relation to the gelation/fusion of PVC pastes, "gel point" is one of a number of terms which have been used to denote the termination of the first sharp rise in viscosity of a PVC paste in the course of gelation on heating. At the gel point the steep rising portion of a plot of viscosity versus temperature terminates in a shoulder (or sometimes a small peak). This feature marks the stage at which the originally liquid paste solidifies to a state in which it begins to acquire some physical strength.

GEL TIME (GELATION TIME)

(1) Of thermosetting powders (e.g. as used for powder coatings). The time, determined under specified test conditions, for a specified amount of the powder to melt and become non-deformable (cf. ISO 8130-6:1992).

(2) Of reactive liquid resin systems. The time for the liquid system to form a gel[1.1,1.5] in specified test conditions. In tests on two-reactant or multi-reactant liquid systems the gel time is normally the period between the mixing of the reactants and a specified manifestation of gelation[1.7]. In systems which require an input of energy (e.g. heating; irradiation) to promote reaction, gel time is usually measured from the moment the energy is applied[1.7]. Several standard methods are available for determining the gel time of thermosetting resins, as follows. *For phenolics*: ISO 9396:1989(E). BS 2782:Part 8:Methods 835 A and 835 D:1980. *For unsaturated polyesters*: ISO 2535:1974(E). BS 2782:Part 8: Methods 835B, C and D:1980. *For epoxy resins*: BS 2782:Part 8:Method 835 C:1980; BS 3815:1964. *For thermosetting resins generally*: BS 2782:Part 8: Methods 835 D:1980 and E:1980 (gelation time at 160°C, under pressure, in impregnated materials used for production of laminates). ASTM D 2471-88 (determination of gel time and exothermic temperature). ASTM D 4473 (measurement of cure progress in thermosetting resins by dynamic mechanical methods). See also *gel (2)* and *gel point (1)*.

GEOMEMBRANE

A geosynthetic, often consisting of one or more layers of polymeric sheeting, of essentially negligible permeability to those fluids and/or other chemical species whose migration the membrane serves to control when it is incorporated in a structure or system (cf. ASTM D 4439-92, D 4833-88, and D 4885-88). Geomembranes are sometimes referred to as "flexible membrane liners", or simply "membranes" or "liners" (cf. also ASTM D 4885-88). See also ASTM Publication PCN 03-435091-38, ASTM Standards on Geosynthetics (listing over 20 ASTM standards dealing with specifications for plastics sheeting for use as geomembranes, and with several properties of, and tests for, geosynthetics of various kinds).

GEOSYNTHETIC

A laminar product, normally polymeric in nature, used in conjunction with a geotechnical engineering material (e.g. sand, soil, rock) as a component of a man-made structure or system. (Cf. ASTM D 4439-92)

GEOTEXTILE

A geosynthetic that is permeable (in contrast with a geomembrane) and that consists entirely of a textile material or materials (cf. ASTM D 4439-92). Some geotextiles are used as support for geomembranes in certain applications.

GLASS TRANSITION (RUBBER-GLASS TRANSITION; GLASS-RUBBER TRANSITION)

The reversible change between a glassy state[1.2] (i.e. hard and relatively brittle condition[1.1,1.5,1.7]) and a rubbery viscous state[1.1,1.5,1.7], which can be undergone - in either direction - by a non-crystalline (amorphous[1.1,1.5,1.7]) polymer, or the amorphous regions of a semi-crystalline polymer (see also ASTM D 4092-90; F 1142-93b; and F 412-94a). Transition from glassy to rubbery state typically occurs with a sufficient increase in temperature (see *glass transition temperature*), or upon introduction into the polymer of such penetrants as plasticisers or swelling agents, which increase the free-volume content. The reverse transition is brought about by cooling, or by removal of plasticising or swelling additives. Major manifestations of glass transition in non-crystalline polymers include marked changes in modulus (see e.g. Tobolsky 1965) and in specific volume. A good account of the nature and effects of glass transition in polymers has been published by Brydson (1972). The glass transition is sometimes referred to as a second order transition, although it does not possess all the thermodynamic features characteristic of such transitions (one of the main terminology standards[1.2] specifically deprecates this usage).

GLASS TRANSITION TEMPERATURE (T_g)

The glass transition undergone by a polymer on heating (or cooling) does not take place at a single, unique temperature but, typically, over a temperature range. It is the temperature at the approximate mid-point of the range that is conventionally quoted as the T_g[1.1,1.2,1.5,1.7]. The actual value of this temperature for a given polymer is not a constant: it is influenced by such factors as the molecular weight (MW) of the polymer, the nature of the material property monitored in the course of T_g determination, the determination method used, and - in dynamic test methods - the rate of testing. Thus, for example, the T_g of non-crystalline polymers can increase substantially with increasing MW in the low to medium MW region but, typically, levels off at high MW's. Test methods involving slow rates of testing, such as measurement of specific volume by dilatometry, give lower T_g values than other, faster-rate methods. Apart from dilatometry, the methods commonly used to determine the T_g of polymers are differential thermal analysis (DTA), differential scanning calorimetry (DSC), and dynamic mechanical analysis (DMA); dielectric analysis (DEA) is also applicable. In a dilatometric determination, volume change is measured as a function of temperature, the data are plotted as a graph of specific volume versus temperature, and the T_g is located as the mid-point of the region in which the plot changes slope. In DTA and DSC determinations carried out in the standard manner (cf. e.g. ASTM D 3418-82

(1988)) the glass-transition region is marked by a base-line shift in the thermal curve. In DMA, the plot of storage modulus versus temperature slopes sharply downwards in the T_g region, whilst the corresponding plot of the loss tangent exhibits a peak. In DEA there are analogous distinct changes in plots of permittivity and loss factor versus temperature.

GLOSS

The capacity of a surface to reflect light: on the subjective level this is perceived as shiny or lustrous appearance of the surface - cf. ASTM E 253-96. For a given surface, this reflecting capacity is often characterised in terms of the degree to which it approaches that of a surface of perfect optical smoothness[1.1,1.5]. In its quantitative aspect gloss is a complex property of a surface, which cannot be completely expressed by a single number (cf. ASTM D 2457-90). The surface gloss of a plastics material or product is usually measured in terms of *specular gloss* (q.v.). Apparatus used in such measurements is known as a glossmeter. Several commercial versions are available, including some capable of accurate gloss measurement on a curved (convex) surface (e.g. the Novo-Curve equipment, Rhopoint Instrumentation Ltd., UK) cf. also ASTM E 430-91 (measurement of gloss of high-gloss surfaces), and the paper by Hammond (1987) on the concept, significance and evaluation of gloss.

GLYPTAL RESINS

See under *alkyd resins.*

GRAFT COPOLYMER

A graft polymer derived from more than one species of monomer. See note to *graft polymer.*

Reproduced, with permission, from ISO 472:1988.

GRAFT POLYMER

A polymer whose molecules have one or more species or block connected to the main chain as side chains, these side chains having constitutional or configurational features different from the constitutional units comprising the main chain, exclusive of junction points.

NOTE - In the graft polymer molecule

```
AAAAAAAAAAAA
  |       |
  Bm      Bn
```

the A-chain, B_m and B_n are regular blocks, the A-chain is the main chain, and B_m and B_n are the side-chain grafts.

The –A– units are junction points and are considered to be part of the main chain. Where A and B are derived from the same monomer, for example:

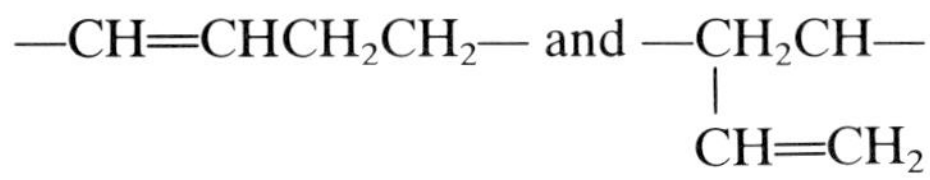

(A) (B)

the polymer is a graft polymer. The graft polymer whose molecules consist of

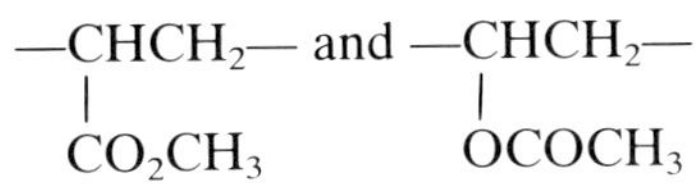

(A) (B)

is a graft copolymer.

See also *graft copolymer.*

Reproduced, with permission, from ISO 472:1988.

GRANULAR MOULDING COMPOUND (GMC)

Term and abbreviation sometimes used specifically in relation to compression, transfer, and injection moulding compounds based on thermosetting resins (especially UP) with fillers and/or reinforcements (cf. e.g. ASTM D 1201-81(1992)). GMC's are available e.g. in the commercial *Durez, Plenco,* and *Polyrite* ranges.

GREY SCALE

A scale used as standard for visual evaluation of small to moderate differences in colour between non-luminous coloured specimens. A colour difference determined in terms of the grey scale is relatable to the equivalent value of CIELAB colour difference (q.v.).

The grey scale comprises nine pairs of neutral-grey colour standards, referenced respectively 1, 1 - 2, 2, 2 - 3, 3, 3 - 4, 4, 4 - 5, and 5. The reference pair 5 consists of two panels of the same grey colour. Each of the other pairs is made up of one panel identical with the reference pair 5 and a lighter-grey panel, the lightness increasing in defined progression from reference pair 4 - 5 to reference pair 1.

The basis of colour-difference evaluation with the scale is identifying the grey standard pair in which the contrast in lightness between the two panels matches the most closely the analogous contrast between the specimens concerned. Methods of using the grey scale to evaluate colour differences are given in ISO 105:A02:1993; BS 1006:A02:1990; and ASTM D 2616-96.

HARDENER; HARDENING AGENT

A curing agent (q.v.) for resins[1.1,1.5] or adhesives[1.5], which itself participates in the curing reaction[1.5].

HARDNESS

Most commonly defined as the ability of a material to resist indentation under strictly specified conditions, including particular, specified combinations of indentor and stress (load). Numerically expressed in units related to arbitrary hardness scales associated with particular test methods. On a given scale, higher numbers correspond to greater hardness, except in the case of the BS softness scale where the reverse applies (the numbers increasing with increasing *softness*). Some methods employ more than one scale to accommodate progressively greater hardness (where mentioned below, these scales are given in order of increasing hardness). In such cases the scales normally overlap to some extent, it being recommended that the lower part of the next, more severe scale should be used in preference to the uppermost part of the preceding "softer" scale.

The determination methods and their corresponding scales in use for plastics and rubbers are: *durometer (Shore) hardness* (scales A and B), applicable in the range soft rubbers and plastics to rigid plastics (ISO 868; BS 2782, Part 3, Method 365B; BS 2719; ASTM D 2240, DIN 5305); *International Rubber Hardness* (IRH), applicable to a wide range of rubbers (ISO 48; BS 903, Part A26; ASTM D 1415); *Rockwell hardness* - scales R, L M and E (ISO 2039, Part 2; BS 2782, Part 3, Method 365C; ASTM D 785) applicable to most plastics, but especially to the more rigid ones, including some reinforced plastics (scale E); the most severe scale K of ASTM D 785 is not widely used; *Barcol hardness* (test with Barcol Impressor Model 934-1) applicable particularly ro reinforced plastics, sometimes used also for other rigid ones (EN 59 and the identical Method 1001 of BS 2782, Part 10; ASTM D 2583); *ball indentation hardness* (ISO 2039, Part 1, the identical Method 365D of BS 2782, Part 3; and DIN 53456). There is no close mutual correspondence between hardness values determined by different methods, except in the following two cases. The values in the upper half of the IRH scale are virtually identical with the corresponding Shore A hardness values, and the standard equipment and conditions for determining the Rockwell R hardness may be used to determine the *Rockwell-α hardness* numbers which correlate with the ball-indentation hardness of ISO 2039, Part 1.

Modulus is an important (though not the sole) factor in the indentation hardness of polymeric materials. For those which, like many vulcanised rubbers, are substantially elastic and isotropic, the relationship between the indenting force (F, newtons), the material's Young's modulus (M, MPa), the radius of the indentor ball (R, mm) and the depth of penetration (P, mm) obeys fairly closely the expression:

$$F/M = 1.9R^2(P/R)^{1.35}.$$

The IRH scale is in fact based on a curve which relates it to Young's modulus (BS 903; Part A26; 1969; ASTM D 1415-88).

Definitions of hardness in terms of scratch resistance or rebound resilience are also relevant in some contexts, e.g. in connection with, respectively, resistance to abrasive wear and determination of modulus by dynamic methods.

HAZE

Cloudy[1.1,1.5,1.7] or turbid[1.7] appearance[1.1,1.5,1.7] or aspect[1.7] caused within or on the surface of a plastic[1.1,1.5] which is otherwise transparent[1.7], by scattering of light at these sites[1.7]. *Haze in reflection*, caused by light scattering at a glossy surface is manifested as an apparent reduction in the image contrast of objects viewed by reflection at the surface (cf. ASTM E 284-89 and - for determination by two specular-gloss measurements - ASTM D 4039-87). In the context of relevant tests, *haze in transmission* - which results in similar blurring of images of objects viewed in transmitted light - is defined and numerically expressed as that percentage of transmitted light which in passing through a specimen deviates from the incident beam by forward scattering exceeding a specified angle (2.5° in most relevant standards). See also BS 2782:Part 5:Method 521A:1992; BS 4618:Section 5.3:1972; ASTM D 1003-61 (1988); and ASTM D 1695 (haze as turbidity in some cellulose derivatives).

HAZEN COLOUR SCALE

See *platinum-cobalt colour scale.*

HEAT AGEING

A general term for procedures which - subject to certain restrictions - can be used to assess likely long-term resistance of a plastic to ageing, on the basis of shorter-term heating tests at elevated temperatures. The main relevant standards are : ISO 2578:1993 (identical with Method 135:1993 of BS 2782:Part 1); IEC Publication 216:1990:Parts 1 to 5; BS 4618:Section 4.6:1974, and ASTM D 3045-92. See also *relative thermal index, temperature index, thermal endurance,* and *thermal stability.*

HEAT CAPACITY (THERMAL CAPACITY)

The quantity of heat necessary to produce a unit change of temperature in a given quantity of material (cf. e.g. BS 4618:Section 3.2:1973). Usual symbol : C. SI units : joule per kelvin (J/K). An ISO source (ISO 31-4:1992) gives a formal definition in terms of small increments of heat (dQ) and temperature (dT), viz. $C = dQ/dT$. See also *specific heat capacity.*

HEAT STABILITY

See *thermal stability.*

HEAT-DISTORTION TEMPERATURE (HDT)

A term originally used (now deprecated by ISO[1.5]) for the *deflection temperature under load* (q.v.).

HEAT-SHRINKABLE ("HEAT-SHRINK") TUBING

Plastics tubing, which on heating at an appropriate temperature shrinks radially so that its diameter is reduced by a predetermined percentage. The shrinkability is imparted by radial stretching, typically in a separate operation after extrusion, at an elevated temperature corresponding to that to be used in the ultimate heat shrinking. In most current commercial versions of the tubing, its constituent polymeric material is crosslinked (chemically or by irradiation) at a stage between extrusion and heat-stretching. This improves the ultimate shrinkability, as well as the mechanical properties and thermal resistance of the tubing. Heat-shrinkable tubing is produced from a number of polymers, notably polyethylene and PVC. Standards dealing with the property requirements and tests for this kind of tubing include ASTM D 2671-91; UL 224-92; CSA C 22.2 No. 198.1-86; and some US-MIL standards.

HEPARINOID RESIN

A polymer with anticoagulant action similar to that of heparin - see *poly(glucosyloxyethyl methacrylate) sulfate.*

HIGH-MODULUS FIBRES

In the context of fibrous reinforcement for or in plastics materials, high-modulus fibres are ones whose modulus exceeds 20 GPa (3×10^6 lbf in^{-2}) - cf. ASTM D 3039-76(1989).

HIGH PRESSURE DECORATIVE LAMINATE (HPDL)

Essentially in line with the definition given in ISO 4586:1995, a high-pressure decorative laminate may be described as a laminate (q.v.) in sheet form, made up of a number of fibrous-sheet layers (which may be paper) impregnated with thermosetting resin(s) and consolidated by bonding under the combined action of heat and pressure (the latter of at least 5 MPa). One or both of the outer layers, which form the surfaces of the finished laminate, may be coloured, and have decorative patterns or designs. In a typical HPDL the resin impregnant of the inner (core) layers is an amino resin (or occasionally a phenolic resin), and that of the surface layers is usually a melamine-formaldehyde resin. ISO 4586:1995 gives HPDL as the abbreviated term for high-pressure decorative laminate, but points out that a relevant European (EN) standard (see BS EN 438 below) favours "HPL".

None of the three main plastics terminology standards[1.1,1.5,1.7] defines HPDL. The ISO standard[1.5] gives a definition of "decorative laminate", which is very similar to the one given here for HPDL, except that it makes no reference to moulding pressure, and does not confine the laminar reinforcement to fibrous sheets (mentioning, in addition to paper and fabric, also film and foil). It may also be noted that the ASTM terminology standard[1.7] includes *high-pressure laminating* in its definition of "high-pressure moulding", characterising both as methods in which the pressure used is greater than 1400 kPa (200 psi).

The following standards are directly relevant to HPDL products and their properties (see also under "laminate" for other standards of some possible relevance) : ISO 4586 (HPDL sheets made from thermosetting resins - Part 1:1995 - classification and specifications; Part 2:1995 - determination of properties). BS EN 438:Parts 1 and 2:1991 (1997) - thermosetting-resin HPDL's.

HIGH-PRESSURE LAMINATING

See under *high-pressure decorative laminate.*

HIGH-PRESSURE MOULDING

See under *high-pressure decorative laminate.*

HINDERED-AMINE LIGHT STABILISERS (HALS)

A group of useful light (UV) stabilisers, particularly highly effective in polyolefins, structurally characterised by the presence of amine groups sterically protected ("hindered") by other adjacent groups. HALS are effective at relatively low concentrations (e.g. small fractions of one percent by weight in polyolefins) : the reason usually cited for this is that, in exercising their function, they are believed to be used up more slowly and less completely (with, possibly, some degree of regeneration) than other light stabilisers. The mechanism of their action is not fully elucidated, but it is variously thought to include termination of free radicals created in the polymer by incident UV-light photons, decomposition of hydroperoxide groups formed in the course of photooxidation of the polymer, and favourable energy transfer. Both low-molecular-weight and high-molecular-weight (polymeric) HALS are available, the latter offering particularly good resistance to migration and extraction, as well as negligible volatility. Examples of commercial representatives of the two kinds of HALS are *Tinuvin 770 DF* and *Tinuvin 765* (low-molecular-weight compounds), and *Tinuvin 622 LD* and *Chimassorb 944 LD* (polymeric compounds). A useful feature of HALS is their effectiveness in thin polyolefin products, such as films, and in surface layers.

HOMOPOLYMER

A polymer made by (derived[1.1,1.5], or resulting[1.7], from) polymerisation of only one kind (one species[1.1,1.5]) of monomer[1.1,1.5] (referred to as "a single monomer" in the ASTM terminology standard[1.7]).

HOSE

In the context of plastics and rubber, this term is applied to some kinds of tubing; thus e.g. garden hose is PVC or rubber tubing for watering gardens, brake hose is tubing conveying compressed air to pneumatic brakes, etc.

HOT BENCH

See under *temperature-gradient bar.*

HOT-SET TEST

A test originally developed for crosslinked polyethylene cable coverings, to provide an indication whether the required minimum degree of crosslinking has been attained. The test is now also occasionally used for other products, e.g. heat-shrink tubing. In the test, a standard dumbbell specimen of the material is suspended, under a specified static load, for a specified period (typically 15 minutes) at a specified elevated temperature. The resulting elongation is measured, and then the recovery after a specified cooling period. Both the elongation and the recovery are influenced by the degree of crosslinking of the material. Relevant standards include IEC Publication 811-2.1 and BS 6469, Section 2.1 (Clause 9).

"HUNDRED PERCENT MODULUS"

The value of the secant modulus (q.v.) at 100% elongation in a tensile test. At this elongation the strain is $(2\ell_0 - \ell_0)/\ell_0 = 1$, where ℓ_0 is the original length of the test specimen, so that the secant modulus is numerically equal to the stress (modulus = stress/strain, i.e. here modulus = stress/1). 100% modulus values (as well as those of "200% modulus" and "300% modulus", i.e. secant moduli at 200% and 300% elongation) are sometimes determined, and quoted, among other tensile properties of those plastics materials (e.g. EVA, some flexible PVC compositions) which, like most elastomers, give continuously-curving stress-strain plots without a well-defined straight initial portion, and thus cannot meaningfully be assigned regular modulus values.

HYDROLYTICALLY DEGRADABLE PLASTIC

A degradable plastic susceptible do degradation by hydrolysis[1.5,1.7]

IGNITION

Initiation of combustion[1.8,1.9,1.12,1.13] (q.v.). See also *spontaneous-ignition temperature.*

IGNITION SOURCE

A source of heat which initiates combustion[1.9] of (i.e. ignites) a combustible material or object[1.8]. In flammability tests, the various ignition sources used (cf. ISO 10093) are almost invariably externally applied[1.8].

IGNITION TEMPERATURE

See *minimum ignition temperature*.

IMPACT STRENGTH (IMPACT RESISTANCE)

(1) In the general, qualitative sense, impact strength or resistance is the ability of a material to withstand shock loading[1.1] (i.e. rapid application of force, resulting in a high rate of deformation) without failure (which would be typically manifested by fracture: - indeed one of the terminology standards[1.2] defines impact resistance as resistance to fracture under shock force).

(2) In more quantitative terms, impact strength or resistance is the energy expended (work done[1.1]) in breaking a defined test specimen under shock loading[1.5] in defined test conditions. In flexural impact strength tests, such as the Izod, Charpy and Dynstat tests, the energy is referred to the cross-section of the test specimen (which, in notched specimens, is the cross-section of the portion of the specimen directly under the tip of the notch)[1.5] : thus the units are those of energy per unit cross-sectional area, viz. $kJ.m^{-2}$ in the SI system or, in other systems, $kgf.cm.cm^{-2}$ and $ft.lbf.in^{-2}$. Also still encountered are the units of ft.lbf per inch of specimen width (notch length). Other standard impact tests, involving different modes of shock loading, widely used where appropriate to the plastics material tested, are the tensile impact tests and falling-weight tests. An advantage of the latter type of test it that suitable versions are applicable directly to plastics products (as distinct from specially prepared test specimens), such as pipes and fittings, extruded profiles, sheeting and films. Impact behaviour of plastics (with special reference to design data) is discussed in BS 4618:Section 1.2:1972. See also *relative impact strength*, and *toughness*.

Methods of determination of impact strength are given in the following standards.

Flexural impact tests

Izod : ISO 180:1993 (identical with BS 2782:Part 3:Method 350:1993). ASTM D 256-93.

Charpy : ISO 789:1993 (identical with BS 2782:Part 3:Method 359:1993). ASTM D 256-93.

Dynstat : DIN 53 435-1983.

Falling weight or dart tests

ISO 6603-1:1985 (identical with BS 2782:Part 3:Method 353A:1991) : falling-dart test for rigid plastics. ISO 6603-2:1989 (identical with BS 2782:Part 3: Method 353B:1991) : instrumented-puncture test for rigid plastics.

ISO 7765-1:1988 : falling-dart test for film and sheeting.

ISO 7765-2:1994 : instrumented puncture test for film and sheeting.

BS 2782:Part 3:Method 352D:1979 : test for thin sheet and film.

ASTM D 3029-94 : test for flat, rigid specimens.

ASTM D 4272-90 : dart impact test for films.

ASTMD 5420-93 : impacted-striker (Gardner impact) test for flat, rigid specimens.

Tensile impact tests

ISO 8256:1990 (identical with BS 2782:Part 3:Methods 354A and B:1991). ASTM D 1822-93.

Determination of brittleness temperature (q.v.) by impact tests

ISO 974:1980. ASTM D 746-95. ASTM D 1790-94.

Miscellaneous impact tests

ASTM D 3420-94 (Pendulum burst impact test).

ASTM D 3763-93 (High speed impact puncture test).

ASTM D 3998-81 (Impact test on plastometer extrudates). *Standard discontinued in 1990.*

ASTM D 4508-93 (Impact test on "chips", i.e. small specimens of different thicknesses).

ASTM D 4226-95 (Test for impact resistance of PVC building products).

ASTM D 4812-93 (Cantilever-beam impact strength of unnotched specimens).

INHIBITOR

A minor additive incorporated in a chemical system to restrain a reaction[1.1,1.5,1.7]. The ASTM definition[1.7] points out that, unlike a catalyst, an inhibitor is used up in the course of discharging its function.

INITIATOR

A minor additive incorporated in a chemical system to set in motion the required reaction (e.g. by supplying free radicals in a polymerisation system[1.1,1.5]). See also *catalyst*.

ASTM 5318-92 provides a test method for the determination of hydrolysable chloride in peroxy esters and dicarbonates used as initiators and chemical intermediates in the production and modification of polymers, including PVC, PE, and PP.

INSULATION RESISTANCE

(1) Defined[1.1,1.5], by reference to an arrangement of two electrodes set into, or in surface contact with, a test specimen, as the ratio of the direct (dc) voltage applied to the electrodes to the total (volume and surface) current flowing between them at a specified time after the application of the voltage. As pointed out in the two main terminology standards which provide a definition[1.1,1.5], insulation resistance is thus dependent upon both the volume and the surface resistance of the test specimen. (Cf. also ASTM D 1711-94). The unit of insulation resistance is the ohm. Standards relevant to methods of determination include BS 2782:Part 2:Method 232:1983 and IEC Publications 167-1964 and 345-1971.

(2) In the general case of insulation on a conductor carrying direct current, insulation resistance is the resistance offered by the insulation to the flow of current resulting from an impressed direct voltage[1.6].

IONOMER

Thermoplastic copolymer of an alpha-olefin with a minor proportion of a carboxylic acid, in which the acid groups are largely or fully neutralised with a base, so that they are in the form of ionic metal salts. In the commercial versions (*Aclyn*; *Bexloy*; *Iotek*; *Surlyn*) the comonomers are typically ethylene and methacrylic acid, and the metals most often sodium and zinc, although potassium and magnesium can also be used.

The ionic groups of an ionomer are arrayed into domains about 10 nm in size, in which the anionic components linked to the metal cations are provided by carboxyl groups of different polymer chains. The domains thus function as ionic cross-links within the polymer structure, imparting additional strength, toughness and stiffness in the solid state. At melt-processing temperatures the domains dissipate so that the ionomer behaves like a true thermoplastic, with - however - melt strength enhanced by some residual ionic linkages. At sheet-thermoforming temperatures the proportion of such remaining linkages is higher, and this imparts to the material particularly good hot-drawing characteristics.

IRREGULAR POLYMER

A polymer whose molecules cannot be described by only one species of constitutional unit in a single sequential arrangement.

NOTE - The polymer whose molecules consist of a random arrangement of the constitutional units

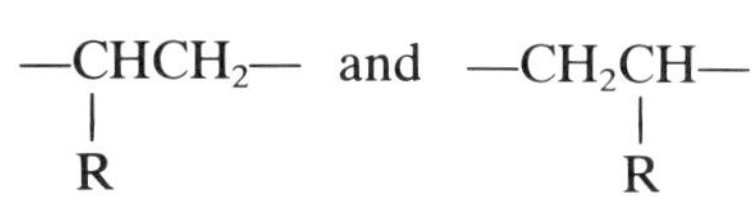

as in the fragment

is an irregular polymer.

See also *regular polymer*.

Reproduced, with permission, from ISO 472:1988.

ISOTACTIC POLYMER

A regular polymer, the molecules of which can be described in terms of only one species of configurational base unit (having chiral or prochiral atoms in the main chain) in a single sequential arrangement.

NOTE - In an isotactic polymer molecule the configurational repeating unit is identical with the configurational base unit. In the polymer $-[CH(CO_2R)CH(CH_3)]_n-$, if only one main-chain stereoisometric site of each constitutional repeating unit is defined, as in

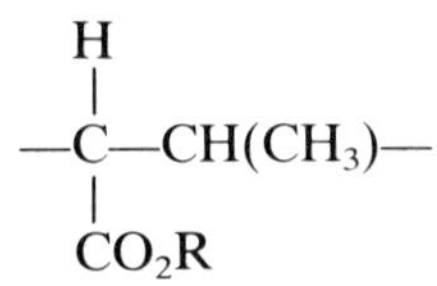

it is a configurational repeating unit and the corresponding polymer

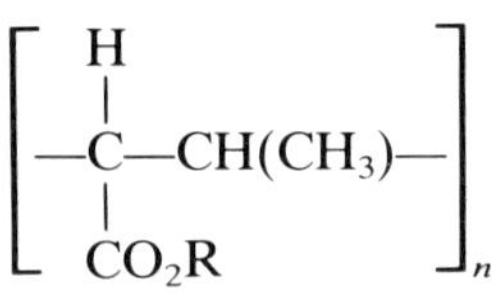

is an isotactic polymer. It is not a stereoregular polymer because the configuration at the stereoisomeric centre $-CH(CH_3)-$ is not defined. The following diisotactic polymers are stereoregular:

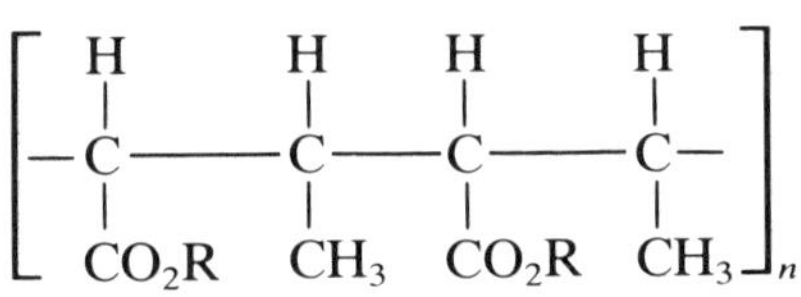

and

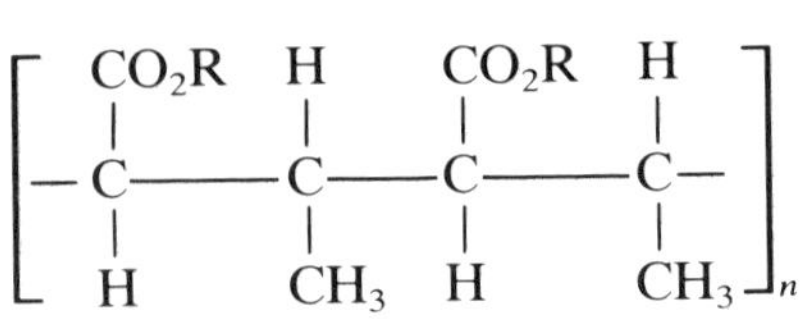

See note to *stereorepeating unit*.

See also *chirality* and *sequential arrangement*.

Reproduced, with permission, from ISO 472:1988.

KAISER EFFECT

Certain fibre-reinforced materials and other composites may suffer a lowering of the yield point, and ultimately fracture, when subjected to repeated stressing well within their elastic limit as established in a conventional test. In most cases such materials produce acoustic emissions - typically in the ultrasonic frequency range of 100 kHz to 2Mhz - under repeated stressing. The emissions can be detected and monitored with a suitable combination of transducers and sound amplifiers and filters. When a specimen of this kind of material is repeatedly stressed by an increasing applied force - and until it suffers structural damage in the process - acoustic emissions will continue occurring (in bursts) only when on its next consecutive application the force has reached and exceeded the previous level. The onset of permanent structural damage is signalled by the occurrence of acoustic emission before the previously applied force level has been reached. This phenomenon is known as the Kaiser effect. Thereafter, as the damage to the specimen develops into fractural failure, acoustic emissions will normally continue without further increase in the applied force.

The ratio of the applied force at which the acoustic emission begins again at each step in the test, to the maximum force applied previously, is termed "felicity ratio". So long as no damage has been suffered by the specimen, the felicity ratio will be equal to or greater than unity, and it will drop below this value when permanent damage has occurred.

KARL FISCHER TEST

A sensitive method of determination of low water contents (between 0% and 2%) in organic liquids and some other materials. Two relevant standards are ASTM E 203-92a and ASTM E 1064-92.

KILOPOND (kp)

An obsolescent term sometimes used (notably in Germany) instead of kilogram - force, i.e. the force represented by one kilogram weight (and thus equal to about 9.8N).

K VALUE

A number indicative of the average molecular weight of a polymer, calculated from the relevant results of measurements of viscosity of a dilute solution of the polymer. Originally introduced by Fikentscher (1932) as an index of molecular weight of cellulose polymers, the K value came to be generally used and quoted primarily for vinyl chloride polymers and copolymers (occasionally also for polyamides).

K value determined from viscosity measurements at 25°C on a solution of a vinyl chloride polymer in cyclohexanone, at a concentration of 0.005g polymer in 1 cm^3 of solution, is sometimes referred to as the Fikentscher K value. In general, of the three parameters, the nature of the solvent has the greatest effect on the numerical result of a determination.

The K value increases with molecular weight. Most commercial vinyl chloride polymers have K values in the range 50 to 80. As an example, PVC resins of weight-average molecular weights of, respectively, about 100 000 and 200 000 would have Fikentscher K values of 55 and 71. For a PVC resin, the Fikentscher K value determined in accordance with DIN 53 726 is related to the viscosity ratio (η_r - see under *viscosity*) by the expression

$$\log n_r = \frac{75K^2 \times c}{10^4 + 1.5 \times 10^3 \times K \times c} + \frac{K \times c}{10}$$

where K is the Fikentscher K value, η_r is the viscosity ratio, and c is the concentration in $g.cm^{-3}$.

LADDER POLYMER (DOUBLE-STRAND POLYMER)

A polymer whose main chain is double-stranded[1.1,1.5], and ladder-like in the sense that the two strands are connected at regular intervals by crosslinks (as the two uprights of a ladder are connected by the rungs). A typical double-stranded ladder polymer is a fused-ring system, the simplest structure being

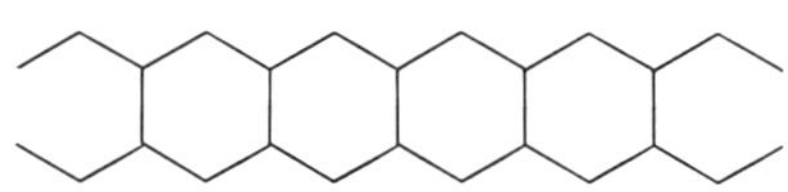

A double-bonded ladder ring structure is sometimes styled an *allene ladder*. Where variation in bonding produces a regular twisted-ladder configuration, the polymer is occasionally referred to as a *spiro-ladder* polymer.

Planar extension of the simple fused-ring ladder structure gives rise to a polymer sometimes termed "parquet" or "sheet" polymer

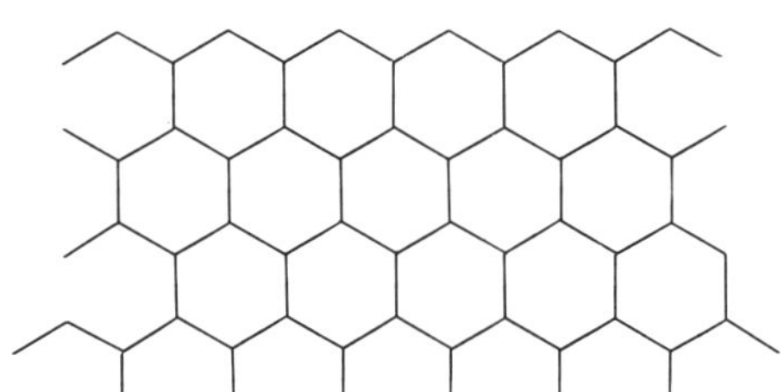

It may be noted that Amendment No 4 (1993) to the ISO terminology standard[1.5] additionally defines a *double-strand chain (ladder chain)* as a macromolecule that can be described by constitutional units joined so as to form an uninterrupted sequence of rings through two atoms. A macromolecule in which an uninterrupted sequence of polycyclic structures is joined through more than two atoms is styled a *multi-strand chain* by the same source. These definitions thus apply respectively to the above-illustrated structures of a single-ladder polymer and simple sheet polymer.

LAMINATE (noun)

All the three main plastics terminology standards[1.1,1.5,1.7] adopt the same general-dictionary definition of this term, which is not specific to plastics laminates. According to this definition, a laminate is the product that results when two or more layers of a material or materials are bonded together. However, in the field of plastics the term is most commonly applied to a particular kind of laminate, viz. a composite consisting principally of a resin, typically thermoset (such as UP, epoxy or vinyl ester resin), and laminar fibrous reinforcement in various forms (e.g. loose fibres, rovings, mats, fabrics). The definition of "laminate" in an ASTM terminology standard for high-modulus reinforcing fibres and their composites (ASTM D 3878-95c) is more closely in line with this usage. In the context of so-called high-pressure decorative laminates (q.v.), the fibrous laminar reinforcement is often paper, and the thermoset resin is typically an aminoplastic.

The material, product and test standards relevant to plastics laminates include the following (see also under *high-pressure decorative laminate*).

Material and product standards: Laminated sheets : ISO 1642:1987 (specification for industrial laminated sheet - withdrawn in 1994); ISO 4586:1995 (high-pressure decorative laminate sheets); BS EN 438:1991(1997) (decorative high-pressure laminates); BS 2572:1990(1997) (phenolic and epoxy resin cotton-fabric-reinforced sheets); BS 3953:1990(1997) (synthetic resin bonded, woven glass fabric laminated sheets); BS 4965:1991(1997) (decorative laminated sheets, boards and panels); BS EN 60893:1995 (multi-part specification for industrial rigid sheet for electrical purposes); BS 7332:1990 (decorative continous-sheet laminates). Printed-circuit bases : BS 4584:Part 103:Section 103.1:1990; BS EN 60249 (multi-part specification:1993-1995). Laminated industrial rods and tubes : BS 6128:Parts 1 to 13. GRP vessels and tanks : BS 4994:1987 (specification for design and construction). Classification of visual defects in GRP laminates : ASTM D 2563-94. Laminates from glass fabrics (woven yarn and woven rovings) : ASTM D 4357-85(1990).

Test standards: Preparation of laminate specimens for tests: ISO 1268:1974 (= BS 2782:Part 9:Methods 920 A to C:1977(1989). Temperature of deflection under load of high-strength laminates : ISO 75-3:1993 (= BS 2782:Part 1:Method 121D:1994). Tensile testing : BS EN 2561:1995 (carbon-fibre reinforced plastics). Heat resistance of thermosetting laminated sheet : BS 2782:Part 1:Method 131D:1978. Determination of inter-layer cohesion in laminated tube : BS 2782:Part 3:Method 346A:1984(1990). Resistance of laminates (inter alia) to chemical reagents : ASTM D 543-95. Determination of acetone-soluble matter in laminated or moulded phenolic products : ASTM D 494-91.

LAMINATED SHEET (as applied to thermosets)

A sheet consisting of superimposed layers of paper, fabric, veneer, or felt (mat) that have been substantially impregnated with a thermosetting or curable resin, and bonded together under pressure, with or without heat, to form a single piece.

NOTE - Other ingredients, for example colouring matter, may be incorporated.

Reproduced, with permission, from ISO 472:1988.

The BS terminology standard[1.1] provides the same definition. The ASTM standard[1.7] does not define the term.

LATEX

See under *dispersion.*

LAY-FLAT FILM OR SHEETING (LAY-FLAT TUBING)

Tubular film or sheeting produced by the process (often referred to as the tubular process) comprising the following principal elements.

1. Extruding a relatively thin-walled tube of a thermoplastic through a ring die.
2. Closing the end of the tube between nip rollers positioned above the die.
3. Expanding the tube radially - by inflating (blowing) with air introduced through the die - to the extent required for the desired final wall thickness (the ultimate film or sheeting thickness). The part of the tube undergoing expansion is known as the "bubble".
4. Winding up the resulting blown, cooled, expanded tubular film or sheeting in the flattened (collapsed) form in which it emerges from the nip rollers.

LEACHING

Essentially extraction by a liquid - especially by percolating water - of substances removable by dissolution from a material (typically one in particulate form). Some of the methods of ASTM C 871 (chemical analysis of thermal insulation materials for leachable chloride, fluoride, silicate, and sodium ions) are applicable to plastics used as thermal insulants.

In the specific context of rubber latex products, leaching is the operation of washing the product in water to remove water-soluble substances[1.2(Add. 20),1.3].

LIMITING OXYGEN INDEX

See *oxygen index.*

LINEAL (noun)

In the plastics context, a term sometimes applied to extruded plastics profiles, especially rigid PVC profiles used for the production of window frames.

LIQUID CRYSTAL (LIQUID-CRYSTALLINE) POLYMER (LCP)

A polymer whose molecular chains contain relatively rigid, rod-like units which can assemble into ordered arrays when the polymer is in the liquid state (melt or solution) and at a temperature within an appropriate range. The units are rod-like in the sense that their axial dimensions are significantly greater than their radial dimensions. In a true crystal constructed from such units, their axes would be lined up in a regular way (i.e. the arrangement would have full orientational order), and the units would be regularly disposed in fixed positions in the crystal lattice (i.e. there would be complete positional order). The arrays into which the units can assemble in the liquid phase of a liquid-crystalline polymer are not true crystals: they are characterised by a high degree of orientational order (long-range in the sense that millions of units can be substantially lined up), but their positional order is only short-range. The direction of axial alignment of the units is termed the "director" (usually denoted by *n*). Various arrangements are possible, the main general kinds being termed smectic, nematic and cholesteric (see figure). The ordered liquid phase of a liquid-crystalline polymer is referred to as the mesophase, and the units in the polymer molecule responsible for the liquid-crystalline properties are called mesogenic units or mesogens (the latter term was first applied to the entire rod-like molecules of non-polymeric liquid-crystalline materials). In a liquid-crystalline polymer, the mesogenic units may be located within the molecular chain (main-chain liquid-crystalline polymers), or they may be pendant to it (side-chain liquid-crystalline polymers). Polymers of the latter kind are discussed in an informative paper by Attard and Williams (1986).

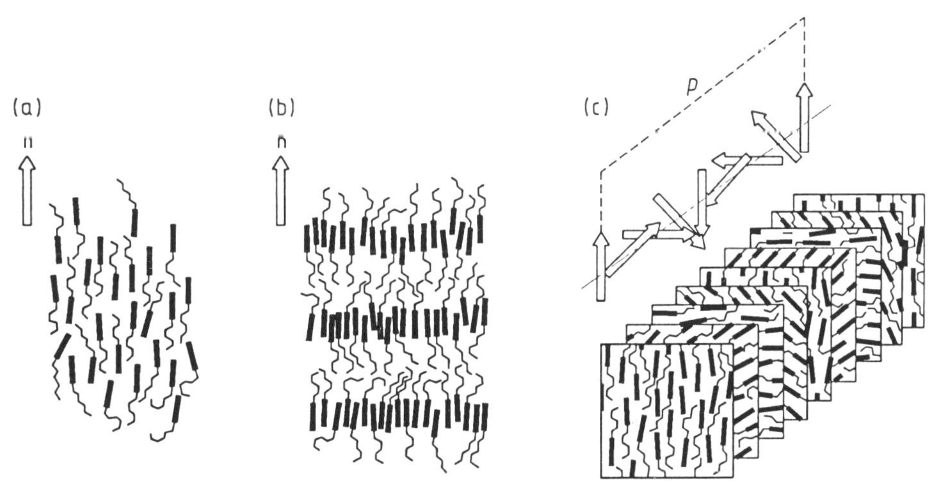

Schematic illustration of the arrangement of simple (non-polymeric) molecules in nematic (a), smetic (b) and cholesteric (c) liquid crystals. After Attard and Williams (1986).
Reproduced by permission from *Chemistry in Britain*, October 1986

Polymers liquid-crystalline in the melt are referred to as thermotropic LCP's. Commercial representatives of this group, which are thermoplastic, melt-processable LCP's, are *HX, Rodron, RTP, Xydar, Vectra* and *Xenite*. LCP's forming the mesophase in solution are called lyotropic. Some commercial aramid (q.v.) polymers used for the production - by a variant of solution spinning - of high-strength, high-modulus fibres, are of this kind. A notable example is poly(p-phenylene terephthalamide) the polymer of *Kevlar* fibres (q.v.), which forms a mesophase in concentrated sulfuric acid, within a defined range of temperature and solution concentration. Because of the complications of the solution systems, and the aggressive nature of the solvents used, the practical applications of commercial lytropic polymers are essentially limited to the production of fibres and some specialist films. The thermotropic polymers offer the advantageous combination of high molecular weight and relatively low melt viscosity in processing and, in the products, high tensile strength and modulus, high heat-distortion temperature, and very good resistance to solvents and UV radiation, coupled with low flammability and smoke generation. The high degree of molecular orientation induced in LCP products by melt flow under shear in processing gives rise to pronounced anisotropy of mechanical properties. This effect can be reduced by incorporation of fillers. ASTM D 5138-90 provides a standard specification for thermotropic liquid-crystalline polymers. The nature and properties of LCP's are the subjects of a number of authoritative publications (cf. e.g. McArdle 1989; Ceferri 1991; Donald and Windle 1992).

LIVE POLYMER

Term coined to indicate reversible changes said to be undergone by certain thermoplastic polyurethanes (cf. e.g. *Isoplast 2500*) during melt processing. The changes are believed to consist in partial depolymerisation to a low-molecular-weight state at the melt temperature - resulting in ease of flow - followed by rapid re-polymerisation on cooling.

LOSS ANGLE (in damping)

δ (rad) : The phase angle ϕ between stress and strain, the tangent of which is the loss factor *d*.

See also *loss factor* (in damping)

Reproduced, with permission, from ISO 472:1988

For a definition of loss angle in the electrical context (i.e. dielectric loss angle) see e.g. ASTM D 1711-91 or IEC Publication 250. A concise ISO definition is also available[1.5].

LOSS FACTOR (IN DAMPING), *d*; LOSS TANGENT; tan delta (dimensionless)

(1) The tangent of the loss angle δ between stress and strain.

(2) The ratio of the loss modulus (loss compliance) to the storage modulus (storage compliance) measured in tension, shear, compression, or longitudinal compression.

$$\tan \sigma_E = \frac{E''}{E'}$$

$$\tan \sigma_G = \frac{G''}{G'}$$

$$\tan \sigma_K = \frac{K''}{K'}$$

$$\tan \sigma_L = \frac{L''}{L'}$$

NOTES : 1. It is given by the quotient $\tan\delta = M''/M' = C''/C'$.
2. Commonly $d = \tan\delta$ is used as a measure of the damping of a system subject to forced vibrations.

Reproduced, with permission, from ISO 472:1988

For a definition of loss factor (loss index) in the electrical context see Ref. 1.5 and/or ASTM D 1711-91.

LOSS MODULUS, M''(Pa) (in damping) [and LOSS COMPLIANCE, C''(Pa^{-1})]

The imaginary part of the complex modulus (compliance).

NOTE - It is a measure of the energy lost (dissipated) during a loading cycle.

See also *complex modulus* and *complex compliance*.

Reproduced, with permission, from ISO 472:1988.

For some methods of measurement of loss moduli of plastics materials see ASTM D 4065-94 and ISO 6721:1994.

LOSS ON IGNITION (IGNITION LOSS)

Term used for the weight lost by a reinforced or filled plastics material (typically a glass-reinforced thermoset resin) in a combustion test - cf. e.g. ISO 1172-1975 and ASTM D 2584-94. For composites consisting only of resin and glass reinforcement, the resin being completely combustible in the test, ignition loss can be equated with the original resin content, and the glass content can be calculated by difference. See also *ash content; ashing*.

LUBRICANT

An additive incorporated at a low level (often about 1% or less) in a plastics material, as an aid in melt processing[1.1,1.5], or to counteract sticking to a mould[1.1]. One of the terminology standards[1.1] refers also to the application of a lubricant to the mould surface for the latter purpose, although a substance employed in this way is more commonly styled a "mould release" (q.v.). The term "internal mould release" is, however, sometimes used for a lubricant additive in a plastics moulding compound. The materials used as lubricants in plastics include metal stearates, waxes of various kinds (including fatty-acid amides and esters, hydrocarbon and low-molecular-weight polyethylene waxes), silicones and fluorocarbons. Some commercial lubricants are listed, under their trade names, in Section 3 of Part 3 of this book.

LUMINOUS FLUX

See under *radiant flux.*

MACROMOLECULE

Two of the three main terminology standards[1.1,1.5] define this term. Both quote the IUPAC definition, according to which a macromolecule is simply a molecule, organic or inorganic, that is very large. The term thus applies, inter alia, to the molecule of a polymer, and is sometimes used in this more specific sense in the context of polymers and plastics.

MAIN CHAIN; BACKBONE

That linear portion of a macromolecule to which all other chains (long or short or both) may be regarded as being pendant; where two or more chains could equally be considered to be the main chain, the one is selected that leads to the simplest geometrical representation of the molecule.

Reproduced, with permission, from ISO 472:1988:Amendment 4:1993.

MARK-HOUWINK EQUATION

Also known as *Mark-Howink-Sakurada (M-H-S) equation,* or *Staudinger-Kühn equation.* See under *viscosity.*

MAR RESISTANCE

In essence, abrasion or wear resistance (q.v.) as measured or expressed in terms of changes in some relevant optical property or properties, brought about by abrasive disturbance of the plastics material's surface. In the mar resistance test of ASTM D 673-88, in which abrasion is effected by the impact of a standard particulate abrasive, the properties determined - before and after the test treatment - are gloss (for transparent, translucent or opaque plastics materials) and haze (for transparent materials).

MELAMINE-FORMALDEHYDE (MF) RESIN

The kind of amino resin (q.v.) that results from polycondensation of melamine with formaldehyde[1.1,1.5]. The mutually identical definitions in the ISO and BS terminology standards[1.1,1.5] additionally include, as an alternative to formaldehyde, "a compound that is capable of providing methylene bridges".

The ISO and ASTM terminology standards also define *melamine plastic(s)*. The ASTM standard[1.7] confines its definition to plastics based on resins which are condensation products of melamine and aldehydes, whilst the ISO standard[1.5] defines a melamine plastic (in the singular) as one based on those amino resins (in the plural) in which melamine is the amine predominant (by mass) among the amine or amide reactants used in the productions of the resins.

Melamine-formaldehyde resins and/or moulding compounds are available in the following commercial ranges. *Ampal; Bakelite; Cymel; Granular; Isomin; Melmex; Melopas; Pantoxyl* (MF-based bonding agents); *Plenco; Resimene; Ultraplas.*

Standards concerned with MF materials include the following.

Specifications for moulding materials: ISO 4896:1990 (melamine/phenolic moulding materials); ASTM D 704-94 (MF moulding compounds); ASTM PS 15-94 (provisional standard for a number of moulding compounds, including MF inter alia).

Determination of extractable formaldehyde in MF materials: ISO 4614:1977 (1985) identical with BS 2782:Part 4:Methods 462A and 462B:1978 (1993).

Specification for rigid laminated sheets based on melamine resins: BS EN 60893-3-3.1995.

MELAMINE/PHENOLIC RESIN

A synthetic resin, thermosetting or in the thermoset state, that is a product of co-condensation of melamine and phenol with an aldehyde (in practice normally formaldehyde). The term is also sometimes applied to a physical mixture of melamine-formaldehyde and phenolic resins. A melamine/phenolic moulding compound consists of a melamine/phenolic resin in the uncured or partly cured state, incorporating appropriate additives, such as curing agents, colorants, fillers, etc. ISO 4896:1990 provides a standard specification for melamine/phenolic moulding materials. Commercial versions are available from some of the suppliers of melamine-formaldehyde and phenolic materials.

MELAMINE PLASTIC(S)

See under *melamine-formaldehyde resin.*

MELT FLOW INDEX (MFI) (MI)

The term originally applied in a number of standard specifications to the numerical expression of the flow rate of polyethylene (in some cases also polypropylene) polymers and compounds, as determined in a standard extrusion-plastometer test. The term was also widely used in the literature and in the plastics industry, and is still encountered, although it has now been abandoned in standards, both terminological and technical, in favour of *melt flow rate* (q.v.) The standards, regular or tentative, which originally dealt with or referred to the MFI include the following: ISO/R 292:1967; ISO/DR 749:1964; BS 2782:1970:Method 105C; BS 3012:1970; BS 3412:1976 (all references to MFI in this standard changed to "melt flow rate (MFR)" by amendment in 1982); ASTM D 1238-65T (this standard discouraged the use of "melt flow index", and promoted "melt flow rate" for any but polyethylene materials). The standard numerical expression of the MFI of a polyolefin material represents the total weight, in grams, of the material extruded (in prescribed batches, termed "cut-offs" in some specifications) in a specific total time (usually 10 minutes) under standard conditions (including those of piston load, temperature, and extrusion-die dimensions) in a specified extrusion plastometer. The MFI test was first developed for the "grading" of polyethylene polymers and plastics, and the MFI values of such materials were sometimes referred to (especially in the UK) as their "grades". Some fluoropolymers were also similarly graded - cf. e.g. Dexter (1953). The MFI value provides some indication of the molecular weight of the polymer (the higher the molecular weight the slower the extrusion in the standard conditions, and hence the lower the MFI), and of the ease of processing, which should increase with increasing MFI. However, in both these respects the indications are only general, as the MFI value is strongly influenced by a number of factors, including polymer chain branching, the composition (if a plastics compound is being tested), polymer stability (both thermal and rheological), and the conditions of measurement, whilst the shear rates in MFI plastometers are substantially different from those typical for industrial injection moulding and extrusion processes.

MELT FLOW RATE (MFR)

The general meaning of the phrase is self-evident. In the field of thermoplastics, the term denotes the fluidity of a polymer melt, determined in a specific extrusion plastometer in prescribed test conditions (cf. ISO 1133:1991; BS 2782:Part 7:Method 720A:1979; ASTM D 1238-90). Most commonly it is the mass of a polymer flowing in a given time that is determined (sometimes referred to as *melt mass-flow rate*, but still abbreviated to MFR); the usual units in which the test results are quoted are grams per 10 minutes. ISO 1133 also provides for determination of *melt volume-flow rate* (MVR) in cubic centimetres per 10 minutes. Other standards concerned with MFR of plastics are: ISO 12086-2:1993 (this standard gives, inter alia, test methods and conditions for determining the MFR's of melt-processable fluoropolymers); ASTM D 3364-90 (determination of MFR's of PVC and some rheologically unstable thermoplastics); ASTM D 3835-90 (determination of the rheological behaviour of reinforced and unreinforced thermoplastics in a capillary rheometer); ASTM D 2116-90 and D 3159-89 (dealing, inter alia, with the determination of MFR's of, respectively, FEP an ETFE fluorocarbon moulding and extrusion materials in a modified version of the extrusion plastometer of ASTM D 1238). See also *melt flow index*.

MELT FRACTURE

Surface roughness and/or distortion of an extrudate, associated with non-laminar flow. The appearance may range from relatively small-scale, uniform roughness (sharkskin), through rippling, ridging ("bambooing"), to such gross manifestations as helical (screw-thread) effects, or complete irregularity of the extrudate. Factors believed to be instrumental in melt fracture are high stresses in the melt where its flow converges at the inlet of the extrusion die, and slippage of the material at the die exit. Melt fracture occurs when the stress(es) involved exceed the rupture stress of the melt. This is the basis of the concept of critical stress for melt fracture, i.e. the shear stress below which the fracture is not observed (at a given melt temperature - critical stress increases with increasing temperature). Critical stress value for polypropylene has been reported as 0.2 MPa, and for poly(ethylene terephthalate) and some polyamides as 1.0 MPa (cf. DiRaddo et al., 1995).

MELTING POINT

See *melting temperature*.

MELTING TEMPERATURE (T_m)

According to the only definition in any of the three major terminology standards[1,5], the melting temperature of a semi-crystalline polymer (q.v.) is the temperature at which the crystalline phase ceases to exist when the polymer is heated in specified test conditions. However, unlike the crystals of simple organic molecules, polymer crystallites do not melt within a temperature range sufficiently narrow to be represented without undue error by a single value (the melting point). The melting of polymer crystallites is not a sharp transition, but occurs progressively over a range of temperatures which can be quite wide. This is because of different stabilities of individual crystallites, caused by differences in their sizes and degrees of perfection (cf. e.g. Sharples 1972). None the less, single temperature values are often quoted as the melting temperatures of semi-crystalline polymers. Apart from the fact that a value so quoted does not reflect the extent of spread of crystallite melting over a temperature range, it may also be associated with one of a number of methods of determination which measure their versions of T_m at different points of the crystallite melting range. Thus the method used to obtain any quoted T_m value should always be stated, if known. Of the standard methods in use, the one providing T_m most closely consonant with the standard definition of melting temperature is that in which a specimen of semi-crystalline polymer is heated at a controlled rate on the hot-stage of a polarising microscope whilst being viewed in transmitted polarised light. The presence in the polymer of the crystalline phase gives rise to optical anisotropy detectable in this mode of viewing, so that the temperature at which the anisotropy disappears coincides fairly closely with that associated with the completion of melting of the crystallites. The standards featuring the polarising microscope method are the mutually identical ISO 3146:1985 (Method B) and BS 2782:Method 125 B:1991. The method was also the subject of ASTM D 2117 (standard discontinued in 1994).

Four other standard methods of determining T_m values (in one case also a melting range) are available. Their main relevant points are summarised in the table. See also *first-order transition*.

MER

See *monomeric unit*.

METALLOCENE CATALYSTS

A family of organometallic catalysts, occasionally also referred to as "single-site" catalysts (q.v.) at present used principally for the production of special versions of polyolefins with a range of "tailored" properties including as required, increased strength and toughness, transparency, elastomeric properties, reduced permeability, and improved melt strength and hot-tack. These catalysts represent a significant advance in polymerisation technology over the conventional Ziegler-Natta catlysts in a number of respects. The Ziegler-Natta catalysts are functionally non-uniform, in the sense that their active sites vary in both activity and accessibility to the monomers undergoing polymerisation, whereas these features are uniform in metallocene catalysts: this makes for a high, controllable degree of uniformity in the polymer. Appropriate metallocene catalysts are particularly suitable for the production of highly stereo regular alpha-olefin polymers, notably polypropylene; they are also more effective than conventional catalysts in polymerising monomers of large molecular size, and in copolymerising ones that differ substantially in type and/or size. The general activity of metallocene catalysts is higher than that of Ziegler-Natta catalysts, and they leave less residues in the polymer. The expressive if rather inaccurate description "metallocene polymers" is sometimes applied to polymers produced with the aid of metallocene catalysts. See also under *polyolefin* and *polyethylene*.

METHACRYLATE/BUTADIENE/STYRENE (MBS[2.1,2.2,2.3,2.11])

These terpolymers are produced by grafting methyl methacylate and styrene onto polybutadiene chains. Their properties are generally similar to those of ABS (q.v.), with the following exceptions. As in the case of MABS (q.v.) tetrapolymers, the presence of methyl methacrylate in an appropriate proportion confers transparency on the polymer, and improves resistance to thermal degradation, whilst the absence of acrylonitrile results in somewhat lower modulus and improved impact strength. The tensile strength and

STANDARD METHODS OF DETERMINATION OF T_m

TEST METHOD	TEST SPECIMEN	STANDARD	APPLICABILITY	TEMPERATURE MEASURED AS MELTING TEMPERATURE	REMARKS
Controlled heating in capillary tube melting apparatus.	Thin shaving, 5 mm long.	ISO 1218:1975, Method A.	Polyamides (especially homopolymers).	Temperature at which sharp edges of the specimen disappear.	Temperature measured referred to as "melting point".
Controlled heating in capillary tube melting apparatus.	Thin shaving, 5 mm long, *or* powdered polymer.	BS 2782:Method 123A:1976(1993).	Synthetic resins with relatively sharp melting points.	Temperature at which sharp edges of the specimen disappear, *or* at which the powder first begins to transmit light through about half its bulk.	Temperature measured referred to as "melting point".
Controlled heating in capillary tube melting apparatus.	Thin film (10μm to 20μm), *or* powdered polymer.	ISO 3146:1985 Method A, identical with BS 2782:Method 125A:1991.	All polymers and their compounds, even if there is no crystalline phase.	Temperature at which the specimen begins to change shape.	Temperature measured referred to as "melting temperature".
Controlled heating on a heating block (hotplate).	Single granule of polymer supporting a tilted cover glass.	ISO 1218:1975, Method B.	Polyamides (especially homopolymers).	Temperature at which the specimen fails to support the cover glass.	Temperature measured referred to as "melting point".
Controlled heating on a heating block (hotplate).	Single granule of polymer supporting a tilted cover glass.	BS 2782:Method 123B:1976(1993).	Polyamides, but method "may be suitable for use with other crystalline polymers".	Temperature at which the specimen just ceases to support the cover glass.	Temperature measured referred to as "melting point".
Controlled heating in the Fisher-Johns melting apparatus.	Particle of moulding powder supporting a tilted cover glass.	ASTM D 789-94.	Polyamides.	Temperature at which the specimen ceases to support the cover glass.	Temperature measured referred to as "melting point".

STANDARD METHODS OF DETERMINATION OF T_m

TEST METHOD	TEST SPECIMEN	STANDARD	APPLICABILITY	TEMPERATURE MEASURED AS MELTING TEMPERATURE	REMARKS
Controlled heating in the Fisher-Johns melting apparatus.	Sliver of polymer shaved from a moulding, placed between two cover glasses.	ASTM D 2116-91a, Section 13.2. *NB* Method not featured in the 1995 edition of the standard.	Fluorinated ethylene-propylene copolymer (FEP).	Temperature at which the specimen changes from solid to melt at the edges.	Temperature measured referred to as "melting point".
Thermal analysis (both DTA and DSC given as alternatives).	Powder, cut film, or fibre of prescribed size and weight.	ISO 3146:1985, Method C, identical with BS 2782:Method 125C:1991.	All polymers containing a crystalline phase, and their compounds.	Temperature at onset of melting (T_o) and peak melting temperature (T_p) on the DTA or DSC curve.	
Thermal analysis (not specified whether DTA or DSC).	Appropriate to thermal analysis apparatus used.	ASTM D 1457-92; Section 3.1.	PTFE.	Peak temperature on the DTA or DSC curve.	
Thermal analysis (DSC).	Appropriate to DSC apparatus used.	ASTM D 2116-91a, Section 13.1 *NB* In the 1995 edition of the standard this method is described in Section 11.3.	FEP.	Peak temperature on the DTA or DSC curve.	Temperature measured referred to as "melting point".

deflection temperature under load of MBS polymers are below those of ABS. Commercial versions of MBS materials are represented in the *Cryolite* range (Röhm GmbH).

METHYL CELLULOSE

A methyl ether of cellulose. Commercial grades, used as water-soluble thickening agents and gums have a degree of substitution of the cellulose hydroxyl groups of about 1.5 to 2.0 (cf. e.g. ASTM D 1695-77(1989)).

METHYL METHACRYLATE/ACRYLONITRILE/ BUTADIENE/STYRENE (MMABS[2.2,2.11])

In polymers of this kind, the four comonomers, methyl methacrylate $\left(\begin{array}{l} CH_2{=}C{-}CH_3 \\ \qquad\; | \\ \quad O{=}C{-}O{-}CH_3 \end{array}\right)$, acrylonitrile $\left(\begin{array}{l} CH_2{=}C{-}H \\ \qquad\; | \\ \qquad CN \end{array}\right)$, butadiene $(CH_2{=}CH{-}CH{=}CH_2)$, and styrene $(CH_2{=}CH)$, are combined in a two-

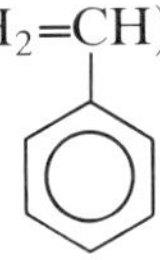

phase, graft tetrapolymer (cf. MBS), which may be regarded as acrylonitrile/butadiene/styrene (ABS) - q.v., with some of the acrylonitrile replaced by methyl methacrylate. The properties of MABS are generally comparable with those of ABS, except that the presence of methyl methacrylate comonomer in an appropriate proportion brings the refractive indices of the two phases sufficiently close to impart transparency to the material; it also improves to some extent resistance to degradation by heat.

MABS materials are available in the *Terlux* (BASF) range. ISO 10366 provides a standard specification for MABS moulding and extrusion materials (Part 1:1993 - Designation system and basis for specification. Part 2:1994 - Preparation of specimens and determination of properties).

MICROBALLOONS

In the plastics context this term is a virtual synonym for "microspheres" (q.v.). It is also sometimes used for the rather larger hollow spheres (typically of polyethylene) floated, as a barrier layer against evaporation, on the surface of liquids in storage tanks.

MICROBIOLOGICAL ATTACK

See under *biostat*.

MICROBIOSTAT; MICROBIOCIDE

See under *biostat*.

MICROCELLULAR PLASTIC (MICROCELLAR FOAM)

In its most general sense, the term is applicable to any cellular plastic (q.v.) with fine cells, and has long been used, for example, in such phrases as "microcellular PVC". In the microcellular plastics and rubbers employed widely in production of certain kinds of shoe and sandal soles, the cell size is indeed small, but the cell population density is relatively low, so that the density of the cellular material remains comparatively high (e.g. in microcellular flexible PVC soling it may be, typically, about 65% of that of the solid plastic).

More recently the term has also been applied specifically to a group of cellular plastics with very small cell size (typically 1 μm to 10 μm) and high cell population density (typically 10^9 to 10^{12} cells per cm^3). This combination of properties results in a very low density of the cellular material (down to 5% of that of the solid polymer in some cases). Such microcellular materials, developed in the late 1980's by the Massachusetts Institute of Technology and some commercial companies in the USA, are produced by blowing with CO_2 of nitrogen. The main advantage claimed for them includes a combination of good mechanical and heat insulation properties. See also *cellular plastic* and *supermicrocellular foam*.

MICROSPHERES

Spherical-particle fillers used in plastics. The microspheres may be solid (commonly spherical glass micro-beads - cf. e.g. *Ballotini; Econ-Glass*) or, more usually, hollow (cf. e.g. *Bubblekup; Ceramcel; Eccospheres; Expancel Holospheres; Q-Cell; Zeeospheres*). The hollow microspheres are produced from glass, ceramics, and some thermosetting resins (phenolic; epoxy). Thermoplastic hollow microspheres are also made (cf. e.g. *Expancel*). Microspheres are supplied in various diameter ranges (ASTM D 2841-94, a standard for the sampling of hollow microspheres, refers to diameters of 1mm or less). Some varieties are available in thin-walled or thick-walled versions. Hollow microspheres are used as weight-reducing fillers in some plastics materials, and in the production of syntactic foams (q.v.), especially those based on thermosetting resins.

MIGRATION

In the general sense of the term, migration is the transfer of a substance from a plastics material (or a rubber) to another solid material when the two are in contact[1.1,1.5]. A typical example is migration of a plasticiser from plasticised PVC (PVC-P). The standard definitions also make the fairly obvious point that migration is usually undesirable[1.1,1.5]. In the context of tests for loss of plasticisers from plastics by migration, the latter is more narrowly defined, in quantitative terms, as the loss of mass, in prescribed test conditions, by a flat specimen closely sandwiched between two layers of a specified material (cf. ISO 177:1988 and the corresponding Method 465C:1990(1993) in Part 4 of BS 2782). The specimen arrangement is reversed in a test for the effect of migration of plasticiser from PVC-P into polyethylene upon the electrical loss tangent of the latter (BS 2782:Part 2:Method 241A:1992). Staining or discolouration can be caused by migration of some components or residues from plastics or rubber materials into contacting bodies. Some examples are: the staining - sometimes accompanied by embrittlement and distortion - of the PVC skins of crash pads in motor car interiors, attributed to migration of amino-catalyst residues from the polyurethane foam filling of the pad (Wilson et al. 1988; Zimmerman and Austin 1988); and staining of contacting materials by certain compounding additives (typically some antioxidants) migrating from rubber compositions (for a relevant general test see ASTM D 925-88(1994), and ASTM D 2151-95 for such migration-staining of PVC). On a strict interpretation of the general definition of migration, the so-called "FDA migration cell" may be considered misnamed, as it is a device for *extraction with liquids* of those components of plastics materials which are susceptible to such extraction in the test concerned (cf. ASTM D 4754-93).

MINIMUM FILM-FORMING TEMPERATURE

In latex (q.v.) technology, the limiting temperature (as determined in a test - cf. e.g. ISO 2115) above which the particles of a layer of latex deposited and dried on a substrate coalesce to form a coherent, continuous film, which will normally be homogeneous and transparent in the absence of pigments or other particulate additives from the original latex. The minimum film-forming temperature is always a few degrees Celsius higher than the white-point temperature (q.v.).

MODULUS OF ELASTICITY (ELASTIC MODULUS)

The ratio of stress (σ) to the corresponding strain (ε) in a material below the proportional limit[1.1,1.5,1.10], i.e. $M = \sigma/\varepsilon$. As the strain is dimensionless (a ratio), the units of modulus are those of stress (Pa in the SI system). The several elastic moduli of interest as properties of plastics materials are named in accordance with the mode of application of stress, viz. modulus of elasticity in tension (tensile modulus; Young's modulus), in shear (shear modulus; modulus or rigidity), in flexure (flexural modulus), and in bulk compression (bulk modulus).

MODULUS OF ELASTICITY IN BULK COMPRESSION (BULK MODULUS)

The ratio of external hydrostatic (isotropic) pressure to the resulting bulk compression (q.v.), below the proportional limit[1.5]. Usual symbol : K. SI unit : Pa. For a material which conforms to the laws of classical elasticity, the bulk modulus is related to Young's modulus by the expression $K = E/3(1-2\mu)$, where E is Young's modulus and μ is Poisson's ratio (q.v.).

MODULUS OF ELASTICITY IN FLEXURE (FLEXURAL MODULUS; TANGENT MODULUS OF ELASTICITY)

The ratio, within the proportional limit, of stress to corresponding strain in bending, calculated by the appropriate theory. Usual symbol : E_B. SI unit : Pa. See also *modulus of rupture in bending*. Methods of determination of the flexural modulus of various plastics materials are given in the following standards. ISO 178:1993 (identical with BS 2782:Part 3:Method 335A:1993); ISO 6721:Part 3:1994 (flexural vibration method); ISO/TR 4137:1978 (alternating flexure method); BS 2782:Part 3:Method 332A:1976(1983) (flexural stiffness of plastics film); ASTM D 747-93 (apparent bending modulus (stiffness) of relatively flexible materials - determination by cantilever beam); ASTM D 790-95a (test relevant inter alia to reinforced plastics and electrical insulating materials).

MODULUS OF ELASTICITY IN SHEAR (SHEAR MODULUS; MODULUS OF RIGIDITY)

The ratio of (nominal) shear stress (σ_{ij}) to the corresponding shear strain (γ), i.e. $G = \sigma_{ij}/\gamma$,[1.5] within the proportional limit of the material. Usual symbol : G. SI unit : Pa. Shear modulus is usually determined by testing in torsion. For a material which conforms to the laws of classical elasticity, the shear modulus is related to Young's modulus by the expression $G = E/2(1+\mu)$, where E is Young's modulus and μ is Poisson's ratio (q.v.). Methods of determination of shear modulus are given in the following standards. ISO 458:Part 1:1985; ISO 6721:Part 2:1994 (replacing ISO 537:1989, withdrawn in 1994); ASTM D 1043-92 (determination of apparent modulus of rigidity (stiffness) of plastics as a function of temperature); ASTM D 4065-94 (tests of dynamic mechanical properties over a wide temperature range).

MODULUS OF ELASTICITY IN TENSION (YOUNG'S MODULUS; ELASTIC MODULUS; TENSILE MODULUS)

The ratio of tensile stress (σ) to the corresponding strain (ε) below the proportional limit of the material[1.5], i.e. $E = \sigma/\varepsilon$. Usual symbol : E. Sl unit : Pa. Some definitions extend the term Young's modulus also to the modulus of elasticity in longitudinal compression[1.10]. For a material which conforms to the laws of classical elasticity, Young's modulus is related to thc bulk and shear moduli respectively by the expressions $E = 3K(1-2\mu)$ and $E = 2G(1+\mu)$, where K is the bulk modulus, G the shear modulus, and μ is Poisson's ratio (q.v.). Methods of determination of Young's modulus of plastics are given in the following standards. ISO 527:1994 (identical with BS 2782:Part 3:Methods 321 and 322:1994); ASTM D 638-95; ASTM D 882-95a (for thin plastics sheeting).

MODULUS OF RUPTURE IN BENDING

The value of maximum tensile or compressive stress (whichever causes failure) in the extreme fibre of a beam loaded to failure in bending, computed from the flexure equation:

$$S_b = Mc/I$$

where:
M = maximum bending moment, computed from the maximum load and the original moment arm,
c = initial distance from the neutral axis to the extreme fiber where failure occurs, and
I = initial moment of inertia of the cross section about the neutral axis.

NOTE 1 - Where the proportional limit in either tension or compression is exceeded, the modulus of rupture in bending is greater than the actual maximum tensile or compressive stress in the extreme fiber, exclusive of the effect of stress concentration near points of load application.

NOTE 2 - If the criterion for failure is other than rupture or attaining the first maximum of load, it should be so stated.

Reprinted, with permission, from the Annual Book of ASTM Standards, copyright American Society for Testing and Materials, 1916 Race Street, Philadelphia, PA 19103.

A less extensive definition, in somewhat different terms, is provided in Ref. 1.1.

MONOMER

A compound (of low molecular weight[1.7] in the typical case) whose molecules can react to form a polymer[1.7], each molecule providing one or more of the polymer's constitutional units[1.1,1.5].

MONOMERIC PLASTICISERS

This term is sufficiently common and well understood not to cause serious confusion by its essential inaccuracy. With a few exceptions (like e.g. the diallyl phthalates), the "monomeric" plasticisers are not monomers in the correct sense of polymerisable compounds. The adjective has been traditionally applied to all plasticisers of relatively low molecular weight (that for many common ester plasticisers lies in the range 300 to 500) in contradistinction to high-molecular-weight plasticisers, which include the group of truly polymeric polyesters with molecular weights up to about 9 000.

See also *polymeric plasticisers* and *plasticisers*.

MONOMERIC UNIT; MER

The largest constitutional unit contributed by a single monomer molecule in a polymerisation process.

Reproduced, with permission, from ISO 472:1988.

MONOSIL PROCESS

A process for the manufacture of chemically crosslinkable polyethylene products, such as cable coverings or heat-shrink tubing. The process is commonly regarded as a one-step operation (in contradistinction to the *Sioplas* process - q.v.) because one passage of the polyethylene feedstock through the processing equipment - typically an extruder with good compounding action - combines the functions of grafting siloxane side-chains onto the polyethylene (for subsequent crosslinking - see *crosslinked polyethylene*) and forming the resulting graft into a product. The feedstock used in the process is a blend of a suitable grade or grades of polyethylene (sometimes with a modifying polymer, e.g. EVA), an initiator (normally a peroxide), vinyltrimethoxy silane, a crosslinking catalyst, stabiliser(s), and any other requisite additives. The product is subsequently immersed in hot water or steam, to hydrolyse the methoxy groups of the grafted side-chains, and to effect crosslinking by condensation of the resulting hydroxyls, in accordance with the following general reaction scheme:

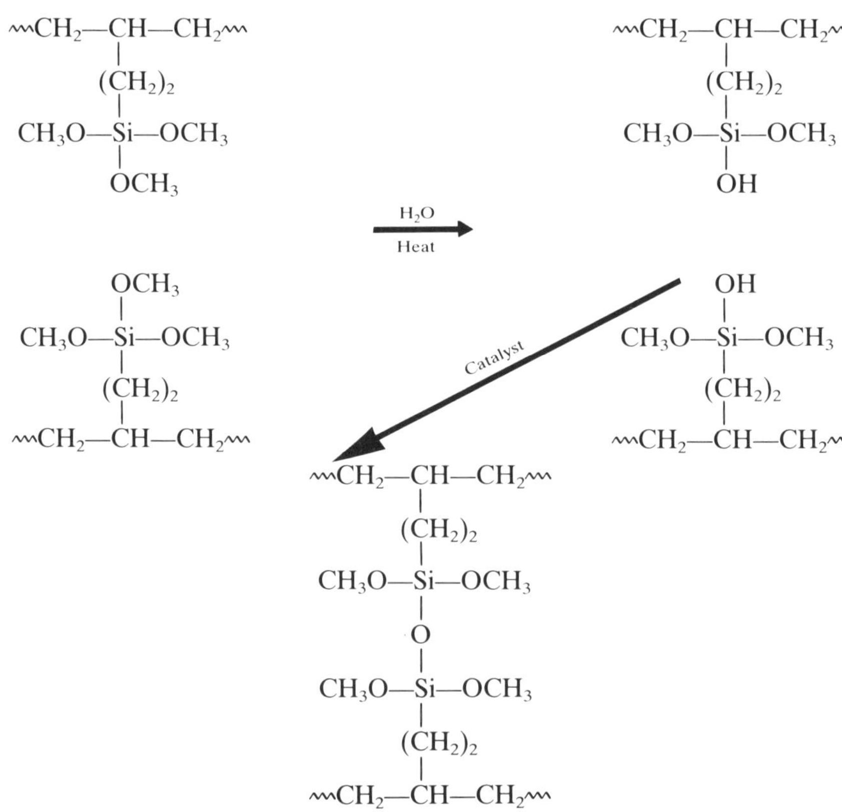

More than one of the methoxy groups on a silicon can be hydrolysed and the hydroxyl group so formed participate in further crosslinking.

MOULD RELEASE

A substance applied to the mould surface to counteract sticking of the moulding and facilitate its removal. Many mould releases are of the silicone or low-molecular-weight fluorocarbon type. Some are based on the materials used also as lubricants (q.v.) in polymer compounds. For trade names of some commercial mould releases see Section 3 of Part 3 of this book.

MOULDING INDEX (ASTM D 731-95)

An expression - in the form of a number with a subscript - of the mouldability of a thermosetting moulding powder in a standard cup mould under specified test conditions. The number is the total minimum force, normally in newtons, required to close the mould in the standardised moulding operation, and the subscript gives the closing time in seconds - e.g. 1686_{15} for the minimum closing force of 1686N resulting in a closing time of 15 s. Other fairly similar standard cup-moulding methods produce expressions of mouldability in terms of "cup flow time" (BS 2782:Method 720B:1979(1986)) and "closing time" (DIN 53 465:1963) : the times are given in seconds in both cases.

MOULDING (MOULD) SHRINKAGE (S_m)

The difference in dimensions between a plastics moulding and its parent mould cavity, measured soon after production, both the mould and the moulding being at room temperature[1.1,1.5]. Moulding shrinkage is the earliest part of the total possible shrinkage S_T of a plastics moulding, in that it may be followed by *post-shrinkage* S_p which may take place during a post-moulding treatment (e.g. annealing), in storage, or in service[1.1,1.5]. Moulding shrinkage is usually expressed in terms of change in linear dimensions, and quoted as a percentage, or as a length per unit length (e.g. mm/mm, in/in). In general, polymers which crystallise to a substantial extent on cooling from the melt exhibit higher moulding shrinkage than non-crystalline polymers. Some fairly representative values are (linear moulding shrinkage, %): PPS 0.2; unplasticised PVC 0.4: PC 0.6; PP 1.8; POM 2.2; PE-LD 3.2; FEP 4.5. Two relevant standards are ASTM D 955-89 and ISO 294:Part 4 (this part still in draft form in 1996).

MULTIPOINT DATA

Data obtained in a number of related tests, in which a given property of a plastics material is determined over a range of test conditions. An example would be the values of Young's modulus measured at each of a number of temperatures within a given range. ISO 11403 (three parts - Part 1:1994, Part

2:1995, Part 3 still in preparation in 1995) deals with the acquisition and presentation of comparable multipoint data for a number of properties of plastics materials. See also *single-point data.*

NATIONAL SANITATION FOUNDATION (USA)

An independent organisation established to serve as a practical focus for many of those concerns of industry, business, official regulatory bodies and the public, which relate to materials, products, services, equipment and procedures that have a bearing on public health and the environment. One of the NSF's main activities in this field is the development of relevant standards and the awarding of a quality mark to materials and products which comply with the standards. An example of a standard of this kind set up for a group of plastics materials is NSF standard 54 for "flexible membrane liners", i.e. plastics sheeting of many kinds used to line water reservoirs and as seepage-barrier linings for the containment of industrial effluents, pollutants, chemicals, wastes, etc.

NATURAL WEATHERING

See *weathering.*

NERVE

The definitions in those major terminology standards in which they are given[1.2,1.3] relate specifically to raw rubber[1.2] unvulcanised rubber[1.2,1.3] or rubber mixes[1.2,1.3]. In that restricted context nerve is defined as elastic resistance to deformation.

NETWORK PARAMETER

An expression of the degree of crosslinking in terms of the notional average molecular weight of a main-chain segment between crosslinks (M_c). If the degree of crosslinking is sufficiently high for the average length of the "free" ends of the molecular chains to be considered negligible, the network parameter is defined by the relationship $M_c = w/q$, where w is the molecular weight of the repeat unit of the polymer chain and q is the crosslink density (q.v.).

NEUTRAL AXIS

See under *section modulus.*

NIBS

See under *gel* (4).

NOMINAL STRESS

The stress at a point calculated, by simple elastic theory, on the basis of the relevant nett cross-section without allowance for any stresses that may arise due to the presence of geometric discontinuities such as holes, grooves, fillets, etc.[1.10]

NORMAL STRESS

The stress, or component of stress, perpendicular to a given plane on which the forces concerned are acting[1.1,1.10]. Such perpendicular stress may be either tensile, i.e. due to a force or forces directed away from the plane, or compressive i.e. due to force(s) directed towards the plane[1.10]. The recommended standard symbols for normal stress[1.1,1.5] are σ and σ_{ii}.

NORMAL TEMPERATURE AND PRESSURE (NTP)

Arbitrarily agreed standard reference conditions of temperature and pressure, viz. the ice-point temperature (273.15 K; 0°C; 32°F) and the pressure of one standard atmosphere (1.01325 bar; 101.325 $kN.m^{-2}$; 760 torr; 760 mm Hg). Volumes of gases measured in different conditions are often recalculated to NTP. See also *standard temperature and pressure.*

NOTCH SENSITIVITY

The degree to which such stress-concentrating features as notches in test specimens, or cracks, scratches, particulate impurities, and sharp profile changes in mouldings, increase the susceptibility of a plastics material to fracture[1.1,1.5]. In general, materials which tend to fail in a ductile mode are less notch-sensitive than those prone to brittle fracture in the same conditions[1.1,1.5].

NOVOLAK[1.1,1.5,1.7] (NOVOLAC[1.7])

A phenolic[1.1,1.5] (phenolic-aldehyde[1.7]) resin (A-stage, i.e. fusible and solvent-soluble): a prepolymer of relatively low molecular weight, which results when a predominant proportion of a phenol is reacted with formaldehyde under acid conditions (phenol to formaldehyde molar ratio greater than one to one[1.1,1.5]; typically about 1.2:1). A novolak normally[1.1,1.5] remains permanently[1.7] thermoplastic, unless it is heated with the requisite amount of a compound (commonly hexamethylenetatramine or formaldehyde) that generates additional methylol groups[1.7] which provide means of crosslinking so that the material ultimately becomes an infusible, insoluble thermoset[1.1,1.5] (cf. *resite*). See also definition in ISO 10082:1991.

NYLON

In common usage, both technical (especially American) and general, this term is a virtual synonym for "polyamide" (q.v.). International (ISO) and British formal terminology favours "polyamide", with some justification as the word reflects the essential structural feature of the polymers concerned, whereas "nylon" is a made-up name. Following the discovery and laboratory development of fibre-forming polyamides by W.H. Carothers working in the chemistry department of Du Pont in the early 1930's, a committee set up by Du Pont considered some 400 names proposed for this group of materials, before finally arriving at "nylon" (cf. Bardsley, 1995). For reasons of company policy, the name was never registered as a trademark. Instead, it was formally announced in the press in October 1938 as a generic term for Du Pont polyamide 66 (which was being made into textile fibres). The name became very popular with the general public, few of whose members would readily understand or recognise the term "polyamide". Even in the more formal technical vocabulary there is no unanimity : differing terminological preferences are reflected in the relevant entries in the main terminology standards. Thus the ASTM standard[1.7] does not define "polyamide", and under "polyamide plastics" refers the reader to its definition of "nylon plastics", whilst the ISO and BS standards both define - in mutually identical terms - "polyamide and "polyamide plastic", the only reference between them to nylon being the BS definition of this term[1.1] (as a generic term for certain synthetic polyamides).

NYLON PLASTICS

Defined in the ASTM terminology standard[1.7] essentially as plastics whose base polymers are synthetic long-chain polyamides in which repeated amide groups are integral constituents of the molecular chain backbone. See also *nylon* and *polyamide.*

OFFSET YIELD POINT

The notional yield point on a stress-strain curve, associated with the offset yield stress [1.5] (see diagram under offset yield stress).

OFFSET YIELD STRESS (OFFSET YIELD STRENGTH)

In a stress-strain plot, the stress value at which the curve departs from linearity by a specified percentage of strain (termed the "offset")[1.1,1.5]. In practice, a fractional offset is usually specified (often 0.2%). The offset value should be stated in a test report in conjunction with the value of the offset yield stress: e.g. "0.2% offset yield stress : XXXMPa", or "yield stress at 0.2% offset : XXX PMa".

The offset yield stress is helpful in the case of materials which give rise to gradually curving stress-strain plots without a definite yield point (cf. also ISO/R 527-1966 (E), ASTM D 638 M-89 and ASTM E 6-89).

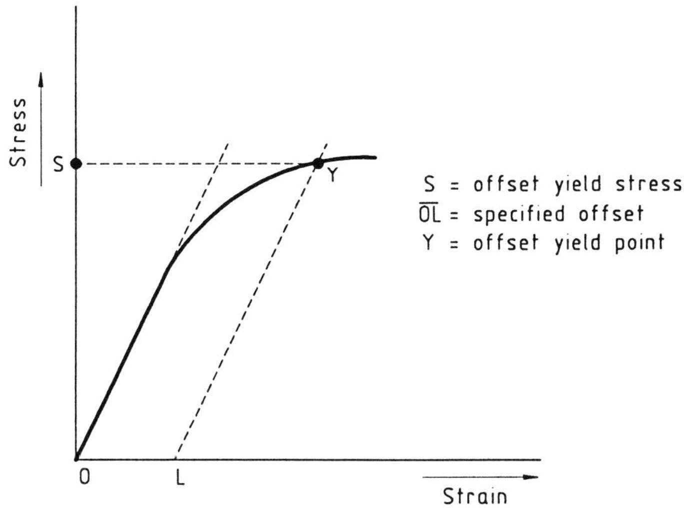

OIL (PLASTICISER) ABSORPTION

The weight in grams of the appropriate oil, needed - in a standard test - to "wet out" 100 grams of a pigment or filler (so that a paste of stipulated consistency is formed). (ISO 787 Part 5 and the equivalent BS 3483 Part 7; ASTM D 1483; DIN 53199 - now abandoned in favour of the ISO standard). The standard test methods, originally evolved for use with pigments for paints (with, typically, linseed oil as the test liquid), are applicable also to plasticiser absorption by pigments and fillers for plastics (ASTM D 281 specifically allows liquids other than a standard oil). An oil absorption test may be used to estimate the plasticiser demand (q.v.) of these additives in a particular compound. The factors reducing the oil absorption, and the plasticiser demand, of fillers and pigments include increasing particle size (which reduces the total surface area for the same volume), good particle packing characteristics, increasing particle crystallinity, and the presence of coatings on the particle surface.

"OIL CANNING"

A term sometimes applied to local distortion - typically in the form of ripples, waves, or bulges - on the surface of some PVC cladding (siding) exposed to solar heating in service. The effect results from recovery of stresses originally induced in the course of cooling after extrusion. Siding made from PVC compositions with high deflection temperature under load (q.v.) is less prone to this fault. Apart from the standard test for resistance of PVC siding to surface distortion (ASTM D 3679-94), thermomechanical analysis (q.v.) can be an aid to evaluating the stress-relaxation behaviour of siding material, and can provide some indication of desirable cooling conditions, or possible annealing temperatures, in production (cf. Changfoot et al., 1976).

OLEFIN

Literally "oil-forming substance" (from the Latin *oleum*, oil and *facio*, I make). An early term for an alkene (q.v.) based on the original finding that gaseous members of the homologous series could add-on chlorine to form oily products.

OLEFIN PLASTICS

Plastics whose base polymers are polymers of an olefin, or copolymers of two or more olefins, or copolymers of an olefin or olefins with other monomers, in which the olefin monomeric component(s) predominate(s) by mass[1.1,1.5,1.7]

NOTE - The definition in the ASTM terminology standard[1.7] apparently also includes copolymers in which the olefinic and non-olefinic components are in equal parts by mass, in that it refers to "the olefins being at least 50 mass %".

A standard classification system for *olefinic thermoplastic elastomers* and their compounds is the subject of ASTM D 5593-94. ASTM D 1603-94 provides a method for the determination of carbon black content in olefin plastics compounds. For other standards concerned with particular plastics materials falling within the very broad category of olefin plastics, see under *polyolefin*, *polyolefin plastics*, *ethylene plastics*, and individual olefin polymers and copolymers.

OLIGOMER

A molecule consisting of a small number of repeated constitutional (monomeric[1.7]) units (a unit comprising one or more species or groups of atoms[1.1,1.5]), or a substance composed of such molecules[1.1,1.5,1.7]. See also *dimer, trimer,* and *tetramer*. An oligomer molecule can be viewed as that of a polymer with a very short molecular chain, but it cannot be formally defined in this way, because the chains are so short that the removal or addition of a few, or even one, of the constitutional units can affect the properties of an oligomeric substance. In contrast, it is a characteristic feature of a true polymer that the shortening or lengthening of its chain molecules by a few constitutional repeating units has no significant effect on the material's bulk properties.

OLIGOMERISATION

The process[1.1,1.5,1.7], or action, of producing an oligomer (q.v.) From a single species of monomer, or from a number (mixture[1.1,1.5,1.7]) of monomers.

OPEN CELL

In a cellular plastic (q.v.) a cell interconnecting with the neighbouring cells, or - if located at the surface of the cellular plastic - open to the exterior[1.1,1.5]. A cell may be open because its walls are incomplete or, as in one type of polyurethane open-cell foam, because they are entirely absent: in the latter case the material has a completely open, filamentary three-dimensional network structure in which the cells are defined - in skeletal outline - only by the lengths of the filaments (sometimes referred to as fibrils) between the junction points of the network.

OPEN-CELL CELLULAR PLASTIC

A cellular plastic (q.v.) containing relatively few, if any, closed cells (so that its cell structure is almost completely interconnecting[1.1,1.5]), or in which open cells predominate[1.7]. See also *open cell.* Methods of determination of the respective percentages of open and closed cells in a cellular plastic are given in the following standards: ISO 4950:1981; BS 4370:Part 2:1993:Method 10; ASTM D 1940 (discontinued in 1970) and ASTM D 2856-94.

ORGANOLEPTIC

In its most general meaning this adjective describes ability to stimulate a sense organ, or to perceive a sensory stimulus. A standard of terminology relating to sensory evaluation of materials and products (ASTM E 253-89a) defines the term as "relating to a property of a sample perceived by the sense organs", but at the same time dismisses it - rather arbitrarily - as obsolete, in favour of "sensing" (defined as "pertaining to the sense organs").

In the context of plastics, the term "organoleptic properties" (sometimes simply "organoleptics") normally refers to the possession of odour or taste by a material, or a material's ability to acquire or impart an odour or taste (specifically when in contact with, or proximity to, other substances). In common usage, a plastics material with "good organoleptics" would be one satisfactorily inert in all the above respects. Standard methods for determining (by a test panel) odour and taste transfer from packaging materials are described in ASTM E 462-84(1989) and BS 3755:1964(1971). Sensory assessments are believed to obey the so-called Weber-Fechner law, according to which sensory reaction is proportional to the logarithm of the stimulus. ASTM standards dealing with sensory evaluation of materials and products are listed in ASTM Publication PCN 03-51808836. Another source of relevant information is ASTM Special Publication STP 434 (PCN 04-434000-36) - Manual on Sensory Testing Methods.

ORGANOSOL

See under *paste (1)*.

ORGANOTIN CATALYSTS

A group of organotin compounds used as catalysts in the production of polyurethane foams, where they are often combined with amine catalysts, e.g. triethylene diamine. Stannous octoate is probably the most widely used member of the group. Some others - e.g. dibutyltin dilaurate or dioctyltin dilaurate - are also employed as catalysts in formulations for polyethylene chemically cross-linked via trimethoxysilane side chains (either grafted onto the polyethylene backbone or introduced by copolymerising ethylene with vinyltrimethoxysilane) - see also *cross-linked polyethylene*.

ORGANOTIN STABILISERS

A group of organotin compounds providing powerful heat stabilisers (some with a measure of light-stabilising action) for PVC compositions. Commercial organotin stabilisers are typically compounds of the general formula I, sometimes used in mixtures with compounds of the formulae II and III.

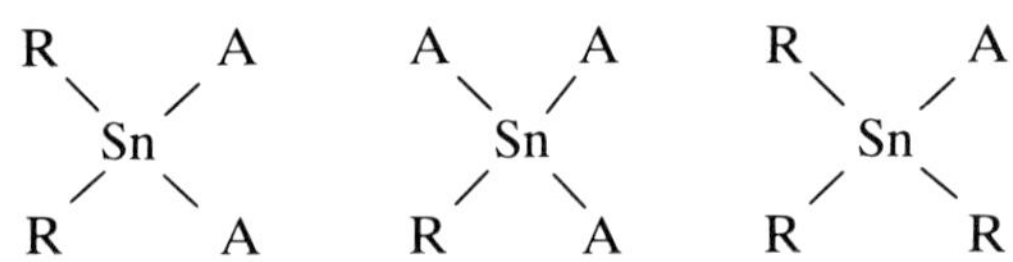

I II III

In these general formulae R represents an alkyl or an ester group, and A is commonly derived either from a mercapto compound (a mercaptide, a mercaptoacid or ester) or from a carboxylic acid or ester (e.g. maleic acid or its esters).

OXIDATIVELY DEGRADABLE PLASTIC

A plastic susceptible to degradation through oxidation[1.5,1.7]. Standards directly relevant in this connection include the following.

ISO 4577:1983 (determination of oxidative stability in air of polypropylene and propylene copolymers).

BS 7816:Part 1:1995 (discussion of techniques for monitoring diffusion-limited oxidation).

ASTM D 2445, discontinued in 1986 (test for thermal-oxidative stability of propylene plastics).

ASTM D 3012-94 (test for thermal-oxidative stability of propylene plastics - biaxial rotator method).

ASTM D 5510-94 (standard practice for heat ageing of oxidatively degradable plastics).

See also *antioxidant*.

OXIDATIVE INDUCTION TIME (OIT)

Sometimes also called *oxidation* induction time. It is a measure of the thermal oxidative stability (q.v.) of a polymer specimen in a test applicable in particular to polyolefin moulding and extrusion materials, and the materials of such polyolefin products as pipes and pipe fittings, mouldings, and sheeting. In the test, a specimen - held isothermally at an elevated temperature in a DTA or DSC apparatus - is exposed to a regulated stream of oxygen, and the time is measured between the first contact with oxygen and the onset of significant degradation. This time (usually expressed in minutes) is the OIT of the specimen's material. The test temperature is at least 200°C, i.e. comparable with polyolefin processing temperatures, so that the specimen is molten during the test.

If the specimen contains an antioxidant, the OIT may be interpreted as the time over which this additive inhibits oxidative degradation in the conditions of the test, and the test is often used to assess the effectiveness of antioxidants, especially in polyethylene. However, the applicability of the test in this context is not universal: there is evidence that it is only the phenolic type of antioxidants whose effects are reflected, or reflected fully, by the OIT; that even at the same concentration of phenolic antioxidant the OIT can vary with differences in the polyethylene compound and sometimes from one DSC apparatus to another; that OIT values do not provide fully reliable indications of the likely stability of polyolefins in service; and that they do not correlate with the stability times obtained in oven-heating tests, which are normally conducted at lower temperatures (with the polymer in the solid state) and in which the effects of stabilisers other than phenolic antioxidants - e.g. HALS (q.v.) - can be very pronounced.

The OIT test is featured in ISO/TR 10837:1991, BS 2782:Part 1:Methods 134A and B:1992, and ASTM D 3895-94. Some manufacturers of polyolefin plastics (e.g. Hoechst) and stabilisers (e.g. Ciba) also operate it in accordance with their own versions of the method.

OXIDATIVE THERMAL STABILITY

See *thermal oxidative stability*.

OXYGEN INDEX (OI)

Sometimes referred to as "limiting oxygen index" (LOI). The minimum percentage volume of oxygen in a mixture of oxygen and nitrogen that is just sufficient to maintain flaming combustion of a material tested, under specified conditions[1.5,1.8,1.9], in the form of strip specimens mounted vertically and ignited at the top so that they burn candle-fashion. (Cf. ISO 4589-1984 or its equivalent BS 2782:Part 1:Method 141:1986, and ASTM D 2863-91).

PARTLY CRYSTALLINE POLYMER

See *semi-crystalline polymer.*

PASTE

(1) In the context of PVC, a liquid or semi-liquid suspension, comprising - as the basic components - fine particles of PVC polymer, or sometimes copolymer (paste grades, made by the emulsion or microsuspension processes) and a plasticiser or plasticisers in which the particles are dispersed. Various additives are also usually present in a PVC paste (notably heat stabilisers), as required for processing and product properties. Suspension-produced PVC polymers are also sometimes incorporated in pastes for particular effects. The special features of the paste, that enable a wide range of plasticised PVC products to be conveniently made by manipulating the paste as a liquid (see e.g. Titow, 1990), are the resistance of the paste-grade polymer particles to swelling by, or solution in, the plasticiser at room temperature, and the elimination of that resistance by heating at suitably elevated temperatures. As the temperature of the paste rises, the plasticiser penetrates into the polymer particles which swell and merge, first loosely and then progressively more fully, until - if the temperature is sufficiently high and the time adequate - the process culminates in complete mutual solution of polymer and plasticiser, with the formation of a homogeneous plasticised-PVC melt - on cooling this solidifies into solid plasticised-PVC material. Overall, the transformation may be regarded as a phase inversion : the conversion of a suspension of particulate polymer in the continuous phase of liquid plasticiser(s), into a molecular-level dispersion (solution) of plasticiser(s) in a continuous polymer matrix, solid at room temperature. Thus the transformation is concluded (before the final cooling step) by the fusion of polymer and plasticiser into a homogeneous plasticised PVC material[1.5]. This has led to the use of "fusion" (or "plastisol fusion"[1.5]) as the general term for the entire transformation process. "Gelation" is also sometimes used, interchangeably with "fusion" in this general sense, especially in the context of industrial paste processing, and also in some publications. However, strictly speaking, gelation of a PVC paste is a relatively early consequence of its heating in processing, when - as a result of absorption of the plasticiser(s) by polymer particles - the originally liquid paste becomes a comparatively weak (but already cohesive) gel.

The term *plastisol*, formally defined[1.1,1.5,1.7] as a (liquid[1.7]) suspension of a finely divided (PVC[1.7]) polymer (or copolymer[1.7]) in a plasticiser, is often used as a synonym for "paste" in the PVC context (in fact "paste" does not even rate a definition in any of the three main terminology standards[1.1,1.5,1.7]). However, the latter term is common usage in the UK and elsewhere (cf. e.g. ISO 4612:1979 and ISO 4575:1985).

Three special kinds of PVC paste are denoted by the following terms.

Organosol: a paste in which the liquid suspending medium is a mixture of plasticiser(s) with a relatively volatile organic liquid or liquids[1.1,1.5,1.7] (e.g. white spirit, toluene, xylene, methyl ethyl ketone). Organosols are useful in applications where low initial paste viscosity is required, bu the final product

is to be relatively hard, with comparatively low plasticiser content. This ultimate condition is achieved by evaporation of the organic liquid in the final stages of processing.
Plastigel: a paste thickened (by the addition of thickening agents) to a gel-like[1.1,1.5] consistency (resembling butter or putty in some cases). Plastigels are used where such consistency is a desirable applicational feature, e.g. in some spread-coating operations, or in paste moulding.
Rigisol: this term was coined for pastes formulated to combine relatively low plasticiser contents (for comparatively hard ultimate products) with fluidity sufficient for normal paste processing (cf. e.g. Titow, 1990).

The following standards give methods of preparation of standard PVC pastes for test purposes : ISO 4612:1972 and ASTM D 1755-92.

Standards relevant to determination of viscosity of PVC pastes include the following.

Determination at high shear rates (with a Severs extrusion viscometer) : ISO 4575:1985, and ASTM D 1823-95.
Determination at low shear rates (with a Brookfield rotational viscometer) : ASTM D 1824-95.
General instructions on rotational viscometry of polymer emulsions and dispersions : BS EN ISO 3219:1995 (identical with BS 2782:Part 7:Method 730B:1994) and ISO 2555:1988.

(2) A standard definition of an *adhesive type paste*[1.16] relates to the kind of composition exemplified by a starch-based paste obtained when a suspension of starch granules in water is heated, and the resulting hydrolysed product finally cooled.

PERFLUORO ALKOXYL ALKANE (PFA)

A regular or modified copolymer of tetrafluoroethylene and a perfluorinated alkyl vinyl ether. The perfluoroalkyl groups in the vinyl ether side-chain are often trifluoromethyl or heptafluoropropyl groups, but in polymers for certain special applications some of these groups may be replaced by ones with ionic functionality. Such polymers have the general formula

$$(CF_2.CF_2)_w—(\underset{\displaystyle OR}{CF_2.CF})_x—(CF_2.CF_2)_y—(\underset{\displaystyle OR_1}{CF_2.CF})_z—$$

where R is typically $–CF_2.CF_2.CF_3$ or $–CF_3$, and R_1 can be for example, $–(CF_2)_nCOOH$; $–(CF_2)_nSO_3H$; $–(CF_2)_nSO_2NH_2$; with n having the values 1, 2, or higher.

Functional polymers of this kind which contain more than about 15 mole percent of the ionic comonomer are used as water-soluble polyelectrolytes. Those with a lower ionic comonomer content are solids, used inter alia as separators in chemical cells. ASTM D 3307-93 is a standard specification for PFA moulding and extrusion materials. Also relevant are ASTM D 4591-93a and ISO/DIS 12086 (still under discussion in 1996). See also *fluoropolymer*.

PERFLUORO(ETHYLENE-PROPYLENE) COPOLYMER; FLUORINATED ETHYLENE-PROPYLENE (FEP) COPOLYMER

A copolymer of tetrafluoroethylene and hexafluoropropylene. In the commercial versions the hexafluoropropylene ranges between about 50% and 90% by weight: this is reflected in the density range (2.14 g cm^{-3} to 2.17 g cm^{-3}). The copolymers are mentioned in ISO/DIS 12086 (still under discussion in 1996), and - by reference to their function as the base polymers of perfluoro(ethylene-propylene) plastics - in two of the major terminology standards[1.1,1.5]. FEP copolymers are semi-crystalline (T_m 285°C to 290°C, depending on the method of determination), the degree of crystallinity varying between about 40% and 70% - according to the cooling and/or annealing conditions. Unlike PTFE, FEP can be processed as a regular thermoplastic (albeit at relatively high temperatures). The service temperature range of FEP products is usually quoted as - 80°C to 205°C (up to about 250°C on short-term exposure without stress). The electrical properties of FEP copolymers, and their resistance to chemicals and to weathering, are comparable to the corresponding properties of PTFE, but their impact resistance is significantly higher. Standard specifications for FEP extrusion and moulding materials, tubes, and sheeting and film are provided, respectively, by ASTM D 2116-91, ASTM D 3296-93, and ASTM D 3368-94. FEP copolymers are included, inter alia, in ASTM D 4591-93 - a test method standard for determination (by DSC) of the temperatures and heats of transition of fluoropolymers. The commercial ranges in which FEP materials are available include *Hostaflon FEP*, *Neoflon*, and *Teflon FEP*.

PERMEABILITY

In the general sense of the term (as distinct from its mathematical expressions), the permeability of a polymeric material is its ability to allow the passage of a permeant (penetrant) through it. In the most typical situation, which is also the usual arrangement in permeability tests, the polymeric material is a laminar barrier (a film, sheet, slab, coating layer on a substrate, or the wall of a container) and the permeant is a gas, vapour, or liquid. The permeant enters (is sorbed into) one face of the barrier, is transported through the thickness of the material by diffusion (q.v.), and passes out (is desorbed) at the opposite face. The standard ISO[1.5] definition of permeability is a shorter, less complete version of the above.

It can readily be seen that the permeability of a plastic to a given permeant should be a function of the chemical nature, morphology, and composition of the plastic, and that the barrier thickness and area involved should also be factors. This is indeed so, and the relevant relationships may be represented, in a general way, by the following expression (applicable in steady-state conditions).

$$J/A = P(F/L) \qquad (1)$$

In this expression J is the flux (flow rate) of the permeant, i.e. the amount (in suitable units, e.g. of mass or volume, or in moles) passing through the barrier in unit time; A is the area through which the passage takes place; F is the driving force actuating the passage, e.g. the difference between the concentrations (Δc), pressures (Δp), temperatures ($\Delta\theta$) or chemical potentials ($\Delta\mu$), of the permeant at the 'upstream' and 'downstream' faces of the barier; L is the barrier thickness; and P is the permeability coefficient which embodies (i.e. whose value will reflect) the effects of chemical and physical interactions in the system permeant/barier material.

As can be seen, F/L is the gradient of the driving force across the barrier. It is self-evident that, for a given value of F, the gradient will increase with decreasing barrier thickness. The various quantities in equation (1) may be expressed in any convenient units, so long as those are mutually consistent; their choice will determine the units (and hence will influence the numerical value) of the permeability coefficient P.

In practical tests of permeability of plastics materials it is the flux of the permeant in steady conditions that is usually measured, but the test results may be expressed in different ways, depending inter alia on the context in which they are to be used. The expressions of permeability most often encountered in practice are *transmission rate, permeance,* and *permeability*. The adoption of this last term for *one particular kind* of numerical expression of permeability is not entirely felicitous, as it can be a gratuitous source of confusion.

The definitions of the three terms (illustrated by reference to the relevant elements of equation (1)) are as follows.

Transmission rate (permeation rate[1,3])

This is the amount of permeant passing in unit time (in steady-state conditions) through a unit area of a plastics lamina (assumed to be parallel-sided) - i.e. J/A in terms of equation (1). Transmission rate is popular in practice as a quantitative expression of permeability of plastics films and sheeting to water vapour: in that context it is usually termed simply "water vapour transmission" (abbreviated to "WVT"). The units of WVT most widely employed are $g.m^{-2}.d^{-1}$, or occasionally their submultiples, e.g. $g.m^{-2}.h^{-1}$ or $\mu g.m^{-2}.s^{-1}$. The SI units are $mol.m^{-2}.s^{-1}$. As can be seen, if a numerical statement of transmission rate is to be meaningfully compared with other expressions of permeability, it will be necessary in most cases to know also the value of the driving force, and the thickness of the plastics barrier (as well as to ascertain that the respective figures relate to the same temperature). In determinations of WVT of plastics materials, the driving force is the difference between the water vapour pressures on the two sides of the plastics barrier : this is usually controlled in a test (and subsequently stated) in terms of the corresponding relative humidities. Tables are available for converting the latter to water vapour pressures (cf. e.g. Weast and Astle 1979-1980). *Gas transmission rate* (GTR) through plastics films and sheeting - of practical interest e.g. in packaging applications - is also often determined (with gas pressure difference as the driving force). The ISO units of GTR are $mol.m^{-2}.s^{-1}$; other frequently encountered units are $ml.m^{-2}.d^{-1}$. In this case the volume (in millilitres) is understood to be the volume at STP (q.v.), so that these units are often written as $ml(STP).m^{-2}.d^{-1}$. The analogous inch-pound units are $ml(STP).100in^{-2}.d^{-1}$. An ASTM standard for determination of permeability of plastics films and sheeting (ASTM D 1434-82(1992)) gives a number of tables of factors for converting gas transmission rates (as well as permeances and "permabilities") from one system of units to another.

Permeance

This is the rate of transmission of a permeant per unit driving force, i.e. J/(A.F) in terms of equation (1). In most practical situations, including permeability tests, the permeant is a gas or vapour, and the driving force is the difference in its pressures on the two sides of the polymeric barier (i.e. $F=\Delta p$: see above). The definition of permeance in an ASTM terminology standard[1.3] is in fact limited to this case. The expression J/(A.F) does not include the barrier thickness: thus in the context of most practical tests on plastics, permeance is the flow of permeant per unit time and unit area under a specified pressure difference through the entire thickness of a specimen. This way of expressing the test result is useful if the specimen is heterogeneous e.g. foam with solid skins). In SI units permeance is often expressed in ng $N^{-1}s^{-1}$; *perm* or *metric perm* are also employed, being, respectively, 1 grain/ (ft^2.h.inHg) and 1g/(m^2.d.mmHg).

Permeability

What is normally quoted, in numerical terms, as the "permeability" of a plastics material is in fact the permeability coefficient P of equation (1), i.e.

$$P=(J/A)/(F/L)=(J.L)/(A.F) \qquad (2)$$

The only definition of permeability (in the specific quantitative sense considered here) to be found in a major terminology standard[1.3] reflects this interpretation, in that the definition is "the permeation rate divided by the pressure gradient of the gas or vapor", permeation rate being defined by the same source as "the flow rate of gas or vapor, under specified conditions, through a prescribed area of a solid body, divided by that area". It is noteworthy that these definitions are confined (like the same standard's definition of permeance - see above) to the case where the permeant is gaseous and the driving force is its pressure difference. For this case, the SI units of P are $g.s^{-1}.m^{-1}.Pa^{-1}$ or $g.m.s^{-1}.N^{-1}$.

NOTE - The dimensions of the terms of equation (2) for the case considered here are J - mass divided by time; A - length squared; F - pressure; and L - length.

$$\text{Thus units of } P=\left(\frac{g}{s}.m\right)\Big/(m^2.Pa)$$

$$=\frac{g.m}{s.m^2.Pa}$$

$$=\underline{g.s^{-1}.m^{-1}.Pa^{-1}}$$

If the pascal is replaced by its equivalent $N.m^{-2}$, the units of P become

$$\frac{g.m.m^2}{s.m^2.N} = g.m.s^{-1}.N^{-1}$$

Various other units of P are also employed, depending on the context : cf. e.g. Yasuda (1966), Comyn (1985), Hopfenberg (1974), ASTM D 1434-82 (1992) and the other relevant test standards listed below.

The basic mechanism of transmission of a permeant through a continuous plastics barier (including closed-cell foams, in which the cell walls constitute individual barriers) is diffusion. In effect, the permeant dissolves in the barrier material on the side where its pressure (or concentration, or chemical activity) is higher, and its molecules travel through the material down the driving-force gradient (clustering can occasionally occur) and come out at the opposite surface. As diffusion is a thermally activated process, the permeability in any plastic/permeant system depends on the temperature in this sense, and also, secondarily, because the free volume content of the plastic will increase with temperature (regularly outside the glass transition temperature, and sharply at the T_g). The chemical structures of the system's components - and hence chemical interactions between them - constitute another important factor, as does the fine structure and morphology of the plastics material. For many plastics materials the rate of diffusion may also increase with the sorbed permeant concentration, especially if the permeant interacts strongly with (i.e. is a good plasticiser, swelling agent or solvent for) the polymer.

For simple gases, and many other penetrants sorbed from the vapour phase, the solubility in the plastics barrier material will - under certain conditions - be proportional to the partial pressure (i.e. Henry's law will be obeyed), and the rate of diffusion will be directly proportional to the concentration gradient (i.e. Fick's law will be obeyed). In such cases the permeability coefficient P of equation (2) may also be defined by the expression

$$P=DS \qquad (3)$$

where D is the diffusion coefficient (assumed to be constant or dependent only on permeant concentration), and S is a solubility factor expressing the equilibrium content of the permeant in the plastics barrier material at the relevant temperature. If S is expressed as a ratio (of the equilibrium concentration of the permeant in the barier material to the permeant concentration 'upstream' of the barier, with both concentrations in the same units), it becomes a dimensionless partition coefficient, and P will have the same dimensions and units as the diffusion coefficient ($length^2.time^{-1}$).

The permeability of porous plastics materials i.e. ones in which the voids are intercommunicating, is governed by the rate of flow of the bulk permeant through the pores. In such a case the chemical and other intrinsic properties of the barrier material are not as important as they are in diffusional transfer.

Standards relevant to the determination of permeability of plastics are listed in the table below.

PERMEANCE

See under permeability.

PERMEATION RATE (TRANSMISSION RATE)

See under permeability.

PERMITIVITY

Term fairly commonly used for *relative permittivity* (q.v.).

PHENOL-FORMALDEHYDE (PF) RESIN

A phenolic resin, the product of polycondensation of phenol and formaldehyde[1.1,1.5].

PERMEABILITY - TEST STANDARDS

PERMEABILITY ASPECT CONCERNED	TEST STANDARD	PLASTICS MATERIAL TESTED	TEST METHOD (GENERAL TYPE)	REMARKS
WATER VAPOUR TRANSMISSION RATE (WVT)	ISO 1663:1981 (under re-vision in 1996)	Rigid cellular plastics	Gravimetric. Test material seals dish con-taining desi-ccant	Standard refers also to permeance and perma-bility
	ISO 2528:1995 BS 2782:Part 8: Method 820A:1966	Sheet materials	Gravimetric. Test material seals dish containing desiccant	
	BS 2782:Part 8: Method 822A:1992	Films	Gravimetric. Sealed film sachet, water-filled	
	BS 3177:1959 (1995)	Flexible sheet materials (especially sheeting for packaging)	Gravimetric. Test material seals dish containing desiccant	
	BS 3837:Part 1: 1986	Expanded polystyrene boards	Gravimetric. Test material seals dish containing desiccant	Standard refers also to permeance and perm-ability
	BS 4370:Part 2: 1993 (identical with ISO 1922: 1981)	Cellular materials	Gravimetric. Test material seals dish containing desiccant	Standard refers also to permeance and perm-ability
	ASTM E 96-94	Materials generally	Gravimetric. Test material seals dish containing - either desiccant or water	Standard refers also to permeance and perm-ability

PERMEABILITY ASPECT CONCERNED	TEST STANDARD	PLASTICS MATERIAL TESTED	TEST METHOD (GENERAL TYPE)	REMARKS
GAS TRANSMISSION RATE (GTR)	ISO 2556:1974 BS 2782:Part 8: Method 821A: 1979 (1993)	Films and thin sheeting	Manometric determination	
	ASTM D 1434-82 (1992)	Films and sheeting	Manometric or volumetric determinations	
	ASTM D 3985-81 (1988)	Films; sheeting; laminates; coextrusions; plastics-coated paper or fabrics	Coulometric determination of oxygen transmission	
TRANSMISSION OF OTHER PERMEANTS	ASTM D 2684-89	Plastics containers	Gravimetric determination of permeability to packaged reagents and proprietary products	

PHENOL-FURFURAL RESIN

A phenolic resin, the product of polycondensation of phenol and furfural[1.1,1.5].

PHENOLIC PLASTICS

Plastics based on phenolic resins[1.1,1.5,1.7] (q.v). See also *phenolic resin compound*. The following standards are directly relevant to phenolic plastics (moulding materials, mouldings, and other products). *Moulding materials and mouldings*. Specifications for: ISO 800; ISO 4896 (melamine/phenolic); BS 771; ASTM D 4617; ASTM PS 15-94 (thermosetting moulding compounds, including phenolics). *Testing*: ISO 119 (=Method 451 E of BS 2782) - determination of free phenols; ISO 120 (=Method 451 D of BS 2782) - determination of free ammonia and ammonium compounds; ISO 59 (=Method 451 A of BS 2782), ISO 308 (=Method 451 B of BS 2782), BS 2782: Method 451 C, ASTM D 494 - determination of acetone-soluble matter; BS 2782: Methods 451 F to J (determination of formaldehyde, sulfates, and chlorides); BS EN ISO 8975 (pH of water extract); BS 2782: Method 451 L (detection of free ammonia). *Phenolic products*. Phenolic adhesives (for timber and plywood): BS 203; BS EN 301; BS 1203; BS 1204. Phenolic foams: BS 3927 (slabs and profiled sections); BS 5946 (punking resistance test). Laminated sheet: BS 2572. Phenolic mouldings for aircraft: BS 2PL 1. Laminated sheet - phenolic/cellulose paper, copper-clad: BS EN 60249; BS 4584. Cork sheet gaskets (phenolic-bonded): BS 4332.

PHENOLIC RESIN

Conventionally defined as a member of the family of thermosetting resins comprising condensation[1.7](polycondensation[1.1,1.5]) products of phenols (including phenol itself[1.1,1.5,1.7]) or ketones[1.1,1.5]. See also *phenolic plastics* and *phenolic resin compound*. However, phenol can also react with some unsaturated hydrocarbons (e.g. acetylene, terpene) to produce phenolic resins that fall outside the group of typical phenolic condensation products (cf. e.g. ISO 10082:1991). Phenolic resins are available in different physical forms, viz as solids (powders, granules, flakes, pellets), liquids, and solutions or dispersions (in solvents or water/solvent mixtures). Examples of commercial material ranges (many of which include also phenolic compounds) are: *Bakelite, Cellobond, Durax, Durez, Fenochem, Fenoform, Fiberite, Genal, Methylon, Moldesite, Nestorite, Plenco, Plyophen, Resinoid, Resinal, Resocel, Rutaphen, Sirfen, Supraplast, Tecolite, Trolitax, Trolon, Valite, Varcum, Vyncolit.* The following standards are concerned with phenolic resins in their various aspects.

Basic standards: BS EN ISO 10082 : 1996 (definitions and test methods), ISO/TR 8244 (basis for classification).
Determination of cure behaviour and reactivity. Gel time: BS EN ISO 9396, BS 2782 Method 835 A, ASTM D 2471. Heats and temperature of reaction (by DSC): ISO 11409; ASTM D 2471. Cure behaviour (by dynamic mechanical procedures): ASTM D 4473. Reactivity (of resols under acid conditions): ISO 8987.
Testing and analysis. Resin powder: ISO 8619 (flow distance); ISO 8620 (sieve analysis). Liquid resins: BS EN ISO 8618 (non-volatile matter); BS EN ISO 8989 (water miscibility); ISO 9371 (viscosity; also of resin solutions). Solid resins: ISO 8974 (residual phenol content); ISO 8975 (pH); ISO 8988 (hexamethylene tetramine content); ISO 9397 (free formaldehyde content); BS EN ISO 9944 (electrical conductivity of resin extracts); ISO 11401 (separation by liquid chromatography).

PHENOLIC RESIN COMPOUND, SINGLE-STAGE[1.7]

A composition based on a phenolic resin, in which the resin still contains reactive groups and can thus undergo further polycondensation on appropriate heating. See also *A-stage* and *resol.*

PHENOLIC RESIN COMPOUND, TWO-STAGE[1.7]

A composition based on a phenolic resin in which the resin is only partly cured but not normally reactive in storage at ordinary temperatures, and which incorporates an additive that can promote further curing of the resin on heating. See also *novolak*.

PHOSPHAZENE POLYMERS

Polyphosphazenes (phosphonitrilic polymers). Polymers with backbone of alternating phosphorus and nitrogen atoms, with two side-groups on the phosphorus. The side-groups may be of various kinds, including perfluoroalkoxy groups (in fluorophosphazene rubbers). See also *Orgaflex* and *Eypel-A*.

PHOTOCONDUCTIVITY

Promotion (increase or development) of electrical conductivity in a material on exposure to light. Materials possessing this property are termed photoconductors[1.4]. (cf. ASTM F 335-94a). Photoconductive plastics materials find application in photocopying.

PHOTODEGRADABLE PLASTIC

A plastic degradable by the action of natural daylight. This is the essence of the definition provided - in mutually identical terms - in the two terminology standards[1.5,1.7] in which it is featured. Thus the standard definition does not extend to plastics susceptible to degradation by photochemically active light from artificial sources. This limitation would not appear to reconcile either with the circumstance that prolonged exposure to fluorescent-tube lighting, quite common in ordinary environments, can result in degradation of some plastics materials, or with the fact that in most standard tests for photodegradability of plastics the material under test is exposed to light from an artificial source (see below).

The following standards deal with tests for photodegradability of plastics.

ASTM D 4647-89 (colour stability of plastics under indoor fluorescent lighting and window-filtered daylight).
ASTM D 5071-91 (standard practice for exposure of photodegradable plastics to xenon-arc radiation with water spray).
ASTM D 5208-91 (standard practice for exposure of photodegradable plastics in fluorescent-UV and condensation apparatus).
ASTM D 5272-92 (standard practice for outdoor-exposure tests on photodegradable plastics).

One of the usual elements in photochemical degradation of a polymeric material is oxidation by atmospheric oxygen. In this connection see *oxidatively degradable plastic*.

A common route to deliberately rendering a plastic photodegradable is to use a polymer suitably modified by incorporation in the molecular chains of UV-sensitive carbonyl groups which serve as sites of chain scission on sufficient exposure to light containing a UV component. Additives susceptible to UV degradation may also be physically incorporated in the material, as may combinations of a pro-oxidant and a photo-activator, whose combined action promotes the breakdown of polymer chains.

PILOT IGNITION

Term used in ISO 10093 synonymously with flash ignition - see *flash point, flash temperature,* and *flash-ignition temperature.*

PIPE

In the plastics context, a long tube, rigid or semi-rigid[1.5], used to convey liquids, gases and, in some cases, particulate solids. In this context, the standard term for a flexible tube is *tubing* (q.v.)

PLASTICISER

A substance incorporated in a plastics material principally to impart to it - and hence to the ultimate product(s) - the properties of flexibility and extensibility[1.1,1.5,1.7], and softness (the German word for "plasticiser" is "weichmacher" - literally "that which makes soft"): the degree of these modifications increases with increasing plasticiser content. Plasticisation also makes for easier melt processing (i.e. better "workability"[1.1,1.5,1.7]), as it lowers the fusion temperature and melt viscosity of the polymer, and provides internal lubrication which reduces the work (shear) heat generated. Two other important typical effects of plasticisation on the properties of a plastics material are an increase in toughness (impact resistance), and a lowering of the T_g (glass transition temperature - q.v.) of the polymer and hence the softening temperature or temperature range of the plastic[1.1,1.5]. The extent of the T_g drop is often used as a semi-quantitative indication of a plasticiser's effectiveness. The drop is associated with an increase in free volume (q.v.) brought about by the introduction of a significant amount of plasticiser into the polymer: this factor is also responsible for an increase in the take-up of most penetrants (gases and liquids) and for their accelrated diffusion through the plastics material.

To exercise properly its property-modifying functions in a plastics material, the plasticiser should be compatible with the base polymer of the plastic (as well as with any additives present) and should, ideally, be of high permanence in the ultimate product, i.e. be resistant to migration (q.v.), extraction, and volatilisation. It is for this reason that plasticisers are, typically, high-boiling liquids of low volatility[1.1,1.5] or, rather less commonly, solids (including, e.g. such plasticising polymeric modifiers for PVC as certain grades of nitrile rubber, chlorinated polyethylene, and EVA copolymers). Applications in PVC account for over 80% of all plasticisers sold: of those the majority are phthalate esters, with phosphate and trimellitate esters also used on a significant scale, and other compounds mainly utilised for special purposes, as e.g. certain esters of aliphatic dibasic acids for good low-temperature properties, saturated polyesters (e.g. polypropylene adipate) of various molecular weights for high permanence, chlorinated paraffins as extenders for primary plasticisers, as well as miscellaneous speciality plasticisers including epoxy compounds, monoesters (e.g. phenoxy ethyl oleate), glycol esters, and others.

Plasticisation of nitrocellulose by camphor - a good plasticiser for this polymer - represents the earliest industrial application of plasticisation. Examples of plasticisers effective in other thermoplastics include the following. Dimethyl phthalate, benzyl butyl phthalate, some aromatic phosphates, and sulfonamides for cellulose acetate (e.g. mouldings); benzyl butyl phthalate, heptyl nonyl adipate and epoxidised soya bean oil for various acrylic coatings; *o,p*-toluene sulfonamide and some of its derivatives for polyamides (used mainly to aid melt-processing, but also in certain coating formulations and flexible products). See also *polymeric plasticisers and monomeric plasticisers.*

For abbreviated terms (letter symbols) used for plasticisers, and for plasticiser trade names see, respectively, Section 3 of Part 2 and Section 2 of Part 3 of this book.

Standard specifications for some individual plasticisers (as chemical materials) can be found, under the names of the compounds concerned, in the catalogues of international and national standards. Several of these specifications are mentioned below, as are standards dealing with the properties of plasticisers and some relevant tests.

Properties in Bulk : Specifications and Tests

Phthalate esters

ISO 1385:1977	Part 1 - General; Part 2 - Measurement of colour after heat treatment; Part 3 - Determination of ash; Part 4 - Determination of acidity : Part 5 - Determination of ester content.
ISO 6795:1983	Determination of alcohol content (phthalic and adipic esters).
BS 4835:1973 (1993)	Methods of test for plasticiser esters.
BS standard specifications for individual phthalate ester plasticisers:	BS 573:1973(1993) - Dibutyl phthalate; BS 574:1973(1993) - Diethyl phthalate; BS 1995:1973(1993) - Di-(2-ethylhexyl)-phthalate; BS 1996: 1973(1993) - Dimethyl phthalate; BS 3647:1973(1993) - Dimethoxyethyl phthalate; BS 4968:1973 - Diisobutyl phthalate, diisooctyl phthalate and diisooctyl sebacate.
ASTM D 1249-92	Standard specification for octyl ortho-phthalate ester plasticisers.

Adipate esters

ISO 2523:1974	List of test methods.
ISO 2524:1974	Measurement of colour after heat treatment.
ISO 2525:1974	Determination of acidity.

ISO 2526:1974	Determination of ash.
ISO 2527:1974	Determination of ester content.

Sebacate esters

BS 2535:1973(1993)	*Standard specification for dibutyl sebacate.*
BS 2536:1973(1993)	*Standard specification for di-(2-ethylphenyl) sebacate.*

Phosphate esters

ISO standards for *tritolyl phosphate* :	ISO 2520:1974 - Methods of test; ISO 2521:1974 - Determination of acidity; ISO 2522:1974 - Determination of free phenols.
BS 1998:1970 (1993)	Standard specification for triphenyl phosphate
BS 1999:1964	Standard specification for tritolyl phosphate (standard discontinued).

General

BS 4835:1973	Methods of test for plasticiser esters.
ASTM D 1045-95	Standard test methods for sampling and testing plasticisers.
ASTM D 3465-75(1990)	Determination of purity of monomeric plasticisers by gas chromatography.
ASTM D 2288-92	Standard test method for weight loss by plasticisers on heating.

Properties in Association with Plastics (Compatibility; Permanence etc.)

Tests for loss of plasticiser (activated carbon method)
ISO 176:1976, identical with Methods 465A and B:1979(1992) of BS 2782; ASTM D 1203-94.

Tests for migration of plasticisers - see under *migration.*

Test for plasticiser/polymer compatibility.

ASTM D 2383-92	Plasticiser compatibility in PVC compounds under humid conditions.
ASTM D 3291-92	Plasticiser compatibility in PVC compounds under compression (loop test).

Test for plasticiser absorption

Absorption by PVC resins: (a) at room temperature : ISO 4608:1984, identical with Method 454 E:1992 of BS 2782; ASTM D 1755-92 (Section 15); ASTM D 3367-75(1990) - absorption under applied centrifugal force; (b) at elevated temperatures : ISO 4574:1978.

Identification and analysis - see under *vinyl chloride plastic.*

PLASTICISER ABSORPTION

See under *oil absorption*.

PLASTICISER DEMAND

That part of the plasticiser in a plasticised compound which is taken up by a particulate additive, such as a filler or pigment, and is thus not directly available to exercise its plasticising function on the polymer of the compound. The plasticiser demand of a particulate additive may be estimated on the basis of its oil absorption (q.v.), and should be compensated for by including the appropriate additional amount of plasticiser in the compound's formulation.

PLASTIGEL

See under *paste (1)*.

PLASTISOL

See under *paste (1)*.

PLATE-OUT

A persistent sticky deposit that may sometimes appear on the working surfaces of PVC-processing equipment. It can occur in any of such principal processes as extrusion, calendering and moulding, as well as in ancillary operations (e.g. milling, embossing). Plate-out deposits are a nuisance in the equipment - necessitating cleaning-up operations involving loss of time and output - and detrimental to product quality: when present e.g. in extrusion dies, pipe-sizing sleeves, or on calender rolls they impair product surface finish and dimensional conformity; plate-out on embossing rolls may blur the emboss, or even obliterate it in severe cases. The deposit may also be transferred to the product causing marks, blemishes or stains on its surface. There is still no complete, detailed understanding of the mechanism of plate-out formation, which appears to be complex and probably varies somewhat with the nature of the PVC formulation concerned and the methods and actual conditions of processing. It is known, however, that plate-out deposits are formed by material exuding from the hot PVC stock in consequence of incompatibility - which may become manifested only in certain processing conditions - of some constituents of the PVC composition with one another and with the PVC resin. The exudate moves out to the surface of the hot stock or melt and thence to the working surfaces of the processing equipment. The process factors which can promote plate-out are local changes in temperature, pressure, and the rate and direction of flow of the stock (melt) in the equipment. The formulation components particularly implicated in plate-out formation are some lubricants (especially metal soaps) and stabilisers: incorrect stabiliser/lubricant balance is often a factor. It is known that lubricants and stabilisers are normally the predominant components of a plate-out deposit, which may also contain some of the other additives, notably fillers and pigments (if those are present in the PVC composition) occluded by the exudate.

PLATINUM-COBALT COLOUR SCALE

A colour scale for clear, slightly coloured liquids (light yellow to yellowish brown). The scale is defined by the colours of specified dilutions of a standard aqueous solution of platinum and cobalt salts (of prescribed light absorbance - see ASTM D 1209-93), each dilution corresponding to a colour standard number within the range 5 (lightest colour) to 500 (darkest colour). In some of its applications the scale may be extended upwards to 2 000 (cf. ASTM E 284-95).

Standard methods of measurement of colour in accordance with the platinum-cobalt scale are given in ISO 2211:1973 (identical with BS 5339:1976) and ASTM D 1209-93. These methods, and the associated descriptions of colour, are applicable inter alia to unsaturated polyester and epoxy resins. Application to plasticisers is specifically covered in sections 19 to 21 of ASTM D 1045-95.

The platinum-cobalt colour scale is also referred to as the *Hazen colour scale*, and its colour standard numbers as the *Hazen units*, after A. Hazen who first formulated the platinum-cobalt stock solution (Hazen 1892) and put forward the idea of a colour scale based upon it for characterisation of colouration in water. In the USA, the water colour rating based on the scale is also sometimes referred to as "APHA color" (the initials being those of the American Public Health Association). Both the "Hazen" and "APHA" nomenclature are deprecated by ASTM D 1209-93, on the grounds that these terms are associated specifically with colour of water.

POISSON'S RATIO

This form of the term (the possessive case) is used in the main ISO and BS terminology standards[1.1,1.5], and in all ASTM standards, as well as in some technical BS standards, e.g. BS 4618: Subsection 1.1.3:1974, and BS 1991:Part 4:1961 (discontinued in 1983). However, one of the older ISO standards concerned with quantities, units and symbols (ISO 31:Part 3:1978) favours the nominative ("Poisson ratio") and recognises also "Poisson number" as an alternative name for this important material property.

When a solid body is being stretched or compressed by a uniaxially applied force of sufficient magnitude, it will undergo deformation (extension or compression) in the direction in which the force is acting. This deformation will also be accompanied by a lateral one: a lateral contraction if the axial deformation is an extension, or a lateral expansion if the axial deformation is compressive. For a body which is homogeneous and isotropic, and in which the applied force produces uniform axial stress within the proportional limit (q.v.), the ratio of lateral strain to the corresponding axial strain is a constant, known as *Poisson's ratio* (in the case of a uniformly anisotropic body the ratio may have more than one value - cf. e.g. ASTM D 4092-90 and ASTM E 28-81).

Poisson's ratio is one of the characteristic, basic properties of a material: for homogeneous, isotropic materials which conform to the laws of classic elasticity, the ratio is related to the three principal elastic moduli by the expressions $E=2G(1+\mu)$ and $E=3K(1-2\mu)$, where E is the modulus of elasticity (Young's modulus), G is the shear modulus (modulus of rigidity), K is the bulk modulus, and μ is Poisson's ratio.

It may be noted that there is no single standard symbol for Poisson's ratio: the symbols used include μ (ISO 31:Part 3:1978, and several ASTM standards - cf. e.g. ASTM E 6-89 and E 132-86), or ν(ISO 31:Part 3:1978 and ASTM D 4092-90), σ (BS 1991:Part 4:1961), and λ (ASTM D 1043-87). The International Union of Pure and Applied Physics (IUPAP) recommends μ and γ[1.5].

A relatively simple standard method is available for the determination, in a tensile test, of Poisson's ratios of materials in which tensile stressing rapidly produces substantial strain with relatively negligible creep - cf. ASTM E 132-86.

For isotropic, viscoelastic materials - a group to which many plastics belong - Poisson's ratio may usefully be determined, as the *creep lateral contraction ratio*, in a uniaxial creep experiment (in tension or compression). This ratio is one of the fundamental parameters charactersing the deformation behaviour of viscoelastic materials (BS 4618: Subsection 1.1.3:1974).

As can readily be shown, an incompressible material has a Poisson's ratio of 0.5. This value is fairly typically approached in practice by the results of determinations on soft rubbers at room temperature (ASTM D 1043-87). For a number of non-crystalline polymers a fairly representative value of Poisson's ratio would be about 0.3.

POLYACETAL

A polymer whose molecular chains contain repeated structural units of the acetal type.[1.5] See *acetal plastics* and *POM*.

POLYAMIDE (PA)

A thermoplastic polymer in whose molecular chains amide-type groups are the repeated structural units[1.1,1.5].

The polyamides which have attained industrial importance as the base materials of polyamide plastics (q.v.) or fibres (or both in some cases - e.g. polyamides 66 and 6) are produced by polymerisation (polycondensation) of either a diamine with a dicarboxylic acid (a salt is usually formed first to secure the acid:amine proportion required), or of an ω-aminoacid (or its lactone). The polyamides based on the two components are sometimes termed "dyadic", and the single-component ones "monadic".

Polyamides in whose molecular chains the amide linkages $(-\underset{H}{\underset{|}{N}}-\underset{O}{\underset{\|}{C}}-)$ of the structural units are separated by aliphatic segments are semi-crystalline polymers. In the crystalline regions the neighbouring chains are strongly held by hydrogen bonding via the amide groups, whilst the aliphatic segments contribute to such flexibility as may be imparted to the polymer by its amorphous regions (in general, the polyamides are relatively rigid materials). Thus the length of the aliphatic segments, which determines the distance between the amide linkages (sometimes expressed in terms of the

($-CH_2-$):($-NH-CO-$) ratio) is an important factor in the morphology and properties of an aliphatic polyamide. The relatively high crystallinity of these polymers is responsible for their lack of transparency. In aromatic polyamides the bulky aromatic groups hinder crystallisation, and thus promote transparency.

The chemical structure and morphology of the polyamides are jointly responsible for their typical properties, which include high strength and stiffness (the strength, tensile and flexural, and the moduli are very substantially increased by glass-fibre reinforcement), high melting temperatures, generally good chemical resistance (effective common solvents are limited to formic acid, glacial acetic acid, concentrated sulfuric acid, phenols and cresols, and there can be some swelling in alcohols), and resistance to abrasion. The moisture and water absorption of polyamides is relatively high, due to the presence of the amide groups, and the water absorbed lowers the mechanical properties (especially the moduli) and the T_g, and affects the electrical properties (which are also influenced by the polarity of the amide groups). The salient property of aliphatic polyamides in processing is their low viscosity at the processing temperatures.

The standard way[2.1,2.2,2.3,2.11] of representing a linear aliphatic polyamide by an abbreviated term (letter symbol) is to use the capital letters PA followed by a number or numbers indicative of the number of carbon atoms in the component (original monomer) unit or units. With a polyamide produced from a diamine and a dicarboxylic acid, the first figure gives the number of carbon atoms in the diamine, and the second figure their number in the acid; the two figures are not separated by any mark of gap. Thus, for example, the abbreviated term for the polyamide produced from hexamethylenediamine (which contains six carbon atoms) and sebacic acid (with ten carbon atoms) is PA 610. In the abbreviated term for a polyamide produced from a single compound (an ω-aminoacid or a lactam) a single number is used to show the number of carbon atoms in the compound. For example, the abbreviated terms for the polyamides produced from ε-coprolactam and 11-aminoundecanoic acid are, respectively PA 6 and PA 11.

In the symbols for copolyamides, i.e. copolymers produced from more than one ω-aminoacid or lactam, or from a combination of more than two amine and acid compounds, an oblique stroke (solidus) is used between the numbers representing the carbon atoms in the polyamide components of the resulting copolyamide. For example, the abbreviated term for a copolymer of ε-caprolactam and ω-dodecanolactam is PA 6/10, that for a copolyamide made from hexamethylenediamine and adipic and sebacic acids is PA 66/610, and that for a copolymer made from hexamethylenediamine and adipic acid with ε-caprolactam is PA 66/6. As the properties of a copolyamide depend not only on the nature of the components but also on their proportions, the relevant ISO standard (ISO 1874-1:1992) recommends that - where such information is of interest - the abbreviated term for a copolyamide should include (in brackets at the end) figures showing the proportions (in terms of percentages by mass) of the amide-forming starting materials. For example, a copolyamide based on 80% of hexamethylenediamine and adipic acid with 20% of ε-caprolactam can be designated PA 66/6 (80/20), and one formed from the same starting materials in reverse proportions would be represented by the abbreviated term PA 66/6 (20/80).

Additional letter symbols are used in abbreviated terms for polyamides and copolyamides with monomer units derived from non-linear aliphatic, aromatic, and certain other compounds. For example, a copolyamide based on laurolactam, isophoronediamine and isophthalic acid is designated PA 12/IPDI, one based on m-xylylenediamine and adipic acid is PA MXD6 (represented by the commercial products *Ixef* and *Reny*), whilst PA NDT and PA INDT are the ISO abbreviated terms for, respectively the polyamides produced from 1,6-diamino-2,2,4-trimethylhexane and terephthalic acid, and 1,6-diamino-2,4,4-trimethylhexane and terephthalic acid. (In practice a mixture of these diamines is used - see Table 3 below).

NOTE - *Trogamid*, a commercial representative of these polyamides (produced by Huels) is sometimes designated by the non-standard abbreviated term PA 6-3-T (in which 6-3 stands for "trimethylhexamethylenediamine").

A full list of letter symbols for non-linear monomer units encountered in polyamides, with the corresponding amine and carboxylic acid compounds from which they originate, is given in Table A.3 of Annex A to ISO 1874-1:1992.

Some data on various aliphatic and semic-aromatic polyamides are listed in tables 1 to 4 below. For fully aromatic polyamides see *aramid*. See also *polyamide plastics*.

POLYAMIDE (PA) PLASTICS

Plastics whose base polymers are polyamides, with molecular chains containing repeated structural units of essentially only the amide type[1.1,1.5]. See also *nylon plastics*, and *polyamide*.

The industrial importance and high volume usage of PA plastics are reflected in the large number and variety of commercial PA polymers and compounds, which include numerous reinforced, filled, lubricated, and other versions. The following are some examples of the names of commercial PA material ranges (see also tables 1 to 3 under *polyamide*, and Section 1 of Part 3 of this book). *Aesno* (PA 12); *Aculon* (PA 6 and 66); *Albis* (various PA compounds); *Alemid* (PA 6 and 66); *Aqualoy* (PA 66 and 612); *Arlen* (modified PA 6); *Baylon* (PA 6 and 66); Besno (PA 11); *Celanese N* (PA 6 and 66); *Compodic* (PA 6 and 12); *Comtuf* (PA 6; 66; and 12); *Crestamid* (semi-aromatic PA); *Denyl* (PA 6 and 66); *Durethan* (PA 6 and 66); *Ekamid* (PA 6 and 66); *Grilamid* (PA 12, and semi-aromatic); *Grilon* (PA 6; 66; and 612); *Grivory*(semi-aromatic PA); *Hiloy* (PA 6; 66; and 612); *Ixef* (PA MXD6); *Jackdaw* (PA 6 and 66); *Jonylon* (PA 6 and 66); *Kelon* (PA 6; 66; and 6/66); *Latamid* (PA 6; 66; 6/66; and 12); *Marnyl* (PA 66 and 6/66); *Minlon* (mineral-filled PA); *Nylamid* (PA 6; 66; and 12); *Nypel* (PA 6); *Nyrim* (PA 6 for RIM); *Orgamide* (PA 6); *Oromid* (recycled PA 6 and 66); *Polynil* (various PA polymers and compounds); *Rilsan* (PA 11 and 12); *Rimplast* (PA 12 for RIM); *Sevamid* (PA 6); *Sniamid* (PA 6 and 66); *Stanyl* (PA 46); *Synlon* (PA 6 and 66); *Technyl* (PA 66 and 6/66); *Texalon* (PA 6; 66; and 610); *Trogamid* (semi-aromatic amorphous PA); *Ube* (PA 6; 66; and 612); *Ultramid* (PA 6; 66; 610; and copolymers); *Verton* (long-fibre-reinforced PA compounds); *Vestamid* (PA 12); *Voloy* (PA 6; 66; and 612); *Zytel* (various PA's and compounds, including PA 11; 66; 610; and 612).

TABLE 1 COMMERCIALLY SIGNIFICANT LINEAR ALIPHATIC POLYAMIDES PRODUCED FROM SINGLE STARTING MATERIALS (AMINOACIDS OR LACTAMS). GENERAL FORMULA : $\{HN\text{–}(CH_2)_x\text{–}CO\}_n$

DESIGNATION; CHEMICAL NAME, AND REPEATED STRUTURAL UNIT	PARENT COMPOUND(S)	T_m °C	T_g (DRY POLYMER)* °C	APPEARANCE AND/OR MORPHOLOGY	EXAMPLES OF COMMERCIAL MATERIAL NAMES	REMARKS
PA 6 Polycaprolactam $\text{–}HN\text{–}(CH_2)_5\text{–}CO\text{–}$	ε-caprolactam (Polymer can also be made from ε-amino-caproic acid)	About 220°	60° to 70°	Translucent; semi-crystalline	Very numerous. See under *polyamide plastics.*	After PA 66, PA 6 is the most important material of PA plastics and fibres. Monomer suitable also for in-mould polymerisation (casting) and RIM.
PA 7 Polyheptanoamide (Polyenantholactam) $\text{–}HN\text{–}(CH_2)_6\text{–}CO\text{–}$	ζ-aminoenanthic acid $H_2N.(CH_2)_6.COOH$	About 200°	About 50°	Translucent; semi-crystalline	*Enant*	Used (originally in the former USSR) as the polymer of some PA fibres.
PA 9 Polypelargonamide $\text{–}HN\text{–}(CH_2)_8\text{–}CO\text{–}$	ω-aminopelargonic acid $H_2N.(CH_2)_8.COOH$	About 210°	About 50°	Translucent; semi-crystalline	*Pelargon* (or *Perlargon*)	Used (originally in the former USSR) as the polymer of some PA fibres.
PA 11 Polyundecanoamide $\text{–}HN\text{–}(CH_2)_{10}\text{–}CO\text{–}$	ω-aminoundecanoic acid $H_2N.(CH_2)_{10}.COOH$	About 185°	About 45°	Translucent; semi-crystalline	*Becno; Besno; Rilsan; Zytel.*	Used as the material of PA plastics and fibres. Relatively low moisture absorption.
PA 12 Polydodecanoamide $\text{–}HN\text{–}(CH_2)_{11}\text{–}CO\text{–}$	Laurolactam (Polymer can also be made from 12-amino-dodecanoic acid)	About 180°	About 40°	Translucent; semi-crystalline	*Aesno; Grilamid; Latamid; Nylamid; Rilsan; Rimplast; Ube; Vestamid.*	Used as material of plastics mouldings and extrusions. Lowest moisture absorbtion among linear aliphatic polyamides. Some grades suitable for in-mould polymerisation and RIM.

*Absorption of moisture can move the T_g below room temperature.

TABLE 2 COMMERCIALLY SIGNIFICANT LINEAR ALIPHATIC POLYAMIDES PRODUCED BY POLYMERISATIONS OF A DIAMINE AND A DICARBOXYLIC ACID. GENERAL FORMULA : $\{HN-(CH_2)_X-NH-CO-(CH_2)_Y-CO\}_n$

DESIGNATION; CHEMICAL NAME, AND VALUES OF X AND Y IN THE GENERAL FORMULA	PARENT COMPOUNDS	T_m °C	DTL* °C	APPEARANCE AND/OR MORPHOLOGY	EXAMPLES OF COMMERCIAL MATERIAL NAMES	REMARKS
PA 46 Poly(tetramethylene adipamide) (x=4; y=4)	1,4-diaminobutane and adipic acid	295°	150° to 170°	Translucent; semi-crystalline	*Stanyl*	Highly crystalline polyamide (due to mutual closeness of amide groups in the polymer chains). Highest DTL obtained with high mould temperatures.
PA 66 Poly(hexamethylene adipamide) (x=6; y=4)	Hexamethylenediamine and adipic acid	263° - 265°	About 75°	Translucent; semi-crystalline	Very numerous. See under *polyamide plastics.*	Most important among polyamides as the material of fibres and plastics. High-use-volume engineering plastic.
PA 69 Poly(hexamethylene azelamide) Poly(hexamethylene nonamide) (x=6; y=7)	Hexamethylenediamine and azelaic acid	About 210°	-	Translucent; semi-crystalline	*Albis 69PA*	Properties generally similar to PA 610; low moisture absorption. Used mainly for injection mouldings and some PA films.
PA 610 Poly(hexamethylene sebacamide) (x=6; y=8)	Hexamethylenediamine and sebacic acid	About 218°	55°	Translucent; semi-crystalline	*Amilan; Texalon; Ultramid; Zytel*	Used for some precision mouldings (including zip fasteners); wire coverings; tubes; brush bristles.
PA 612 Poly(hexamethylene dodecanoamide) (x=6; y=10)	Hexamethylenediamine and dodecanoic acid (1,10-decanedicarboxylic acid)	About 215°	65°	Translucent; semi-crystalline	*Aqualoy; Aquanyl; Comtuf; Grilon; Hiloy; Voloy; Zytel*	Applications similar to those of PA 610.

*Deflection temperature under load (1.80MPa : ISO 75-2:1993). Dry polymer.

TABLE 3 COMMERCIALLY SIGNIFICANT POLYAMIDES WITH AROMATIC COMPONENTS

DESIGNATION; CHEMICAL NAME, AND/OR REPEATING UNIT	PARENT COMPOUNDS	T_m °C	APPEARANCE AND/OR MORPHOLOGY	EXAMPLES OF COMMERCIAL MATERIAL NAMES	REMARKS
PA MXD6 Poly(*m*-xylylene adipamide) 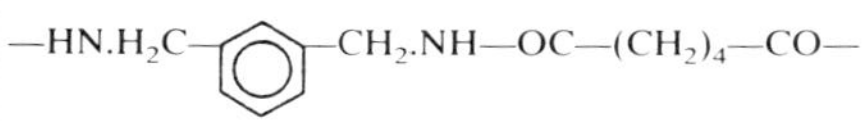	*m*-xylylenediamine and adipic acid.	235° - 240°	Partly crystalline	*Ixef*	Most commercial grades have glass-fibre reinforcement. They are stronger than many other polyamides similarly reinforced. The chemical resistance of Ixef is better than that of many aliphatic polyamides.
PA ND-INDT (PA 6-3-T) Poly(trimethylhexamethylene terephthalamide).	1,6-diamino-2,2,4-trimethylhexane (ND)** and 1,6-diamino-2,4,4-trimethylhexane (IND)** (a mixture), and terephthalic acid.	- T_g about 150°	Amorphous, transparent	*Trogamid T*	High DTL; soluble in some solvents not effective with aliphatic polyamides.
PA 6/6T Copolyamide of ε-caprolactam and terephthalic acid.	ε-caprolactam and terephthalic acid.	295°	Partly crystalline	*Ultramid T*	High DTL (especially with glass-fibre reinforcement) and maximum temperature for continous service. Relatively low moisture absorption.
PA 12/MACMI Copolyamide of laurolactam, 3,3'-dimethyl-4,4'-diaminodicyclohexylmethane (MACM) and isophthalic acid.	Laurolactam, 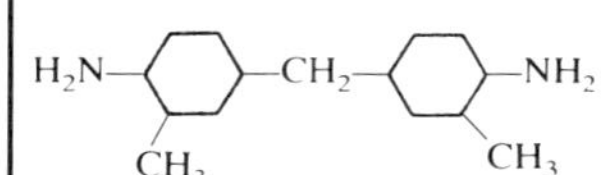 and isophthalic acid.	- T_g about 155°	Amorphous, transparent	*Grilamid TR 55*	High DTL and lowest moisture absorption among transparent polyamides.
PA 6/AMNB*T Copolyamide of bisaminomethyl norbornanes (mixture of isomers), ε-caprolactam and terephthalic acid.	ε-caprolactam, mixture of isomeric bisaminomethyl norbornanes, and terephthalic acid.	- T_g about 150°	Amorphous, transparent	*Hostamid LP* (No full-scale commercial exploitation)	Properties similar to those of Trogamid T.

*Aminomethyl norbornane (a non-standard letter symbol; there is no standard abbreviated term - cf. E.g. ISO 1874-1:1992).
**Initials of *nonyldiamine* and *isononyldiamine* (the simplified versions of the names of the two chemicals concerned).

TABLE 4 MISCELLANEOUS POLYAMIDES OF RELATIVELY MINOR INTEREST

DESIGNATION; CHEMICAL NAME, AND/OR REPEATING UNIT	PARENT COMPOUND(S)	T_m °C	MORPHOLOGY	EXAMPLES OF COMMERCIAL MATERIAL NAMES	REMARKS
PA 1 N-substituted polyurea: –N–CO– | R	An isocyanate (R.NCO)	Depends on the substituent group	Translucent, semi-crystalline	-	Not used in plastics.
PA 2 Polyglycine $–HN–CH_2–CO–$	Glycine : $H_2N.CH_2.COOH$	-	Translucent, semi-crystalline	-	May be regarded as the simplest polypeptide (q.v.). Not used in plastics.
PA 3 Poly(β–alanine) $–HN–(CH_2)_2–CO–$	Acrylamide : $CH_2=CH$ | $CO.NH_2$	Melts with decomposition at 320°C to 325°C	Translucent, semi-crystalline	-	Direct polymerisation of the aminoacid β-alanine not feasible. Polymers (prepared from acrylamide) not used in plastics.
PA 4 Poly(γ-aminobutyramide) $–HN–(CH_2)_3–CO–$	Pyrrolidone: 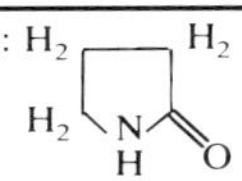	260°C to 265°C	Translucent, semi-crystalline	-	Produced not from γ-aminobutyric acid, but by ring-opening polymerisation of pyrrolidone. Fibre-forming.
PA 5 Poly(δ-aminovaleramide) $–HN–(CH_2)_4–CO–$	Piperidone: 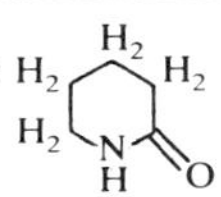	255°C to 260°C	Translucent, semi-crystalline	-	Produced not from δ-aminovaleric acid, but by ring-opening polymerisation of piperidone. Polymers not used in plastics.
PA 8 Polycaprylamide; Polycapryllactam $–HN–(CH_2)_7–CO–$	ω-aminocaproic acid	About 200°C	Translucent, semi-crystalline	-	No significant industrial application in plastics or fibres.

TABLE 4 CONCLUDED

DESIGNATION; CHEMICAL NAME, AND/OR REPEATING UNIT	PARENT COMPOUND(S)	T_m °C	MORPHOLOGY	EXAMPLES OF COMMERCIAL MATERIAL NAMES	REMARKS
PA 10 Polycapramide; Polycaprinolactam $-HN-(CH_2)_9-CO-$	ω-aminocapric acid	About 185°C	Translucent, semi-crystalline	-	No significant industrial application in plastics or fibres.
PA 13 Polytridecanoamide $-HN-(CH_2)_{12}-CO-$	ω-aminotridecanoic acid	About 180°C	Translucent, semi-crystalline	-	No significant industrial application in plastics or fibres.
PA 62 Poly(hexamethylene oxamide) $-HN-(CH_2)_6-NH-CO-CO-$	Hexamethylene-diamine and oxalyl chloride	About 320°C (Decomposes at the melting temperature)	Translucent, semi-crystalline	-	No significant industrial application in plastics or fibres.
PA 91 Poly(nonamethylene urea) $-HN-(CH_2)_9-NH-CO-$	Nonamethylene diamine and urea	About 240°C	Translucent, semi-crystalline	*Urylon* (Toyo Koatsu, Japan)	Polycondensation of parent compounds proceeds with elimination of ammonia. Polymer melt-spun into filaments. Properties comparable with those of PA 11.
PA 1010 Poly(decamethylene sebacamide) $-HN-(CH_2)_{10}-NH-CO-(CH_2)_8-CO-$	Decamethylene diamine and sebacic acid	About 200°C	Translucent, semi-crystalline	-	No significant industrial application in plastics or fibres.

Standards relevant to polyamide plastics include the following.

Material specifications

ISO 1874-1:1992	Polyamide moulding and extrusion materials - Designation.
ASTM D 4066-94b	Nylon injection moulding and extrusion materials.

Material property tests

Test specimens : ISO 1874-2:1987 (PA moulding and extrusion materials - Preparation of test specimens and determination of properties); ISO 1110:1985 (accelerated conditioning of PA test specimens); ASTM D 3641-93 (injection moulding of test specimens from thermoplastic moulding and extrusion compounds, including PA).

Solution viscosity: ISO 307:1984 (determination of viscosity number of polyamides), identical with BS 2782:Part 7: Method 733A:1991; ASTM D 789-94 (tests on polyamides; including relative viscosity inter alia);
ASTM D 2857-93 (determination of dilute-solution viscosity of polymers, including polyamides inter alia).
Determination of *melting temperature* (q.v.) : ISO 1218:1975 (Methods A and B); BS 2782:Part 1:Method 123 B: 1976(1993); ASTM D 789-94 (tests on polyamides, including "melting point" inter alia).
Moisture content : ISO 960:1988; ASTM D 789-94 (tests on polyamides, including moisture content inter alia).
Determination of ash : ISO 3451-4:1986, identical with BS 2782:Part 4:Method 470 D:1991.
Mouldings and products : Specifications
BS 7029:Part 2:1993 (some requirements for thermoplastic mouldings, including PA); BS 5409 - Nylon tubing types 11 and 12; Part 1:1976 (1977) -tubing for pneumatic installations. Part 2:1978 - tubing for use in the automobile industry.

POLYAMIDE-IMIDE (PAI)

A non-crystalline polymer, glassy within its service temperature range (T_g above 275°C), basically thermoplastic, but in the finished (postcured) products the molecular weight is too high for any remelting to be practicable. The molecular weight of the polymer as supplied in the moulding materials is substantially lower, resulting in melt viscosity manageable in processing, but still high. The melt rheology, and the reactivity of the polymer, necessitate both operational care and special equipment features to secure uniform, void-free mouldings, and to ensure that excessive pressure and overheating do not cause further polymerisation in the moulding machine. Manufacture of PAI products is in effect a two-step operation, in that the mouldings require postcuring (typically for several days at 260°C) to complete polymerisation, whereby they are finally stabilised and attain their ultimate strength and other properties. In view of these complications, relatively simple products are often machined from rods, billets, tubes and the like.

The polyamide-imides forming the only widely available commercial PAI plastics range (*Torlon*) are based on trimellitic anhydride and aromatic diamines, so that in the resulting molecular chains the structural units joined by alternating imide and amide linkages are aromatic in character. This conjunction of structural features confers on the polymer very high levels of strength, toughness, and dimensional and thermal stability. PAI has an exceptionally wide service temperature range, given by Amoco Polymers, the makers of *Torlon*, as between –190°C and 260°C, within which it is resistant to ageing. It possesses good natural resistance to wear and weathering, and its flammability is low.

Its limitations include relatively high moisture absorption (which can affect its otherwise very good electrical properties, and cause distortion on rapid heating), susceptibility to attack by strong alkalis and by oxidising acids, and very dark, almost black, natural colour. PAI parts find uses in aircraft, motorcars, office machines, and electrical and electronic equipment. PAI fibres and films, and PAI-based lacquers are also available.

ASTM D 5204-91 provides a standard classification system for PAI moulding and extrusion materials.

POLYANILINE (PANI)

A polymer of aniline, possessing high inherent electrical conductivity. This material has been known for many years, but its intractability in the usual polymer-processing methods (it is insoluble and infusible) has hindered commercial exploitation until recent times. It is now available (from Ormecon Chemie, a subsidiary of the German company Zipperling Kessler & Co.) as fine-particle dispersions in water, solvents, or lacquer formulations for application as conductive or antistatic surface coatings (cf. e.g. *Incoblend*), and coatings for corrosion protection (*Corrpassiv, Correpair*). Blends have also been produced (by Neste OY, Finland) of thermoplastic and thermosetting polymers with PANI, or with the more easily processable complexes of PANI with alkylbenzene sulfonic acids. The properties of conductive dispersions of PANI are discussed in a paper by Wessling (1995), and those of some conductive, soluble copolymers of aniline by Kathirgamanathan et al. (1991).

POLYARYLATE (PAR)

A thermoplastic polyester, entirely aromatic in the sense that, structurally, it is a polycondensation product of an aromatic dicarboxylic acid or acids and a dihydric phenolic compound. Thus each of the carboxyl and hydroxyl groups of the original monomers is linked directly to an aromatic nucleus[1,7]. The first polyarylate to reach the market (*U-Polymer*, produced by Unitika Co., Japan) was a terpolymer of terephthalic and isophthalic acids with bisphenol A, and some of the currently available materials are of similar composition. Commercial polyarylates resemble bisphenol-A polycarbonate in properties, in that they are tough, transparent, high-modulus materials with good electrical properties and naturally low flammability (lower than that of polycarbonate). Their weathering resistance is somewhat superior to that of polycarbonate, as is their heat-deflection temperature and maximum temperature for continuous service in the absence of stress (respectively about 160°C to 170°C at 1.84MPa, and 150°C). Like polycarbonate, polyarylates are susceptible to a number of solvents (especially chlorinated hydrocarbons) and to environmental stress cracking. Their resistance to long-term exposure to hot water is also limited. Polyarylates are available commercially under the trade names *Ardel* (material similar to the original *U-Polymer*), *Arylon*, and *Durel.*

POLYBENZIMIDAZOLE (PBI)

A polymer resulting from polycondensation of an aromatic tetramine and a dicarboxylic acid or ester (usually aromatic in commercial PBI resins). These resins are high-performance plastics materials, similar to polyimides (q.v.) in their resistance to high temperature and chemicals (albeit less resistant than aromatic polyimides to oxidation at elevated temperatures). PBI resins are usually made into plastics parts by direct-forming techniques, but some grades are suitable for moulding and extrusion (see *Celazole*). PBI fibres are also produced (from low-molecular-weight polymer heat-treated after spinning to render the fibres insoluble and infusible).

POLYBUTYLENE (POLYBUTENE) (PB)

A polymer (homopolymer[1.7]) of butylene[1.1.1.5] (butene-1 : $CH_3{-}CH_2{-}CH{=}CH_2$), with the repeating unit

$$-\underset{\underset{\displaystyle CH_3}{|}}{\underset{\displaystyle CH_2}{\underset{|}{CH}}}-CH_2-$$

The polymer is semicrystalline, generally isotactic , with a T_m of about 125°C and T_g of 75°C to 78°C. The crystal form changes within a period of the order of a week after melt processing, with consequent increase in stiffness, hardness and yield stress. In general, the properties of PB resemble those of PE-LD, but it is more resistant to distortion at elevated temperature, to abrasion, and to environmental stress cracking. An early industrial version of the polymer was *Vestolen BT* (Huels).

POLYBUTYLENE (POLYBUTENE) (PB) PLASTICS

Those butylene plastics (q.v.) which are based essentially on polybutylene (i.e. on homopolymers of butylene)[1.1.1.5,1.7]. commercial PB plastics materials, used for extrusion, blow moulding, injection moulding, and wire and cable covering, are found in the following ranges: *Duraflex; Indopol; Parapol.*

Two major standards provide specifications for PB materials, viz.

ISO 8986	Polybutylene (PB) moulding and extrusion materials.
Part 1:1993	Designation system and basis for specification.
Part 2:1995	Preparation of test specimens and determination of properties.

ASTM D 2581-91	Polybutylene plastics - specification for moulding and extrusion materials.

An ASTM standard for determination of carbon black content of polyolefins, ASTM D 1603-94, is applicable to PB inter alia.

POLY(BUTYLENE NAPHTHALATE); POLY(BUTYLENE-NAPHTHALENE-2,6-DICARBOXYLATE) (PBN)

Thermoplastic polyester based on dimethylnaphthalene-2,6-dicarboxylic acid and butane-1,4-diol. Properties similar to those of poly(ethylene naphthalate) (q.v.).

POLY(BUTYLENE TEREPHTHALATE) (PBT); POLY(TETRAMETHYLENE TEREPHTHALATE) (PTMT)

A thermoplastic polyester (q.v.) derived from butylene glycol and terepthalic acid; produced either by the polycondensation of these two components or by ester-exchange polymerisation of butylene glycol and dimethyl terephthalate[1.5]. The polymer is semicrystalline, with a T_m of about 220°C , and the rather low T_g of 50° to 60°C. Some of its other properties are mentioned under *thermoplastic polyester plastics*.

PBT polymers and compounds are represented in the following commercial ranges (see also Section 1 of Part 3 of this book.

Arnite; Beetle; Celstan; Comtuf; Crastin; Deniter; Ekadur; Gafite; Grilpet; Hostadur; Later; Lupox; Orgater; Pocan; Raditer; Retlon; Rynite; Sniater; Texapol; Thermex; Tribit; Ultradur; Valox; Vampter; Vestodur; Voloy.

The following standards are relevant to PBT materials.

ISO 7792	Poly(alkylene terephthalate) moulding and extrusion materials.
Part 1:1995	Designation and basis for specification.
Part 2:1988	Preparation of test specimens and determination of properties.
ISO 3451-2:1984	Poly(alkylene terephthalates) - Determination of ash.
ASTM PS 2-94	Provisional specification for moulding thermoplastics, including PBT inter alia.

POLYCARBONATE (PC)

A member of the general family of polyesters[1.7], in that the repeating unit in its molecular chain contains the ester group $(-\underset{\underset{O}{\|}}{C}-O-)$. It is also more closely classifiable as a polyester of the carbonate type[1.1,1.5,1.7], since the ester groups of the repeating unit are structurally derived from carbonic acid $(H-O-\underset{\underset{O}{\|}}{C}-O-H)$. Indeed, one of the main methods of preparation of a polycarbonate is polycondensation, through ester exchange, of a dicarbonate with a diol. For example, bisphenol-A polycarbonate - virtually the only polycarbonate in wide industrial use - with the repeating unit

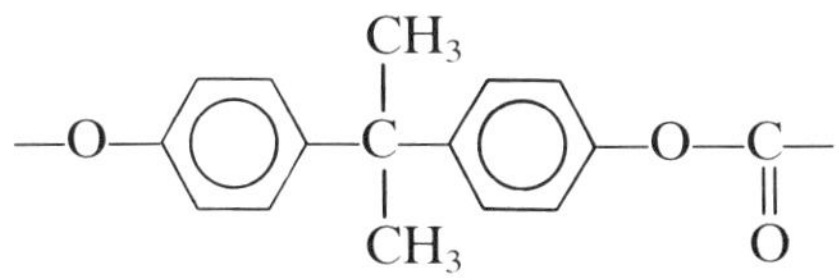

can be made by such transesterification of diphenyl carbonate with bisphenol-A (phenol being generated as a byproduct).

NOTE - Bisphenol-A is 2,2-bis(4-hydroxyphenyl)propane, i.e.

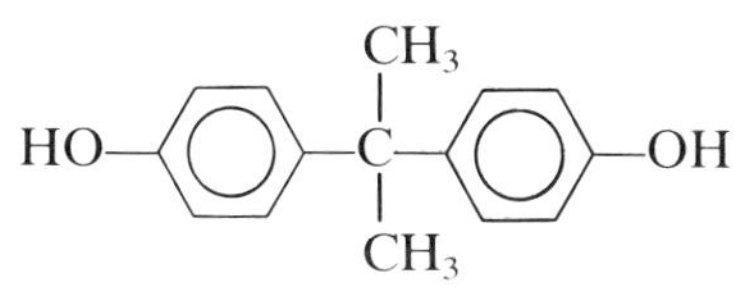

produced by acid-catalysed condensation of acetone with two molecules of phenol. In the common name the "A" stands for "acetone". Similarly, bisphenol-F is the condensation product of formaldehyde with two molecules of phenol. However, in bisphenol-C the letter relates to the phenolic component, the compound being obtained by condensation of two molecules of *ortho*-cresol with acetone. Polycarbonates based on bisphenol-F have been prepared, as have those based on bisphenol-C. However, neither has attained industrial importance.

Another major method of production of bisphenol-A polycarbonate is by reaction of the bisphenol with phosgene ($Cl{-}\underset{\underset{O}{\|}}{C}{-}Cl$) in a solvent or/and strong aqueous alkali. Because of its position as the only polycarbonate of substantial industrial significance, bisphenol-A polycarbonate is often referred to simply as "polycarbonate". The polymer is high-softening (T_g 142°C to 148°C depending on grade and method of determination), and normally non-crystalline when cooled relatively rapidly from the melt (as in moulding and extrusion), which makes for clarity. Crystallisation (up to about 30% crystalline material content) can be brought about by heating for long periods well above the T_g, or by absorption of some solvents (cf. e.g. Titow et al., 1974). In either case the treatment creates enough extra free volume to enable the relatively stiff, large-segmented molecules of the polymer to line-up into regular arrays.

The good mechanical properties (high strength and modulus) of bisphenol-A polycarbonate, and their maintenance over a wide temperature range (typically quoted as –150°C to +135°C), place the polymer and the materials based upon it in the category of engineering plastics. The polymer is tough : in conjunction with its clarity, this property promotes its wide use in safety screens and safety glazing. The toughness is due primarily to the high energy of crack initiation in bisphenol-A polycarbonate. However, the energy of crack propagation is relatively low (lower than that of ABS - cf. e.g. Vincent, 1974) and this makes for notch sensitivity.

Ordinary glazing applications of bisphenol-A polycarbonate include its use, in the form of double-skin or triple-skin sheets (q.v.), as roofing for hot-houses and the like.

The electrical properties of bisphenol-A polycarbonate are good, and relatively unaffected by absorption of moisture. This makes the polymer applicable as the material of electrical components of various kinds, including its use in the form of film as the dielectric in multi-layer capacitors.

The flammability of bisphenol-A polycarbonate is naturally low, and can be further improved by incorporation of flame-retardant additives, or by the use of a halogenated bisphenol-A as comonomer in the production of the polymer.

Bisphenol-A polycarbonate is soluble in, or swollen by, a number of chlorinated-hydrocarbon and ketone solvents, and is susceptible to environmental stress cracking in the presence of such solvents or their vapour. It is noteworthy that ethylene oxide, used as a sterilising agent inter alia for hospital equipment (which may include bisphenol-A polycarbonate compon-ents) is also a stress-cracking agent for the polymer.

Books by Christopher and Fox (1962) and Schnell (1964) - the chemists who independently originated polycarbonates respectively in the USA and Germany - provide much basic information on these polymers.

See also *polycarbonate plastics.*

POLYCARBONATE PLASTICS

Plastics whose base polymers are polycarbonates, with molecular chains containing repeated structural units of essentially only the carbonate type[1.1,1.5,1.7]. (see also *polycarbonate*). In virtually all commercial polycarbonate plastics falling strictly within this definition, the base polymer is bisphenol-A polycarbonate. In the less formal general technical terminology, plastics based on polyestercarbonate and on blends of bisphenol-A polycarbonate with other polymers (in which the polycarbonate predominates) are also included in the group of polycarbonate plastics.

A *polyestercarbonate* contains both aromatic carbonate and ester groups in the molecular chains. The latter groups are typically introduced by including a suitable derivative of terephthalic or isophthalic acid in the polymerisation system (e.g. the acid chloride in addition to bisphenol-A and phosgene - see under *polycarbonate*). An example of the resulting chain segment is

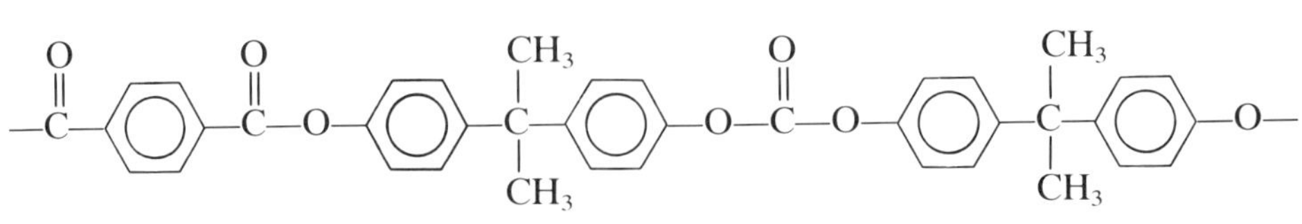

The main effect of the presence of the ester groups in the polymer is to increase significantly the Vicat softening temperature and the deflection temperature under load. In blends of bisphenol-A polycarbonate with polyesters (PBT or PET) effect of this kind is relatively minor; but the polyester component of the blend contributes significant improvements in stiffness and chemical resistance. However, the impact resistance of such blends is normally lower than that of the polycarbonate alone. In some grades of commercial blends this effect is counteracted by incorporation of elastomeric impact modifiers.

In blends of bisphenol-A polycarbonate with acrylic polymers, the latter act as impact modifiers providing notched-specimen impact strength substantially higher than that of the polycarbonate alone.

At sub-zero temperatures, the notched-specimen impact strength of typical blends of bisphenol-A polycarbonate with ABS is better than that of either of the two individual components, whilst the polycarbonate raises the deflection temperature under load of the blend above that of the ABS alone.

The commercial ranges representing polycarbonate plastics and their base polymers include the following (see also Section 1 of Part 3 of this book).

Polycarbonate polymers and compounds

Calibre; Carbotex; Celstran; Denicarb; Ekalon; Electrafil; Iupilon; Karlex; Latilon; Lexan; Lisa; Makrolon; Naxell; Panlite; Radilux; Selon; Sinvet; Trirex; Vampcarb; Xantar.

Polyestercarbonate polymers and compounds

Apec; Inspire; Lexan PPC.

Blends

(PC+ASA) : *Terblend*
(PC+Acrylic) : *Cyrex; Koblend*
(PC+Polyester) : *Carbotex; Ektar; Latiblend; Macroblend; Radiblend; Sabre; Stanuloy; Stapron E; Xenoy.*

For blends with ABS see under *ABS plastics.*

The following standards are relevant to polycarbonate plastics.

ISO 1628-4:1986	Determination of viscosity number and limiting viscosity number of polycarbonate moulding and extrusion materials.
ISO 7391	Plastics - Polycarbonate moulding and extrusion materials.
Part 1:1987	Designation.
Part 2:1986	Preparation of test specimens and determination of properties.
ISO 11963:1995	Polycarbonate sheets - types, dimensions and characteristics.

BS 7029:Part 2:1993	Method of specifying thermoplastic injection mouldings. Additional requirements for polycarbonate, polyamide and polyethylene mouldings.
ASTM D 1238-95	Test for flow rates of thermoplastics in an extrusion plastometer (relevant to polycarbonate inter alia).
ASTM D 2857-93	Determination of dilute-solution viscosity of polymers (relevant to polycarbonate inter alia). Standard similar to ISO 1628-4.
ASTM D 3641-93	Injection moulding of test specimens of thermoplastic moulding and extrusion materials (relevant to polycarbonate inter alia).
ASTM D 3935-95	Standard specification for polycarbonate unfilled and reinforced materials (for injection moulding, blow moulding, and extrusion). Standard similar to ISO 7391.

POLYCHLOROTRIFLUOROETHYLENE (PCTFE)

A polymer of chlorotrifluoroethylene[1.1,1.5], produced by the same method as PTFE (free-radical polymerisation of the monomer in aqueous suspension) although the polymerisation proceeds less readily. Like PTFE, PCTFE is supplied in the form of granules, powder, or dispersion. In comparison with those of PTFE, the properties of PCTFE are modified by the presence of the polar, relatively large chlorine atom in the repeat unit of the molecular chain. The bulk of the chlorine hinders the close, uniform lining-up of the chain molecules necessary for crystallisation, although it does not prevent it altogether. Thus the degree of crystallinity of PCTFE is substantially lower than that of PTFE, as is the crystallite melting point (T_m 214°C to 216°C, depending on the method of determination). The relatively low crystallinity, and reduction of crystallite size effected by rapid cooling in processing, make it possible to produce films and thin-walled mouldings of good transparency. Another consequence of the presence of the chlorine atoms in the molecular chains is that their polar nature increases intermolecular attraction, so that the strength, modulus and hardness of PCTFE are greater than those of PTFE. PCTFE is processed at melt temperatures between 270°C and 300°C, by extrusion, and injection and compression moulding, into such products as chemical-resistant tubes, valves seals and container closures, films and mouldings (especially for packaging medical and pharmaceutical goods), printed-circuit boards and electrical insulation. Whilst the electrical properties and chemical resistance of PCTFE are generally good, in keeping with its fluoropolymer nature, they are inferior to those of PTFE: at elevated temperatures PCTFE is susceptible to attack by some strong acids, and to swelling by ketone and ester solvents. The useful-service temperature range of the polymer is –100°C to 150°C (up to 180°C on short exposure in the absence of stress). Prolonged exposure to intense heat can cause substantial embrittlement. The weathering resistance of PCTFE is good, as is its flame resistance, in the sense that it is not ignitable even on extended contact with a small flame. The polymer is featured in the forthcoming ISO standard for fluoropolymer materials (ISO/DIS 12086 - still under discussion in 1996). It is also included among the fluoropolymers whose temperatures and heats of transition can be determined by the DSC test methods of ASTM D 4591-93. Standard specifications for, respectively, PCTFE plastics and extruded film and sheeting are provided by ASTM 1430-91 and ASTM D 3595-91. Commercial PCTFE materials are available under the trade names *Halar* (300/500 series), *Kel-F, Neoflon, Voltalef.*

POLY(CYCLOHEXYLENEDIMETHYLENE TEREPHTHALATE) (PCT)

Polymer produced, by ester exchange, from 1,4-dimethylolcyclohexane and dimethyl terephthalate:

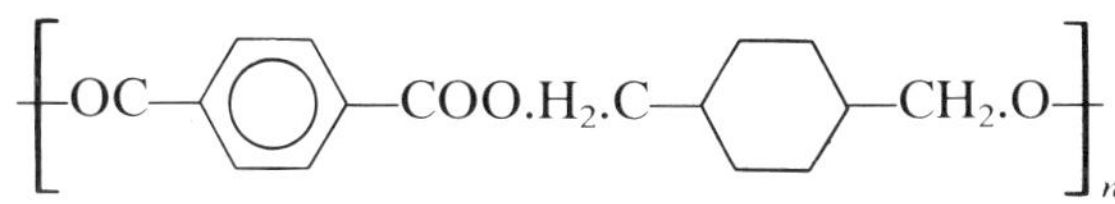

Superior to polyethylene terephthalate in weathering resistance, and less water-absorbent. See *Ektar, Valox.*

POLYESTER

A polymer in whose molecular chains the repeated structural units are, or contain, ester groups[1.1,1.5,1.7]. The ASTM terminology standard notes that polyesters which are derived - either by actual preparation or structurally - from monohydroxy-monocarboxylic acids or through combination of a diol with a dicarboxylic acid are linear and thermoplastic[1.7] (as distinct from thermosetting *unsaturated polyesters* - q.v.). Two examples of widely used thermoplastic polyesters are polyethylene terephthalate (PET) and polybutylene terephthalate (PBT). See also *polyarylate.*

POLYESTERCARBONATE (PEC)

See under *polycarbonate plastics.*

POLYESTER PLASTIC

A plastic base on a polyester (q.v.) or polyesters in whose molecular chains virtually all the repeated structural units are, or contain, ester groups[1.1,1.5]. The ISO and BS terminology standards[1.1,1.5] deprecate the use, occasionally encountered, of "alkyd plastic" for "polyester plastic", whereas the ASTM standard[1.7] regards these terms as synonymous (and defines *alkyd resin* - q.v.).

POLYETHERETHERKETONE (PEEK)

An aromatic linear polyether, with

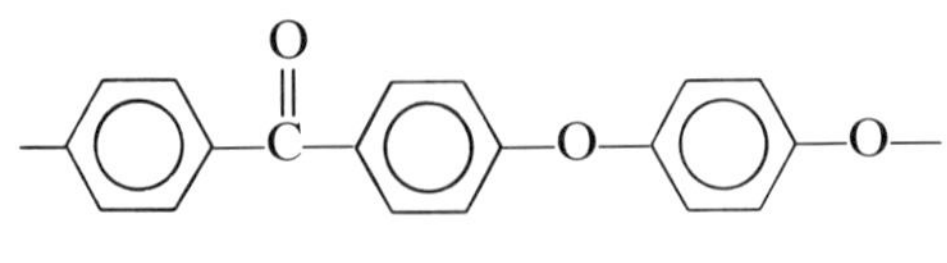

as the repeated structural unit in the chain[1,5]. Melt-processable (at high temperatures - typically between 370°C and 400°C). Highly heat-resistant in service (up to 320°C). Used inter alia in wire coatings and cable coverings; also as the material of *Zyex* (q.v.) filaments and fibres.

POLYETHERIMIDE (PEI)

A non-crystalline polymer, glassy within its service temperature range (T_g about 215°C), typically obtained from a substituted phthalimide and bisphenol-A.

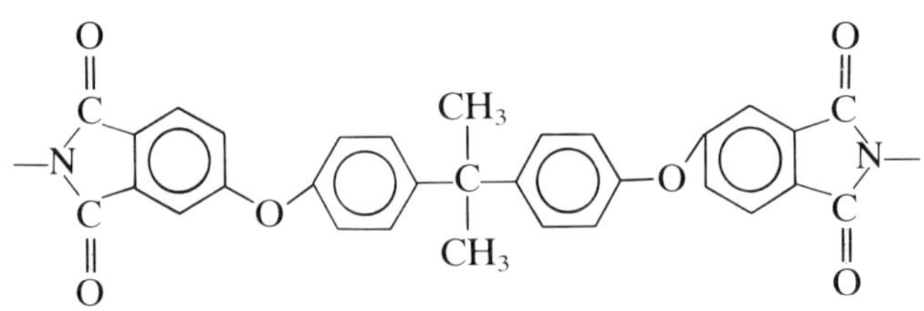

bisphenol-A-residue

The commercial grades of PEI have high maximum temperature for continuous service (170°C, in the absence of stress and aggressive environments), naturally high strength and stiffness (improvable further by glass and carbon fibre reinforcement), high dimensional stability, with low coefficient of linear thermal expansion, good electrical properties, very good resistance to weathering and high-energy radiation, and low flammability and smoke generation. They are resistant to many chemicals, but susceptible to attack by some ketones, esters, and chlorinated hydrocarbons. Their resistance to environmental stress cracking is limited. Unmodified PEI polymers are transparent, with a pronounced amber tinge.

Unlike PAI, PEI commercial materials are processed relatively easily by injection moulding and (some) by extrusion.

Main outlets include some under-the-bonnet applications in motorcars, industrial bearing sleeves, threaded closures, and mechanical couplers, printed circuit boards, electrical connectors, housings and sockets, and some items of sterilisable medical equipment. PEI fibres have been produced for use as reinforcements in special composites.

A range of commercial PEI polymers and compounds is available under the trade name *Ultem*. A carbon-fibre-filled grade is also made (by DSM) within the general *Electrafil* range.

Two standards relavant to PEI materials are ASTM D 5205-93 (standard classification system for PEI materials) and ASTM PS 2-94 (provisional specification for thermoplastic moulding compounds for electrical and electronic components, including PEI-based moulding compounds).

POLYETHERSULFONE (PES)

In its widest meaning the term should include both aliphatic and aromatic polyethersulfones. However, only the aromatic polymers - the polyarylethersulfones (PAESU) - are significant as plastics materials. Within this practical limitation, and in the strict sense of the term, a PES may be defined[1.5] as a polymer whose molecular chains are made up of the repeating unit

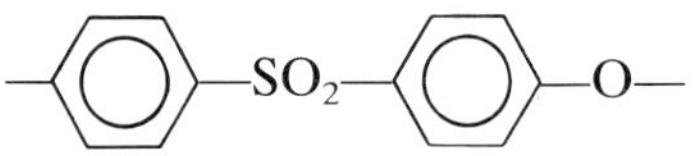

(cf. *Victrex PES*). In general usage the term also covers other polyethersulfone polymers, with such structural repeating units as

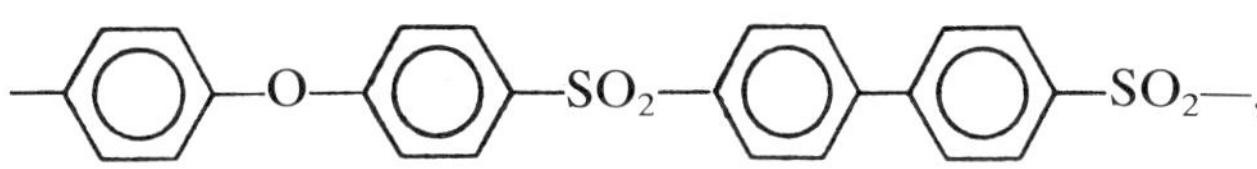

or

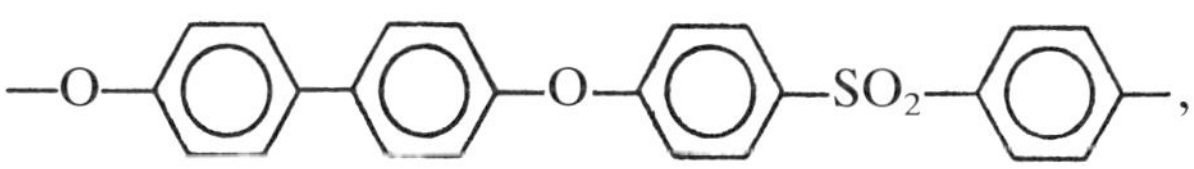

(cf. respectively *Astrel*, and *Radel A*). Commercial PES materials include also *Lubricomp J*, and *Ultrason E*.

POLYETHYLENE (PE)

Formally defined as a polymer[1.1,1.5] (homopolymer[1.7]) of ethylene.

NOTE - In accordance with IUPAC recommendations, the ISO and BS definitions[1.1,1.5] include polyethene and ethene as alternative, systematic terms for, respectively, *polyethylene* and *ethylene*.

The polymers of ethylene categorised in practice under the generic name of polyethylene from a large family, comprising several groups distinguishable on the basis of material properties and processing characteristics, and produced by a number of processes. Among the commercial polymers, each of the groups contains various grades of its constituent members. Not strictly in line with the standard definition (but in full accord with the general usage of the term "polyethylene"), included in the general family are not only homopolymers of ethylene, but also two groups of ethylene copolymers, each copolymer containing a relatively small proportion of an alkene-1 (alpha-olefin) comonomer (butene-1, hexene-1, octene-1, or 4-methylpentene-1). The two groups are known, respectively as linear low-density polyethylene (PE-LLD; LLDPE) and very-low-density polyethylene (PE-VLD; VLDPE). These groups differ essentially in the alkene-1 comonomer contents of their constituent polymers (higher in PE-VLD than in PE-LLD). Spanning some of the groups, whilst by virtue of their origin also making up a group of their own, are the polyethylenes produced with the aid of metallocene catalysts (q.v.; see also under *polyolefin* for some commercial representatives of this group).

All polyethylenes are semicrystalline polymers. The degree of crystallinity varies among the different groups, and this influences the corresponding polymer densities, which provide the basis for the most widely applicable general classification of polyethylenes (illustrated schematically in the diagram). The numerical limits shown in the diagram for LDPE, MDPE, HDPE, LLDPE and LMDPE density ranges are those given in the ASTM terminology standard[1.7].

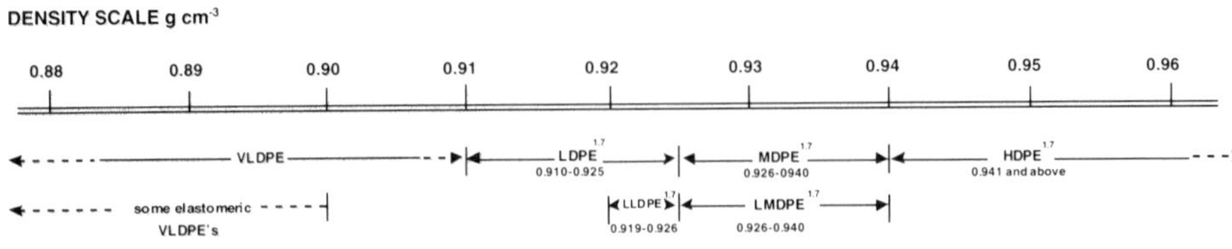

With regard to their molecular weights (MW), commercial polyethylenes range from the waxy, low-MW polymers(MW between about 1000 and 10 000) used as lubricants (notably in some PVC formulations, and in various masterbatches where they aid dispersion of the pigments, fillers, etc. in the polymer carrier) to ultra-high-MW polyethylene (q.v.) with MW between about 1 000 000 and 5 000 000. Use of appropriate metallocene catalysts makes possible production of PE-LLD with controlled, narrow MW distribution for improved material properties (although the resulting increase in melt viscosity makes for more difficult processing) and uniform alkene-1 comonomer content (with conventional catalysts the content decreases with increasing MW of the polymer fractions in a production batch). Polyethylenes are also available with bimodal MW distribution (controlled, balanced content of high-MW and lower-MW fractions). In such materials the high-MW component provides strength and toughness of the ultimate product, whilst the lower-MW component promotes ease of melt processing. Two examples of the materials are bimodal PE-HD made by the Mitsui slurry process (e.g. *Formolene E 905*, Formosa Plastics, USA) and bimodal PE-LLD produced by the *Borstar* process (Borealis, Finland).

The molecular chains of conventional, historically earliest, PE-LD made in high-pressure reactors are branched to a substantial exent, whilst those of PE-HD - produced by catalytic, low-pressure processes - are more linear. The chains of PE-LLD have linear polyethylene backbones with side-chains typically formed by the alkene-1 comonomers. PE-VLD has a similar structure, but with more extensive side-chain branching. See also *polyolefin, polyolefin plastic, ethylene plastics*, and *polyethylene plastics*.

Standards relevant to polyethylene polymers (as distinct from polyethylene plastics) include the following.

ASTM D 1996-92	Determination (by liquid chromatography) of phenolic antioxidants and erucamide slip additives in LDPE.
ASTM D 2839-93	Preparation of PE strands from the melt index test for determination of density.
ASTM D 3895-94	Determination (by DSC) of oxidation induction time of polyolefins.

ASTM D4883-89	Determination of density of polyethylene by an ultrasonic technique.
ASTM D 5017-91	Determination of the composition of LLDPE (with reference to the contents of propene, butene-1, hexene-1, octene-1, and 4-methylpentene-1).
ASTM D 5226-92	Standard practice for dissolving polymer materials - Section A1.2.37 deals with polyethylene.
ASTM D 5524-94	Determination (by liquid chromatography) of phenolic antioxidants in HDPE.

POLY(ETHYLENE GLYCOL)

See under *polyoxyethylene.*

POLY(ETHYLENE KETONE)

An aliphatic polyketone, with the repeating unit in the chain made up of ethylene and carbonyl groups. The first of the *Carilon* aliphatic polyketone range, introduced by Shell Chemicals in 1996. The polymer is reported to be polyethyleneketone of high molecular weight - produced from ethylene and carbon monoxide feedstocks - with ethylene and carbonyl units regularly alternating in the molecular chain. Among the disclosed characteristics are: a high melting point (220°C), short-exposure temperature resistance up to 180°C, good mechanical properties, high elongation at yield (25%), good impact resistance, high heat-deflection temperature under load (of 30% glass-filled grade), good resistance to friction and wear, good general chemical resistance at room temperature (but somewhat limited at elevated temperatures), and sensitivity to UV radiation in the absence of stabilisers.

POLY(ETHYLENE NAPHTHALATE); POLY(ETHYLENE-NAPHTHALENE-2,6-DICARBOXYLATE) (PEN)

Thermoplastic copolymer of dimethylnaphthalene-2,6-dicarboxylic acid and ethylene glycol, with better barrier properties and radiation resistance than its terephthalate analogue (PET), and with higher continuous-service temperature (155°C as against about 130°C for PET). Use as material for food packaging films (e.g. *Kaladex,* ICI Films), hot-fill containers, video and audio tapes, capacitor films, photographic films for advanced camera systems (e.g. *Advantix* - Kodak), and insulating tapes. Originally promoted by Teijin Ltd. in Japan, Du Pont in the USA, and ICI Films in the UK. See also *Cleartuf HP.* Polymer more recently available from Amoco Polymers and Shell (*Hipertuf*).

POLY(ETHYLENE OXIDE)

See *polyoxyethylene*

POLYETHYLENE (POLYETHENE) (PE) PLASTIC

An ethylene (ethene) plastic based on polymers made with ethylene (ethene) as essentially the sole monomer.

Reproduced, with permission, from ISO 472:1988.

NOTE - This definition is also given in the BS terminology standard[1.1]. In a comment accompanying its own, virtually identical definition, the ASTM terminology standard[1.7] interprets the qualifying word "essentially" as meaning that the monomeric constituents of the base polymers must be at least 85% ethylene and at least 95% total olefins.

The following are some examples of trade names of the very numerous commercial polyethylene plastics (many of the ranges include polyethylene polymers alongside polyethylene plastics). See also Section 1 of Part 3 of this book.

Alathon; *Alcudia*; *Alphathene* (recycled materials); *Aquathene*; *Arcel* (expandable beads); *Attane*; *Clearflex*; *Ecothene*; *Eltex*; *Eraclene*; *Escorene*; *Ferrene*; *Finathene*; *Flexirene*; *Fortiflex*; *Hicond*; *Histat*; *Hostalen*; *Hostalloy*; *Innovex*; *Ipethene*; *Lacqtene*; *Lupolen*; *Marlex*; *Microthene* (rotational-moulding compounds); *Montell*; *Neopolen* (foam); *Novapol*; *Petrothene*; *Plexar*; *Polycyre* (crosslinkable); *Rexene*; *Rigidex*; *Saxolen* (sheeting); *Saxorec* (recycled material and sheeting); *Shu-Link* (crosslinkable); *Sclair*; *Si-Link* (crosslinkable); *Stamylan*; *Tenite*; *Trocellen* (crosslinked foam); *Ucar*; *Visico* (crosslinkable).

Various aspects of polyethylene plastics are covered by the following standards.

Moulding and extrusion materials specifications

PE plastics - general : BS ISO 1872:Part 1:1993; BS 3412:1992; ASTM D 1248-84 (1989); ASTM D 4976-89; ASTM D 3981-89 (MDPE for general and packaging applications).
PE plastics - recycled materials : ASTM D 5203-93 (HDPE); ASTM D 5491-94 (recycled material - film sources).
PE-UHMW plastics : BS ISO 11542:Part 1:1994; ASTM D 4020-92.

Film and sheeting specifications

BS EN 6076:1996 (tubular PE film for use as protective covering for burried iron pipes and fittings); BS EN 7344:1990(1997) (reeled PE-LD film for general purposes); BS 3012:1972(1982) - obsolescent in 1997 (low and intermediate density PE sheet for general purposes); BS 4646:1972(1982) - obsolescent in 1997 (PE-HD sheet for general purposes); BS 7476:1991(1997) (PE film for covering silage); ASTM D 2103-92 (PE film and sheeting); ASTM D 4397-91 (PE sheeting for construction, industrial and agricultural applications); ASTM D 4635-91 (LDPE films for general use and packaging applications); ASTM D 4801-88 (low, medium, and high density PE sheeting).

Product specifications

Pipes and pipe fittings : various standards, too numerous to list in detail, including ISO 3458; ISO 3459; ISO 3501; ISO 3503; ISO 3607; ISO 3663; ISO 4056; ISO 4059; ISO 4437; ISO 4451; ISO 8283-2; ISO 8770; ISO 8772;

ISO 8779; ISO 8796; ISO 9625; BS 864-5; BS 3284; BS 5114; BS 5255; BS 6437; BS 6572; BS 7281; BS 7336; BS CP 312-3. See also ISO Standards Handbook 28:1986 - Pipes, fittings and valves - Plastics products, and ASTM Book 08.04 - Plastic pipe and building products.

Miscellaneous products

ISO 5834:Parts 1 and 2:1985	PE-UHMW implants for surgery.
ISO 1639:1990	PE ropes.
BS EN 700:1995	PE fibre ropes for general service.
BS 2919:1968(1982) - obsolescent	PE rod (low and intermediate density).
BS 4645:1970(1982)	PE-HD rod.
BS 7029	Thermoplastics injections mouldings.
Part 1:1993	General requirements.
Part 2:1993	Additional requirements (for PE mouldings inter alia).
BS 6515:1984	PE damp-proof courses for masonry.
BS 7523:1991	Preformed cellular PE insulation for pipes.
ASTM D 1998-95	PE upright storage tanks.

Cable coverings:

BS EN 60811-4-1:1995 (identical with IEC 811-4-1:1985)	Test methods for polyethylene and polypopylene compounds for cable insulation and sheathing.
BS 6234:1987	Specification for polyethylene insulation and sheath of electric cables.
BS 6469:Part 4:Section 4.2:1992 (Identical with IEC 811-4-2:1990)	Miscellaneous tests for PE and PP cable compounds (including test method for copper-catalysed oxidative degradation).
BS 6622:1991	Specification for cables with extruded crosslinked PE and E/P rubber insulation.

Testing standards

Carbon black content and dispersion : BS 2782:Part 8:Method 823A and B:1978(1983) (assessment of carbon black dispersion in PE - microscope method); ASTM D 3015-72 (1990) (determination of carbon black dispersion in plastics compounds, including PE compounds - wire and cable grade); ASTM D 4218-91 (determination of carbon black content of PE compounds - muffle furnace method).

Resistance to environmental stress cracking (ESC) : BS 2782:Part 11:Method 1109A:1989 (resistance to ESC of PE pipes and fittings); ASTM D 2561-91 (resistance to ESC of blow-moulded PE containers).

Resistance to thermal stress cracking : ASTM D 2951-71 (1988).

Preparation of test specimens : ISO 1872-2:1989 (PE and ethylene copolymer thermoplastics - preparation of test specimens and determination of properties); ASTM D 1928-90 (compression-moulded PE test sheets).

Thermal stability : ISO/TR 10837:1991 (determination of thermal stability of PE for use in gas pipes and fittings); ASTM D 3826-91 (degradation end point of PE sheeting).

Miscellaneous tests : ISO 7214:1985 (methods of test for cellular PE plastics); ISO 10147:1994 (estimation of degree of crosslinking in pipes and fittings made from crosslinked PE); BS 2782:Part 2:Method 241 A:1984 (determination of the effect of PVC compound on the loss tangent of PE); BS 2782:Part 4:Method 452C:1979(1986) (determination of butyl rubber content of PE-LD compounds); ASTM D 2741-91 (test for susceptibility of PE bottles to accumulation of soot - indicative of presence of static electricity charge (q.v.) on the bottle surface).

POLY(ETHYLENE SUCCINATE)

A polycondensation product of ethylene glycol and succinic acid, with the repeated structural unit in the molecular chain

$$—OCH_2.CH_2O.\underset{\underset{O}{\|}}{C}.CH_2.CH_2.\underset{\underset{O}{\|}}{C}—$$

This aliphatic polyester is claimed to be fully biodegradable (ultimately to water and carbon dioxide), and processable on ordinary plastics-processing equipment into biodegradable products. It can also be used as a biodegradability-promoting additive in some plastics. Examples of relevant commercial products are *Skygreen* and *Bionolle*. Whilst the name could be abbreviated to "PES", it should be noted that this abbreviation is the standard letter symbol for poly(ether sulfone)[2.1,2.2,2.3].

POLY(ETHYLENE TEREPHTHALATE)(PET)

A thermoplastic polyester (q.v.) derived from ethylene glycol and terephthalic acid; produced either by the polycondensation of these two components[1.1,1.5] or by ester-exchange polymerisation of ethylene glycol and dimethyl terephthalate[1.5]. Depending on the processing conditions, PET mouldings can be essentially amorphous, with good clarity, or partly crystalline (up to about 40% crystalline phase content). Clear films are produced by extrusion. Typical T_g (amorphous polymer) and T_m values are, respectively, about 45°C and 255°C. Some other properties of PET are mentioned under *thermoplastic polyester plastics*.

PET polymers and compounds are represented in the following commercial ranges.

Arnite; Beetle; Comtuf; Ertalyte; Eslon; Kopet; Malon; Mamax; Melinar PET; Melinex (films) *Mylar* (films); *Petra; Raditer; Rynite; Stanuloy; Tenite; Tripet; Valox; Venpet; Vivypak* (bottle grades); *Valoy*.

The following standards are relevant to PET materials.

Material specifications

ISO 7792	Poly(alkylene terephthalate) moulding and extrusion materials.
Part 1:1995	Designation system and basis for specification.
Part 2:1988	Preparation of test specimens and determination of properties.

ASTM PS 2-94	Provisional specification for moulding thermoplastics, including PET inter alia.

Tests

ISO 3451-2:1984	Poly(alkylene terephthalates) - Determination of ash.
ASTM D 3641-93	Injection moulding of test specimens from thermoplastic moulding and extrusion materials.
ASTM D 4603-91	Determination of inherent viscosity of PET.
ASTM D 5814-97	Test for contamination of recycled PET (flakes or chips).

Products

Films : BS 4584:Part 15:1978 (1991) (adhesive-coated PET films used as cover sheets for printed circuits); BS 7290:Part 3 : Section 3.2:1995, equivalent to IEC 674-3-2:1992 (specification for biaxially oriented PET films for electrical insulation); BS EN 60249-2-8:1995 (specification for flexible, copper-clad PET film); ASTM D 5047-90 (specification for biaxially oriented PET film and sheeting).
Capacitors (various aspects) : BS QC 300100; BS QC 300 400; BS CE CC 32200; BS 9076 N001.
Electric cables : British Aero G 189 and G 227 (Tersil electric cables for aircraft).
Bottles : ASTM D 4509-85(1990)) (test for gas-space acetaldehyde content in freshly blown PET bottles).

POLY(GLUCOSYLOXYETHYL METHACRYLATE) SULFATE

An ester of poly(methacrylic acid) in which the pendant ester groups contain sulfated glucose units. The similarity of these units to the sulfated disaccharide ones in heparin, the naturally occurring blood anticoagulant, is reported to impart to the polymer (described as a "heparinoid" resin) the ability to inhibit blood coagulation. It is suggested that antithrombogenic polymers of this kind could adhere more easily and firmly to damaged blood vessels and to the surfaces of surgical implants than heparin whose attachment is neither strong nor permanent.

POLYIMIDE (PI)

A family of polymers whose repeating units - which may contain aliphatic, aliphatic and aromatic, or only aromatic groups - are joined by imide linkages. Fully aromatic polyimides are the materials of some commercial high-performance plastics of superior resistance to chemicals and heat. An example is *Vespel*, a polyimide produced by the condensation of pyromellitic dianhydride and 4,4'-diaminodiphenyl ether in a two stage process. Plastics products made from polyimides of this kind are typically manufactured by direct forming, as the polymer resulting from the second stage of the process is normally infusible and insoluble. Some examples of commercial PI materials are *Kapton* (film); *Upilex* (formable thermosetting sheeting); *Kinel* (thermosetting moulding compounds); *Aurum* (thermoplastic moulding compounds).

Aside from their use as plastics materials, some polyimides also find applications as the materials of special fibres, adhesives, and surface finishes.

The only definition of polyimide to be found so far in a major plastics-terminology standard in English[1.1] centres on the joining of the polymer's structural units by imide groupings, and is virtually confined to thermoplastic polyimides.

The following standards deal with polyimide products (mainly in electrical and electronic applications).

Polyimide films and sheets: BS 4584:Part 15:1978(1991) (adhesive-coated PI cover sheets for flexible printed circuits); BS 60893:Part 3:Section 7:1995 (rigid laminated sheets based on PI resin); BS EN 60249:Part 2:Sections 13 and 15:1994 (flexible, copper-clad PI films); BS EN 60674:Part 3:Sections 4 to 6:1995 (PI films used for electrical insulation).

Various PI-enamelled wires: BS 6811:Part 4:Section 4.7:1993 (equivalent to IEC 317-18:1990) - rectangular copper wire; Part 5:Section 5.3:1993 (equivalent to IEC 317-25:1990)-round aluminium wire; BS EN 60317:Part 15 - round aluminium wire; BS EN 60317:Part 15 - round aluminium wire enamelled with polyester imide; BS EN 60317:Parts 7, 8, 13, 36, 37 and 38 round copper winding wires.

POLYISOBUTYLENE (POLY-2-METHYLPROPENE) (PIB)

A homopolymer of isobutylene (2-methylpropene)[1.1,1.5] :

$$\begin{array}{c} CH_3 \\ | \\ C{=}CH_2. \\ | \\ CH_3 \end{array}$$

According to the molecular weight, PIB polymers are thick, oily liquids or rubbery solids of varying degrees of tackiness. The main plastics applications are as films for use as damp-proof coursing in buildings and corrosion-protective wrappings. Lower molecular weight grades have been used as plasticisers in polyethylene cling film. Non-plastics applications included the use of PIB as a component of hot-melt adhesives and as chewing-gum base. Various grades of PIB are marketed under the trade names *Oppanol; Parapol;* and *Vistanex*.

POLYMER

A substance composed of molecules characterised by the multiple repetition of one or more species of atoms or groups of atoms (constitutional units) linked to each other in amounts sufficient to provide a set of properties that do not vary markedly with the addition or removal of one or a few constitutional units.

Reproduced, with permission, from ISO 472:1988.

POLYMER-FIBRE OPTICS (PFO)

Fibres used in fibre-optics systems. A typical structure of such fibres (strictly speaking continuous filaments) is an acrylic polymer light-transmitting core and fluoropolymer sheath (termed "cladding"). The exact nature of the proprietary polymers used is not disclosed by the producers. The composite polymer fibres are less efficient light transmitters than their glass counterparts in glass-based systems, but are more resistant to breaking and vibration damage, lighter and cheaper.

POLYMERIC PLASTICISERS

The majority of commercial polymeric plasticisers are saturated polyesters resulting from the reaction of a diol with a dicarboxylic acid. They differ basically from "monomeric" plasticisers (q.v.) in that their molecules are chains consisting of repeated constitutional units, and their molecular weight can be varied by controlling the chain length. For example, in commercial poly(propylene adipate) plasticisers the molecular weight may range between 2 000 and 9 000. It is common practice to include a third reactant in the production of a polyester plasticiser, to modify the product by terminating (end-stopping) the molecular chains. The terminating agent is usually either a monohydric alcohol or a monocarboxylic acid, although isocyanates and even diazomethane have been used to modify terminal hydroxyl and acid groups.

Thus there are three possible general structures for polymeric plasticiers:

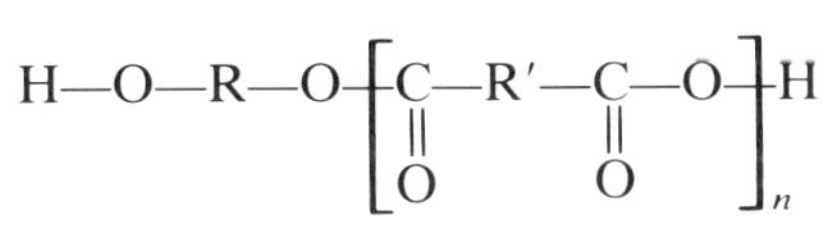

non-terminated

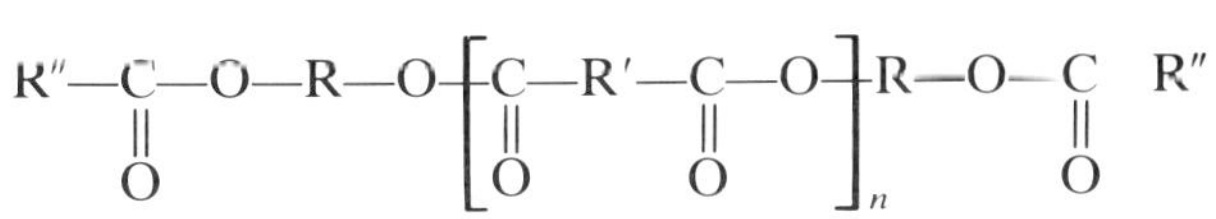

acid-terminated

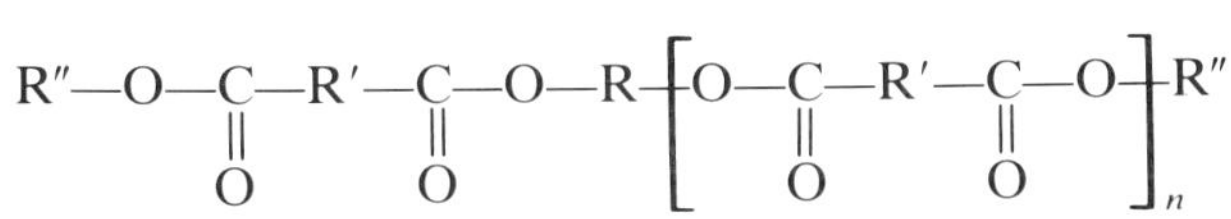

alcohol-terminated

There are therefore two main elements which can be varied in polymeric plasticisers, viz. their constituents, including chain stoppers, and their degree of polymerisation (and hence the molecular weight). Since both these factors are important in determining the resulting properties, the scope for variation is extensive. It also follows that the polymeric plasticisers are not simple, readily definable chemicals, as are most of the "monomeric" plasticisers, and it is for that reason that they are marketed under trade names rather than the chemical names indicative of their constitution.

The following are the most common raw materials used in the production of polyester plasticisers.

Dicarboxylic acids : adipic; azelaic; phthalic; sebacic.

Diols : 1,2-propanediol; 1,3-butanediol; 1,4-butanediol; 1,6-hexanediol; neopentyl glycol; diethylene glycol.

Monocarboxylic acids : acetic; caproic; caprylic; lauric; myristic; palmtic; stearic; pelargonic; benozoic.
Monohydric alcohols : isooctanol; 2-ethylhexanol; isodecanol; and some mixed-alcohol stocks, e.g. *Linevol 79* and *911* (Shell Chemicals).

POLY(METHYL METHACRYLATE) (PMMA)

Polymer obtained by polymerisation of methyl methacrylate[1.1,1.5]

$$CH_2{=}\underset{\displaystyle COOCH_3}{\underset{|}{C}}{-}CH_3$$

POLY(METHYL METHACRYLATE) (PMMA) PLASTICS

Acrylic plastics (q.v.) whose base polymers are homopolymers of methyl methacrylate[1.1,1.5]. PMMA plastics are available in the form of moulding and extrusion materials, and as cast products - especially sheets, but also rods, tubes and other simple shapes - in which the polymer may be partly cross-linked. Two of the properties of PMMA plastics, particularly important in such large-scale applications as aircraft glazing, vehicle number plates, sign and display panels, are the natural transparency, clarity, and weathering resistance of PMMA. The weathering resistance can be improved still further by incorporation of HALS (q.v.) or other selected UV stabilisers. The combination of high clarity with high refractive index value (1.49) of PMMA promotes conductance of light by total internal reflection, utilised in light-guide filaments and rods. Production of sanitaryware, such as baths, sinks and hand-basins (typically by thermoforming partly crosslinked sheet) is another substantial outlet for PMMA plastics, in which the high surface gloss and good scratch resistance of the base polymer are advantageous properties. Where slight scratching or marring has occurred, the surface can often be restored by polishing. Except for some special products, transparency and clarity are not primary requirements for sanitaryware, and the PMMA materials used may contain white pigments, calorants and fillers (including marble-effect fillers in some sink lines).

PMMA plastics share their parent polymer's resistance to hot and cold water (the presence of fillers can modify the naturally relatively low water absorption and permeability to water of PMMA), to aqueous solutions of surface-active agents, weak acids and alkalis, and to oils, greases, and non-polar solvents. They are susceptible to attack by strong acids and alkalis, by such solvents as aromatic and chlorinated hydrocarbons, ketones, esters and ethers (solvent susceptibility is utilised in solvent bonding of PMMA plastics parts), and are prone to environmental stress cracking in the presence of such solvents or their vapours. Their maximum temperature for continous service is below 100°C, and about 100°C, on the average, for short-term exposure (in both cases in the absence of stress and agressive environment), with the average DTL also somewhat below 100°C (slightly above for cast, partly crosslinked materials).

Whilst the formal, standard definition of the term "PMMA plastics" is properly restricted to plastics based on methyl methacrylate homopolymers, in common usage high-clarity copolymers of methyl methacrylate and acrylonitrile are also often included in that group. The copolymers are tougher than PMMA, but have a yellow tinge; they are less stable to UV radiation, but exhibit lower water absorption, and are more resistant to solvents (in particular to esters and chlorinated hydrocarbons). Their flammability is comparable with that of PMMA.

PMMA plastics are available in the following commercial ranges (see also Section 1 of Part 3 of this book).

Acrylite; *Acrypoly*; *Acryrex*; *Acrysteel*; *Altuglas*; *Corian* ("cultured marble" grades); *Degalan*; *Diakon*; *Gardlite*; *Kalmax T*; *Lucryl*; *Modar*; *Optix*; *Oroglas*; *Paraglas*; *Parapet*; *Perspex*; *Resarit*; *Sumipex*; *Vedril*; *Vestiform*.

The following standards are relevant to PMMA plastics.

BS ISO 8257	PMMA moulding and extrusion materials.
Part 1:1987(1997)	Designation
Part 2:1990(1997)	Preparation of specimens and determination of properties.
ISO 5833:1992	Acrylic (PMMA-based) resin cements for surgical implants.
ISO 7823	PMMA sheets : types, dimensions and characteristics.
Part 1:1991	Cast sheets.
Part 2:1989	Melt-calendered extruded sheets.
ISO 12017:1995	PMMA double- and triple-skin sheets - Test methods.
BS 7253:Part 1:1993 (technically equivalent to ISO 5833)	Specification for acrylic (PMMA-based) cement for surgical implants.
ASTM D 788-95	Standard classification system for PMMA moulding and extrusion compounds.
ASTM D 4802-94	Specification for PMMA acrylic plastic sheet. This standard replaces ASTM D 1547 (the now discontinued specification for extruded acrylic sheet).
ASTM D 5436-93	Specification for cast PMMA rods, tubes, and other shapes.
ASTM D 2857-93	Determination of dilute-solution viscosity of polymers (including methacrylates inter alia).
ASTM D 3641-93	Injection moulding test specimens from thermoplastic injection-moulding and extrusion materials.

POLY(METHYL METHACRYLIMIDE)[2.2]; POLY(N-METHYL METHACRYLIMIDE)[2.11]; POLYMETHYL-METHACRYLIMIDE[1.7] (PMMI)

A thermoplastic polymer resulting from reaction of poly(methyl methacrylate) (PMMA) with a monomethyl amine[1.7]. One version is a copolymer whose molecular chains are built up of blocks of methyl methacrylate and N-methyl-dimethyl-glutarimide. PMMI resembles PMMA but is superior to it in strength, rigidity, hardness, clarity, and barrier properties. It is also higher-softening, with a somewhat higher maximum temperature for continuous service, and lower coefficient of linear thermal expansion. Like PMMA, PMMI is susceptible to many organic solvents and to environmental stress cracking, and is fairly readily flammable. A commercial copolymer, *Kamax* (AtoHaas) finds applications in packaging (including coextruded and coinjected multi-layer container materials) and in instrument and automotive glazing.

POLY(4-METHYLPENTENE-1) (PMP)

A polymer of 4-methylpentene-1[1.5]

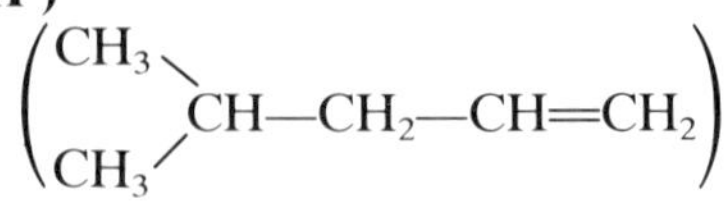

As a polyolefin, PMP shares with the two high-profile members of the group, polyethylene and polypropylene, excellent electrical properties, good resistance to chemicals, and low density (at 0.83 g cm^3 in fact the lowest of any commercial polyolefin). It is also similar in its possession of semi-crystalline morphology, albeit with a lower typical crystalline material content (between 40% and 65% in commercial PMP polymer, depending on the cooling treatment in processing) and relatively very high crystalline melting temperature (homopolymer T_m 245°C). At the same time, it has some interesting untypical properties, consequent mainly upon the presence of a particularly large side group in the repeating unit of its molecular chain:

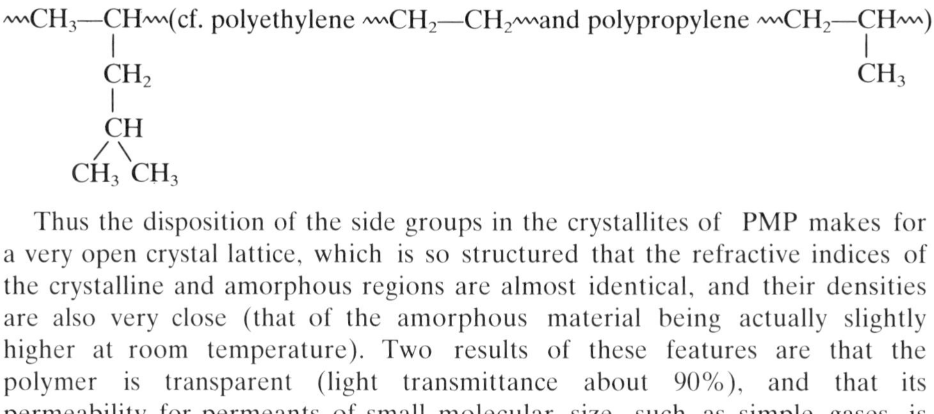

Thus the disposition of the side groups in the crystallites of PMP makes for a very open crystal lattice, which is so structured that the refractive indices of the crystalline and amorphous regions are almost identical, and their densities are also very close (that of the amorphous material being actually slightly higher at room temperature). Two results of these features are that the polymer is transparent (light transmittance about 90%), and that its permeability for permeants of small molecular size, such as simple gases, is relatively high; this is attributable to the permeants' ability to traverse the crystalline regions, which are not accessible in this way in other semi-crystalline polymers.

The tertiary carbon atoms bearing the large side groups in the repeat units of the molecular chains of PMP make the polymer vulnerable to oxidative and thermal degradation. The vulnerability is greater than that of polypropylene, in which the tertiary carbons are linked to methyl groups. Effective stabilisation is therefore essential, especially as the high T_m of PMP necessitates correspondingly high melt temperature in processing (typically about 270°C to 280°C).

The development of PMP was initiated by BP (at an early stage assisted by the Yarsley Research Laboratories in the evaluation of some material properties). It was taken over and completed by ICI, who eventually transferred the technology and material stocks to Mitsui in 1973. The current commercial range is marketed by Mitsui under the trade name *TPX*. In the course of development and early use of the homopolymer it was found that , especially in thick mouldings, transparency was impaired by a haze, attributed to formation of micro-voids at the boundaries between the crystalline (spherulitic) and amorphous regions. The resulting partial opacification was particularly undesirable in thick panels of PMP, otherwise especially suitable as windows in some chemical reactors. Whilst the haziness could be prevented by an appropriate moulding and cooling procedure (one was developed by the Yarsley Research Laboratories who acted as ICI agents for the production of thick mouldings until the take-over by Mitsui), another way was to introduce into the polymer a minor proportion of an alkene-1 comonomer (hexene-1 having been found particularly useful). The copolymererisation also has the effects of lowering the degree of crystallinity (which decreases with increasing molecular size of the comonomer and with its increasing proportion in the polymer) and the T_m. The lower T_m permits some reduction of the melt-processing temperatures.

PMP resembles polypropylene in some of its mechanical properties with, however, very much lower elongation of break (about 15% as against 200% to 600% for polypropylene). Unlike polypropylene, PMP is glassy at room temperature, which makes it more brittle (the T_g of PMP is variously quoted at values between 40°C and 60°C). Apart from its susceptibility to thermal and oxidative degradation in the absence of effective stabilisation, PMP is also prone to photochemical degradation, and its weathering resistance is relatively poor. It is subject to environmental stress cracking and crazing (q.v.) by agents of the kind that similarly affect low-density polyethylene, and exhibits the relatively easy flammability of a polyolefin.

Unlike polyethylene and polypropylene, PMP is a speciality material with applications where its high T_m and the associated high softening temperature, its transparency and chemical resistance are particularly useful. Such applications are e.g. in some transparent components of chemical installations (e.g pipes, and the inspection windows already mentioned), transparent electrical components and covers, and heat-resistant films for special packaging purposes.

POLYOL; POLYALCOHOL; POLYHYDRIC ALCOHOL

An alcohol having many hydroxyl groups.

NOTE 1 In cellular-plastics usage, the term *polyol* includes compounds containing alcoholic hydroxyl groups such as polyethers, glycols, polyesters and castor oil used in urethane foams.

Reproduced, with permission, from ISO 472:1988:Amendment 3:1993.

Polyols for use in polyurethane systems are available under the following table trade names: *Alcupol; Arcol; Celltrek; Glendion; Multranol; Pluracol; Poly-G; Qadrol; Tercarol; Thanol; Voranol;* and *Vulcollan.*

The following standards deal with various aspects of polyols.

ASTM D 4273-94 : determination of hydroxyl content of polyols.
ASTM D 4662-93 : determination of acid/alkalinity numbers of polyols.
ASTM D 4669-93 : determination of the specific gravity of polyols.

ASTM D 4670-93 : determination of suspended matter in polyols.
ASTM D 4671-93 : determination of unsaturation in polyols.
ASTM D 4672-91 : determination of water content of polyols.
ASTM D 4875-94 : determination of polymerised ethylene oxide content of polyols.
ASTM D 4878-93 : determination of viscosity of polyols.

POLYOLEFIN

A polymer produced by polymerisation of one or more olefins[1.1,1.5,1.7].

Polyolefins of the greatest commercial importance are polyethylene and polypropylene. Other polyolefins significant in this sense are copolymers of ethylene with propylene (including EPDM), other propylene copolymers, copolymers of ethylene with certain alpha-olefins, e.g. butene, hexene or octene (some of these copolymers are the various versions of linear low-density polyethylene (PE-LLD) and very-low-density polyethylene (PE-VLD)), polybutylene, polyisobutylene, and poly-4-methylpentene-1.

A recent development in the field of polyolefins is the production of special versions of a number of olefin polymers and copolymers with the aid of metallocene catalysts (q.v.). This polymerisation technology enables polymers to be produced with structure and properties "tailored" for various applications and impoved processing behaviour. The main application areas (with others being targeted) are packaging films, wire and cable coverings, and fibres for non-woven fabrics. Further applications are also being developed for metallocene-produced polyolefin "plastomers": in this context the term is intended to denote polymers with properties intermediate between those of plastics and elastomers. One such application is their use as impact modifiers in other polymers.

The metallocene-polymerised polyolefins which are already (in 1997) either in commercial or pilot-scale production are listed in the table.

POLYMER TYPE	TRADE NAME	PRODUCER	REMARKS
Polypropylene	ACHIEVE HOSTACEN LUFLEXEN	Exxon (USA) Hoechst (Germany) BASF (Germany)	
Polyethylene	-	BP Chemicals (UK)	
Low-density polyethylene	EXCEED	Exxon (USA)	Joint venture in Europe with DSM
	-	Union Carbide (USA)	
"Plastomer"	EXACT	Exxon (USA)	Joint venture in Europe with DSM
	AFFINITY	Dow (USA)	
Elastomer	ENGAGE	Dow (USA)	Joint venture in Europe with Du Pont
Cyclo-olefin copolymer	TOPAS	Hoechst (Germany)	Joint venture with Mitsui Sekka

For trade names of various polyolefins and polyolefin plastics see under the individual polyolefin materials. See also Section 1 of Part 3 of this book.

The following standards are concerned with polyolefins in general (as distinct from individual polyolefin materials) See also under *polyolefin plastics*, *olefin plastics*, and *ethylene plastics*.

ASTM D 3895-94	Determination of oxidative induction time (of PE and PP).
ASTM D 5227-92	Determination of low-molecular-weight (hexane-extractable) material in polyolefins.
ASTM D 5576-94	Determination of type and distribution of branch chains in polyolefins (by Fourrier-transform infra-red spectrophotometry)

Standards dealing with some ethylene/propylene copolymers classifiable under the heading of polyolefins include the following:

EPDM terpolymers: ISO 4097:1991, identical with BS 6063:1992 (methods of evaluation of general-purpose EPDM); ASTM D 3568-95 (test methods); ASTM D 3900-95 (determination of ethylene units in EPDM).
Ethylene/propylene copolymers (E/P[2.1,2.3]; EPM[2.2], EPM rubber[2.4,2.5,2.6]) : BS 6014:1991(1996) (specification for EPM compounds); BS 5467:1989 (cables with thermosetting insulation - EPM rubber or crosslinked polyethylene); BS 6724:1990 (cables as in BS 5467, but with low smoke emission); BS EN 2428:1995 to 2432:1995 (EPM/EPDM with IRHD hardness of 50 to 90).

POLYOLEFIN PLASTICS

Those olefin plastics (q.v.) which are based on polymers or copolymers containing essentially only olefin monomeric units in the molecular chains[1.1,1.5,1.7] (cf. e.g. polyethylene, polypropylene, polybutylene, EPDM, ethylene/propylene copolymers, etc.).

The following standards are concerned with materials and products generally polyolefin-based, i.e. not associated specifically with individual polyolefins (for standards of the latter kinds see under particular polymer/plastic names). Cf. also *olefin plastics, polyolefin, ethylene plastics, polyethylene plastics*, and *polypropylene plastics*.

Flexible cellular polyolefin plastics materials: ASTM D 4819-88 (specification); ASTM D 3575-93 (testing).
Polyolefin pipes and fittings: ISO 4433:1984 (determination of resistance to chemical fluids); ISO 6964:1986 (determination of carbon black content); ISO 8584-1:1990 (determination of basic stress and chemical resistance factor); ISO 9356:1989 (resistance to internal pressure).
Blow-moulded polyolefin containers: BS 4839:Part 1:1972 and Part 2:1974.
Polyolefin hot-water bottles: BS 6728-2:1993
Polyolefin-covered, copper-conductor cables: BS 3573:1990.
Cold-water storage cisterns (polyolefin or olefin copolymer): BS 4213:1991.
Heat-shrinkable polyolefin sleeving (dual wall) for electrical insulation purposes: BS EN 60684-3-246:1995.
Polyolefin agricultural twines: ISO 4167:1979.

Miscellaneous test methods
Determination of carbon black in PE, PP and PB plastics: ASTM D 1603-94; BS 2782:Part 4:Method 452 B:1993.
Rigidity test for polyolefin film and sheeting: ASTM D 2923-94.
Identification and determination of antioxidants and UV absorbers in polyolefin compounds: BS 2782:Part 4:Methods 434 A to D:1975(1993).
Determination of water-extractables in polyolefin compounds: BS 2782:Part 4:Methods 452 D to F:1978(1992) - pH of water extract (Method D); water-soluble sulfates (Method E); water-soluble chlorides (Method F).
Determination of viscosity number of polyethylenes and polypropylenes: ISO 1628-3:1991, identical with BS 2782:Part 7:Method 732C:1991.
Assessment of pigment dispersion in polyolefin pipes and fittings: BS 2782:Part 11:Method 1106A:1983(1988).

POLYOXYETHYLENE; POLY(ETHYLENE OXIDE) (PEOX)[1.5,2.1,2.2,2.11]

A polymer produced by polymerisation of ethylene oxide[1.5] 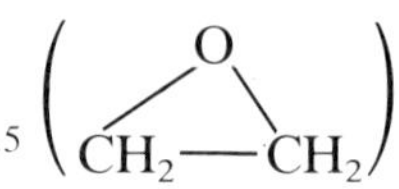 or by polycondensation of ethylene glycol (hence in some cases referred to as polyethylene glycol), with the repeating unit $-CH_2-CH_2-O-$. Polymers of relatively low molecular weight (up to about 600) are liquids of varying viscosity, used mainly as non-ionic surface-active agents. Higher molecular weight materials (up to about 10 000) are hydrophilic, water-soluble waxes, (e.g. *Carbowax*) with main applications is some cosmetic and pharmaceutical preparations. It is these low and medium molecular weight materials that are occasionally termed polyethylene glycols (especially where the molecular weight is low enough for the terminal hydroxyl groups to exert some influence on the properties). Polymers with much higher molecular weights (of the order of a million) are crystalline materials (T_m 65°C to 67°C), strong and tough in the dry state (the strength is greatly reduced by moisture absorption at high ambient humidity). The commercial versions (e.g. *Polyox*) can be processed by injection moulding, extrusion and calendering, although they remain water-soluble. The main uses of PEOX are as films, sachets and the like for water-soluble packaging applications, and as thickening agents for aqueous solutions.

The following standards are relevant to various PEOX materials.

BS 6829: Part 4 (six sections)	Ethylene oxide adducts.
ASTM D 4875-94	Determination of polymerised ethylene oxide content in polyether polyols.
ISO 1065: 1991	Non-ionic surface-active agents obtained from ethylene oxide - Determination of cloud point.
ISO 2268:1972	Non-ionic surface-active agents - Determination of polyethylene glycols.
ISO 2270: 1989	Non-ionic surface-active agents - Polyethoxylated derivatives - Iodometric determination of oxyethylene groups.

POLYOXYMETHYLENE (POM); POLYFORMALDEHYDE

Structurally the simplest polyacetal (q.v.), with molecular chains made up of repeated oxymethylene units[1.1,1.5,1.7]($-OCH_2-$).

POLYOXYMETHYLENE PLASTICS

Acetal plastics whose base polymers are essentially of the polyoxymethylene (q.v.) type[1.1,1.5,1.7].

POLYPEPTIDE

A compound formed from aminoacid units linked by amide groups. The term is normally used in connection with proteins. In that context the amide linkages are usually termed peptide linkages or groups, and the polypeptide molecules of a particular protein are often composed of several different aminoacid units.

POLY(PHENYLENE ETHER) (PPE)

The alternative name and letter symbol for poly(phenylene oxide) (PPO) (q.v. for more detailed definition). This alternative is promoted in the USA because of a trademark restriction on the use of "PPO" in that country; it is also featured in standard ISO and BS terminology, in which "PPE" is now the favoured letter symbol for the polymer concerned. PPE materials (including blends with polystyrene and polyamides) are covered by ASTM D 4349-93. It may be noted that this standard makes passing reference to an earlier one (ASTM D 2874) now discontinued, in which these materials were still referred to as "poly(phenylene oxide) plastics". PPE moulding and extrusion materials are also dealt with in a forthcoming ISO standard (ISO/CD 15103 Parts 1 and 2, still in draft form in 1996). Commercial versions of these materials include the following products or ranges: *Ashlene (200 series); Dimension; Emarex; Emiclear; Iupiace; Laril; Lastil; Luranyl; Noryl and Noryl GTX; Prevex; Tarnoform; Vestoblend; Vestoran; Xyron.*

POLY(PHENYLENE OXIDE) (PPO)

In the strict sense of the term, a polymer whose molecular chains are made up of the repeating unit 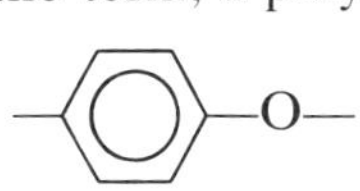 [1.5]. This polymer has a high T_g (about 245°C) and good stability to heat, but it is not sufficiently readily fusible, or soluble in solvents, for industrial processing in melt or solution to be practicable. An advance in these respects was made by the introduction (by General Electric) of a closely related polymer, methyl-substituted on the benzene ring of the repeating units, viz, poly(2,6-dimethyl-1,4-phenylene oxide); and it was this polymer which became generally known as poly-(phenylene oxide). A further improvement, in processability and resistance to oxidative thermal degradation, was achieved by blending this material with polystyrene. The original commercial version of this blend (Noryl) was first marketed by General Electric as "modified PPO". The expression is now also applied to such other, later modifications as blends with polyamides, and grafts of styrene onto poly (2,6-dimethyl-1,4-phenylene oxide). Modified PPO is usually included among engineering plastics with good mechanical properties and weathering resistance, available in flame-retarded grades. Its chemical resistance is moderate (it is susceptible to attack by several solvents and concentrated acids), and it is not fully resistant long-term to oxidative thermal degradation. The name poly(phenylene oxide) and its letter symbol PPO have been replaced in American ASTM standards and, to a considerable extent, in industrial terminology in that country by the alternative name poly(phenylene ether) and the letter symbol PPE[2.2] (q.v.), because in the USA "PPO" is a registered trademark for a different product. "PPE" has also been adopted as the standard letter symbol by the relevant ISO and BS letter-symbol standards[2.1,2.3], but in the ISO terminology standard (second edition, current in 1997[1.5], with a new edition in preparation) - the only one of the three main such standards to define the polymer - the definition appears under the heading "poly(phenylene oxide) (PPO)". Included in the definition is a note on the position in the USA: this presentation is reasonable as both the full name and the abbreviated term for PPO are in use in Europe and the UK. For references to technical standards relevant to PPO, and trade names of some PPO materials see *poly(phenylene ether)*.

POLY(PHENYLENE SULFIDE) (PPS)

An aromatic polysulfide with the constitutional repeating unit 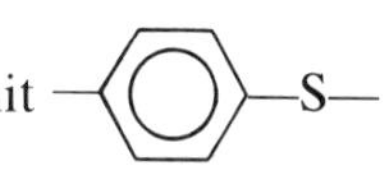 [1.5].

High linearity of this polymer's chains promotes crystallinity, which can reach high degrees, depending on cooling and product annealing conditions. Crystalline melting point 288°C; T_g 93°C. Some commercial grades are cross-linkable. Incorporation of fillers and reinforcements counteracts PPS's inherent tendency to brittleness. A specification for reinforced and filled PPS moulding and extrusion compounds is provided in ASTM D 4067-93; some compounds are also covered by MIL-P-46174(MR). An ASTM provisional standard (PS 2 - 94), which deals with a number of thermoplastic moulding compounds for electrical and electronic components, includes a specification for flame-resistant 40% glass-reinforced PPS. Commercial versions of PPS include *Alton; Cels; Compodic FPPS; IPC; Larton, Lubricomp, Stat-Kon; Primef;* and *Ryton*.

POLY(PHENYLENE SULFONE) (PPSU)

A polysulfone (q.v.) in whose molecular chains phenylene sulfone

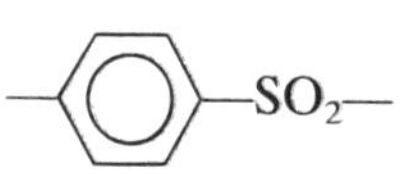 is the constitutional repeating unit[1.5]. The polymer is not used as a commercial thermoplastic because of its very high melting point (over 500°C) and the fact that melting is accompanied by decomposition. The

Radel R materials (q.v.) - a line of polysulfone injection moulding compounds, are sometimes designated by the abbreviation PPSU, but in this case the letters stand for "polyphenylsulfone", the manufacturer's description of these materials.

POLY(p-PHENYLENEVINYLENE) (PPV)

A photoluminescent polymer. PPV and some of its derivatives can be stimulated to produce amplified laser-like emission.

POLYPHOSPHAZENE

See *phosphazene polymers.*

POLYPHTHALAMIDE (PPA)

Member of the family of polyamides containing terephthalic or/and isophthalic acid residues in the repeat units making up the polymer chains. The commercial polyphthalamides (cf. e.g. *Amodel*) are a group of semi-crystalline polymers with softening temperatures acceptably high for some engineering applications (T_g range 120°C - 135°C), low water absorption, and good chemical resistance. ASTM D 5336-93 provides a specification for PPA injection moulding materials.

POLYPROPYLENE (PP)

The product of polymerisation of propylene[1.1,1.5,1.7] ($CH_3.CH{=}CH_2$) used as the only monomer[1.7].

NOTE - The ISO and BS standards[1.1,1.5] mention polypropene as the alternative term.

POLYPROPYLENE (PP) PLASTICS

The standard definitions[1.1,1.5,1.7] distinguish between *polypropylene plastics*, i.e. plastics based entirely on polypropylene (q.v.), and the wider category of *propylene plastics* whose base polymers may be either polypropylene or copolymers in which propylene is the major monomeric component, predominant by mass.

NOTE - The ISO and BS definitions[1.1,1.5] include propene and polypropene as alternative terms for propylene and polypropylene.

In terms of industrial consumption, propylene plastics, and within that general group the polypropylene plastics, are among the most widely used plastics. The high consumption is reflected in the large numbers of commercial materials, both polymers and compounds. Only some examples are given below, out of the greater number listed in Section 1 of Part 3 of this book.

Accpro; Acctuf; Achieve (polymers produced with metallocene catalysts); *Adflex; Adpro; Albis; Amoco; Arpro* (expandable beads); *Artholene; Beta PP; Buplen; Capilene; Cefor; Celstran; Corton; Denilen; Eltex P; Epalex; Escorene; Hostalen; Hostacom; Isplen; Latene; Marlex; Montell; Nortuf; Novolen; Polyfort; Polytrope; Pro-Fax; Propadex* (recycled); *Refax* (recycled); *Rexflex; Rototron* (for rotational moulding); *Snialene; Stamylan P; Stamyroid* (modified, amorphous) *Taffen; Tenite; Technoprene* (glass-reinforced); *Top-alloy; Typlax; Valtec; Vamplen (flame-retarded); Vestolen EM; Vipprop RE* (recycled).

The following standards deal with with propylene and/or polypropylene plastics materials or products.

Material specifications

BS EN ISO 1873:Part 1:1995	PP moulding and extrusion materials - Designation system and basis for specification.
ASTM D 4101-94	Standard specification for propylene plastic injection and extrusion materials.

Material property tests

Test specimens : ISO 1873:1990 (PP and propylene copolymer thermoplastics - Preparation of test specimens and determination of properties); ASTM D 3641-93 (Injection moulding test specimens of thermoplastic moulding and extrusion materials).

Tests for resistance to degradation : ISO 4577:1983 (PP and propylene copolymer thermoplastics - Determination of thermal oxidative stability by an oven method); ASTM D 3012-94 (Determination of thermal oxidative stability of propylene plastics); ASTM 3826-91 (Determination of the degradation end point of PE and PP plastics); ASTM D 3895-94 (Oxidative induction time : DSC test for polyolefins).

Miscellaneous tests : ISO 1628-3:1991 (Determination of viscosity number and limiting viscosity number of PE and PP); ISO 9113:1986 (Determination of the isotactic index of PP and propylene copolymers); BS 6469:Part 4 (Tests specific to PE and PP insulating and sheathing materials for electric cables : Section 4.1 of Part 4 replaced by BS EN 60811-4-1:1995); ASTM D 1603-94 (Test for carbon black in olefin plastics); ASTM D 5492-94 (Test for xylene-soluble matter in propylene plastics).

Product specifications and test standards

Films : ASTM D 2673-94a (Oriented PP films); ASTM D 2578-94 (Test for wetting tension of PP and PE films); ASTM D 4272-90 (Dart impact test for plastics films).

Ropes and cords : ISO 1346:1990 (Specification for PP ropes); BS 4881:1993 (Sepcification for PP film cords, lines and twines).

Miscellaneous products : BS EN 277:1989 (PP sacks for food-aid transport); BS 6782-2:1993 (Specification for PP and PE hot-water bottles); ASTM D 5419-93 (PP and PS threaded closures for containers - test for environmental stress cracking).

Pipes and pipe fittings. Specifications : ISO 8773:1991 (PP pipes and fittings for buried drainage and sewage); BS 4991:19974(1982) (propylene copolymer pressure pipe); BS 5254:1976 (PP waste pipe and fittings). *Tests* : ISO 3212:1975 (bursting test); ISO 3213:1975 (reduction of permissible stress as a function of time and temperature); ISO 3477:1981 (determination and specification of density); ISO 3480:1976 (maximum permissible longitudinal reversion). *Dimensions and systems* : ISO 3609:1977 (tolerances on dimensions); ISO 7279:1984 (sockets for fusion); ISO 7671:1991 (PP pipe fittings and systems); ISO 8242:1989 (dimensions of PP values for pipes under pressure). See also ISO Standards Handbook 28:1986 - Pipes, fittings and valves - Plastics products, and ASTM Book 08.04 - Plastic pipe and building products.

POLYSTYRENE (PS)

A polymer[1.1,1.5] (homopolymer[1.7]) of styrene, with the repeated structural unit $-HC-CH_2-$. Normally amorphous (and hence transparent), with a T_g of

90°C to 100°C, depending on grade.

Semi-crystalline, isotactic versions of the polymer (with a T_m in excess of 200°C) can be produced with the aid of metallocene catalysts (q.v.).

Polystyrene was the first fully synthetic, large-use thermoplastic. After World War II it held for some time the position of the workhorse of the thermoplastics moulding industry, which it took over from cellulose acetate, the leading pre-war thermoplastic moulding material. Polystyrene offers the combination of high clarity and surface gloss, hardness, rigidity, good electrical properties, low water absorption, and easy processability. Its advantageous properties have made it a popular commodity plastic, both in the solid and expanded forms (expanded polystyrene being widely used for thermal insulation, and as a packaging material). However, even with glass-fibre reinforcement (first practiced in the USA by A. Bradt - cf. e.g. Titow and Lanham, 1975) polystyrene does not qualify as an engineering plastics material. Its disadvantageous properties include relative brittleness, susceptibility to many organic solvents (including aromatic and chlorinated hydrocarbons, esters, ethers, and ketones) and to environmental stress cracking, and flammability (particularly pronounced in the expanded form).

See also *polystyrene plastics* and *toughened polystyrene*.

POLYSTYRENE (PS) PLASTICS

Those styrene plastics (q.v.) whose base polymers are homopolymers of styrene[1.1,1.5]. See also *polystyrene, toughened polystyrene,* and *styrene-rubber plastic.*

Commercial polystyrene materials are available in the following product ranges (see also Section 1 of Part 3 of this book). *Aim; Amoco; API; Bapolan; Bustren; Denistyr; Dylene; Edistir; Extir (expandable PS); Ferroflo; Fina; Gedexcel* (expanded PS); *Huntsman PS; Lacqrene; Lastirol; Luran; Lustran; Novacor; Polyman; Poly Star; Polystyrol; Rigipore* (expandable PS); *Saxerol* (PS sheet); *Scopyrol; Solarene; Styradex* (reprocessed materials); Styrocell (expandable PS); *Styrodur; Styrolux; Styron; Styropor* (expandable PS); *Tyril; Vestypor* (expandable PS); *Vestyron.*

Major standards relevant to polystyrene plastics include the following.

Material specifications

BS ISO 1622 (replaces BS 1493:1987)	Polystyrene moulding and extrusion materials.
Part 1:1994	Designation system and basis for specification.
Part 2:1995	Preparation of test specimens and determination of properties.
BS ISO 2897 (Replaces BS 3126:1987)	Impact resistant polystyrene (PS-I) moulding and extrusion materials.
Part 1:1990	Designation.
Part 2:1994	Preparation of test specimens and determination of properties.
BS 3241:1960 (1997)	Specification for toughened polystyrene for sheet extrusion.
ASTM D 4549-85 (1994)	Specification for polystyrene moulding and extrusion materials.

Material property tests

ISO 2561:1974 (identical with Method 453A:1978 (1993) of BS 2782)	Determination of residual styrene monomer content in polystyrene(by gas chromatography).
ISO 6235:1995	Determination of block polystyrene content in styrenic elastomers (ozonolysis).
ASTM D 5296-92	Determination of molecular weight and molecular weight distribution of polystyrene (by size-exclusion chromatography).

Product specifications and test standards

BS 3837	Expanded polystyrene boards.
Part 1:1986(1996)	Specification for boards manufactured from expandable beads.
Part 2:1990(1996)	Specification for extruded boards.
BS 6203:1991 (1996)	Guide to fire characteristics and fire performance of expanded polystyrene materials in building applications.
ASTM D 5419-93	Environmental stress cracking test for threaded plastics closures (including polystyrene closures inter alia).

POLYSULFONE (PSU)

Generically, the term is applicable to any polymer in which the constitutional repeating unit is $-R.SO_2-$, where R may be an aliphatic or an aromatic unit; however only the aromatic polysulfones, i.e. polyarylsulfones, are significant as plastics materials. The term is also often used specifically for bisphenol-A polysulfone, a polymer whose structural repeat unit combines bisphenol-A and diphenylsulfone residues. A commercial version of this material is *Udel*.

Other polyarylsulfones marketed as "*polysulfones*" include materials in the following ranges : *Fiberfil J; Lasuf; Lubricomp G; Mindel M; Ultrason S*. Examples of commercially available polysulfone alloys are *Mindel B, A* and *S*, and *Ucardel*.

POLYTETRAFLUOROETHYLENE (PTFE)

A polymer of tetrafluoroethylene[1.1,1.5], produced - as a relatively fine powder - by free-radical polymerisation of the monomer in aqueous dispersion. PTFE is a polymer of high molecular weight (number-average molecular weight in excess of 10^6), characterised by complete linearity of the molecular chains (virtually no chain branching), a high degree of crystallinity (over 90%), and high crystalline melting point, normally quoted as 327°C (e.g. Brydson 1995). It is commercially the most important and most widely used of fluoroplastics (q.v.), offering an impressive combination of properties, including very high thermal and oxidate stability, high maximum temperature for continuous service (short-term 300°C, long term 250°C; in air, in the absence of stress), exceptional chemical inertness and resistance to solvents, excellent electrical insulation properties and weathering resistance, very low coefficient of friction, low-adhesion ("non-stick") surface, and low flammability. These properties qualify PTFE for a wide range of specialist applications, some of which are reflected in a number of national and international product standards. Although it can be repeatedly melted by heating and resolidified by cooling, PTFE does not entirely fall within the standard definition of a thermoplastic (q.v.) in that its high melting point and melt viscosity make it unsuitable for melt processing by conventional moulding and extrusion techniques. The special methods whereby it is converted into products include preforming followed by sintering (at about 370°C), extrusion of powder (with continuous sintering), and extrusion of a paste of the polymer in a hydrocarbon carrier. PTFE films can also be made by skiving, i.e. continuous peeling from the outer surface of a tube. The trade names of commercial material ranges in which PTFE polymers and compounds are represented include: *Algoflon; Duroid; Fluon; Fluorocomp; Goretex; Halon TFE; Hostaflon TF; Kel-F; Neoflon; Polycomp; Polyflon; Teflon.*

The following standards deal with PTFE polymers, compounds and products.

Material standards

Moulding and extrusion materials : ISO 12086:Part 1*; BS 6564:Part 1:Sections 1.1 and 1.2:1989; ASTM D 1457-92; D 4894-94; D 4895-94. Filled moulding and extrusion compounds : BS 6564:Part 3:1990; ASTM D 4745-91.
Aqueous dispersions : ISO 12086*; BS 7485:1991; ASTM D 4441-91.

Product standards

General : ISO 1300*; Parts 1 and 2 (PTFE semi-finished products); BS 6564:Part 2:1991 (specification for fabricated, unfilled products). Pipes, pipe fittings and flanges : ASTM F 423-93. Tubing, hoses and hose assemblies (aerospace and aircraft) : ISO 7313-1984; 8829:1990; 8913:1994; 9528:1994;

9938:1994; 10502:1992. BS Aero M56; M57. Miscellaneous : BS Aero G210 (PTFE-insulated wires); BS 6974:1991 (unsintered tape for coarse-thread sealing); BS 7786:1995 (unsintered tape); ASTM D 1710-91 (basic shapes); D 3293-91 (moulded sheet); D3294-91 (moulded sheets and basic shapes); D3295-91 (tubing); D3297-93 (parts : tolerances for moulding and machining); D3308-91 (skived tape); D3369-91 (cast film); D4969-91 (PTFE-coated glass fabric).

Test standards

ISO 12086:Part 2* (preparation of test specimens and determination of properties); ISO 7258:1984 (determination of density and relative density of PTFE aerospace tubing); BS 2782:Method 327A:1993 (determination of tensile strength and elongation of PTFE products); ASTM D 4591-93 (determination of the temperature and heat of transition).

*All parts of ISO 12086 and ISO 13000 still under discussion in 1996.

POLY(TETRAFLUOROETHYLENE)-CO-[PERFLUORO-(METHYL VINYLETHER)]

See *PERFLUORO ALKOXYL ALKANE (PFA)*.

POLYURETHANE (PUR)

A polymer, which may be thermoplastic or thermoset, characterised by the presence of the urethane group $-\underset{\text{H}}{\underset{|}{\text{N}}}-\underset{\text{O}}{\underset{\|}{\text{C}}}-\text{O}-$ or groups in the repeated structural unit of its molecular chain (so that the unit can be described as of the urethane type[1.1,1.5]). Polyurethanes are polymeric products of reactions between diisocyanates (mainly) or polyisocyanates and dihydric or polyhydric hydroxyl compounds (diols or polyols). In its definition, the ASTM terminology standard[1.7] refers to the isocyanate reactant specifically as "an organic diisocynate". However, whilst organic diisocyanates - especially the aromatic TDI and MDI - are most widely used in the production of polyurethanes, isocyanates with functionalities higher than two are also employed (e.g. modified polymeric MDI).

NOTE - TDI and MDI are the abbreviated terms for, respectively, tolylene diisocyanate (the material used in practice is a mixture of isomers) and 4,4'-diphenylmethane diisocyanate. Other isocyante reactants include 1,5-naphthylene diisocyanate (NDI), and the aliphatic 1,6-hexamethylene diisocyanate (HDI).

The ASTM terminology standard additionally draws attention to the use of "urethane" for "polyurethane"[1.7] (as e.g. in *urethane plastics* - q.v.), and to the great variety of forms of polyurethane materials, viz. thermoplastic or thermoset, rigid or flexible, cellular or solid[1.7].

The following standards deal with isocyanates used in the production of polyurethanes (for standards relevant to the polyol components of polyurethanes see under *polyol*).

Specification for TDI : ASTM D 1786-90
Determination of amounts of 2,4- and 2,6- isomers in industrial TDI : ISO/TR 9372:1993; ASTM D 4660-90.
Freezing point of TDI : ASTM D 4664-93.
Determination of acidity of isocyanates : ASTM D 4667-93 (acidity in TDI); ASTM D 4876-93; ASTM D 5629-94.
Determination of isocyanate content : ASTM D 5155-91.
Determination of specific gravity : ASTM D 4659-93.
Determination of viscosity : ASTM D 4889-93.
Determination of colour : ASTM D 4877-93.
Determination of chlorine content : ASTM D 4661-93; ASTM D 4663-93; ASTM D 5523-94.

POLY(VINYL ACETAL)

(1) In the generic sense of the term, any member of the class of polymers (typically amorphous) that are the products of reactions of poly(vinyl alcohol) (PVAL) with aldehydes, in which acetal groups are formed[1.1, 1.5].

NOTE - The PVAL itself is produced by conversion of the acid groups of a polyvinyl ester (normally poly(vinyl acetate)) to hydroxyl groups[1.1,1.5] - see *poly(vinyl alcohol).* In the production of some poly(vinyl acetals), the starting material is poly(vinyl acetate) and its conversion to the polyacetal, although proceeding via PVAL, is carried on directly, without intermediate isolation of the PVAL. Such direct conversion is usual e.g. in the manufacture of poly (vinyl formal) - q.v.

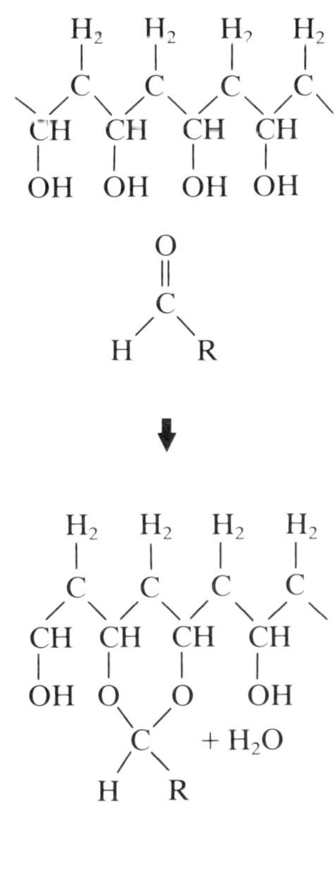

In practice, the PVAL, whether generated in situ in a direct conversion of poly(vinyl acetate) to a poly(vinyl acetal), or used as the starting material, will normally contain some unconverted acetate groups, whilst the acetal-forming reaction of the PVAL with an aldehyde, which involves pairs of adjacent hydroxyl groups, proceeds randomly, so that some hydroxyls remain unpaired, and thus unreacted. Hence a commercial poly(vinyl acetal) will normally contain, in addition to the acetal groups, also some acetate groups and some hydroxyl groups.

(2) In the specific sense, the poly(vinyl acetal) is the product of reaction of PVAL with acetaldehyde CH_3CHO; i.e. R is CH_3 in the general reaction scheme under (1) above. This material has no significant applications in plastics. Its main use is as part of the polymeric base of some wire enamels: - cf. BS EN 60317, Parts 1: 1995; 12: 1995; 17: 1996; and 18: 1996 (specifications for various enamelled copper wires), and Part 14: 1995 (specification for enamelled aluminium wire).

NOTE - It may be noted that none of the main standards for abbreviated terms (letter symbols)[2.1,2.2,2.3,2.11], and none of the three terminology standars[1.1,1.5,1.7], provides a letter symbol for poly(vinyl acetal), either in its generic or specific meaning (the symbol *PVAC*, which might have been applicable, stands for *poly(vinyl acetate)*).

POLY(VINYL ALCOHOL) (PVAL[1.1,1.5,2.11]; PVOH[2.2])

A polymer with the repeated structural unit $-CH_2-\underset{\displaystyle OH}{\underset{|}{CH}}-$, produced not from vinyl alcohol monomer (which cannot be prepared in the free state, as it tautomerises to acetaldehyde : $CH_2{=}\underset{\displaystyle OH}{\underset{|}{CH}} \rightarrow CH_3-\underset{\displaystyle O}{\underset{\|}{C}}-H$, but by hydrolysis of a poly(vinyl ester)[1.1,1.5,1.7]. The ester normally used is poly(vinyl acetate), and the hydrolysis is effected by its reaction not with water but with methanol (most commonly) or ethanol, so that the reaction is more accurately described as alcoholysis; the products are PVAL and methyl (or ethyl) acetate. Although on a strict definition of the term, poly(vinyl alcohol) should be the polymer obtained by essentially complete hydrolysis of the original polymeric ester[1.7], commercial grades of PVAL are made with varying degrees of conversion, inter alia to vary the ease of solubility in water, which its hydroxyl groups confer on the polymer along with ready moisture absorption from humid atmospheres. In the dry state, when it is not plasticised by a significant moisture content, PVAL is strong and tough. It is resistant to many solvents, in particular to aliphatic hydrocarbons, and to oils. The main applications of PVAL are associated with its water solubility. They include its use as a thickener and non-ionic surface active agent in aqueous solutions (one example of such application is its use in ophthalmic "artificial tear" solutions), as the material of soluble packages (e.g. for additives in textile processing), and, in the form of filaments, as removable scaffolding yarns in specialty fabrics. Textile fibres spun from PVAL solutions can be insolubilised by treatment with formaldehyde to convert the polymer to poly(vinyl formal).

POLY(VINYL BUTYRAL) (PVB[1.1,1.5,2.12.2,2.3,2.11])

A poly(vinyl acetal) produced by reacting poly(vinyl alcohol) with butyraldehyde[1.1, 1.5] ($CH_3(CH_2)_2CHO$); i.e. R is $CH_3(CH_2)_2$ in the general reaction scheme under poly(vinyl acetal) **(1)** - q.v.).

Films of high-purity PVB (lightly plasticised with a suitable ester plasticiser) are used as bonding inter-layers between glass sheets in the production of safety glass (cf. e.g. *Butvar; Saflex, Trosifol*) . This important application, for which PVB is especially suited by virtue of the combination it offers of toughness, clarity, excellent adhesion to glass, and resistance to degradation by UV light, is virtually its only utilisation as a plastics material. Its other uses are as a component of some adhesives and surface coatings.

POLY(VINYL CARBAZOLE)[2.2,2.11]; POLYVINYLCARBAZOLE[1.1,1.5,2.1]; (PVK)

A polymer of N-vinyl carbazole[1.1,1.5], with the repeating structural unit.

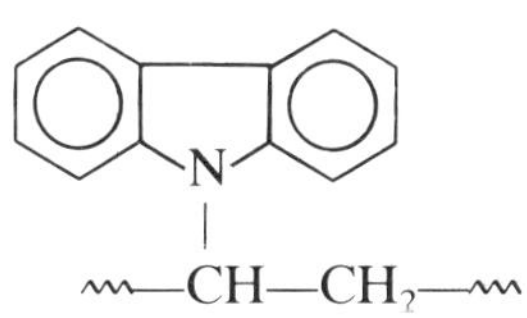

PVK is a strong, stiff (relatively brittle), high-softening material (T_g variously quoted as 173°C and 208°C), with good dielectric properties and low flammability. It is normally non-crystalline (and hence transparent), but some crystallinity can develop with molecular orientation, reducing the original transparency. The material is resistant to many solvents, but susceptible to aromatic and chlorinated hydrocarbons, some hydrocarbon fuels, and certain ketones (e.g. tetrahydrofuran). Its resistance to environmental stress cracking is good. The inherent photoconductivity of PVK (which has been utilised in some photocopying applications) can be further enhanced by chemical-complex formation.

POLY(VINYL CHLORIDE)(PVC)

A polymer (homopolymer[1.7]) of vinyl chloride[1.1,1.5] (CH_2=CHCl). The vinyl chloride repeating units in the polymer chain are linked essentially "head-to-tail", i.e. $-CH_2-CHCl-CH_2-CHCl-$, with very few head-to-head junctions ($-CH_2-CHCl-CHCl-CH_2-$). The end-groups most common in commercial homopolymers are 1,2-dichloroalkane ($\sim CH_2-CHCl-CH_2Cl$) and 1-chloro-2-alkene ($\sim CH_2-CH=CH-CH_2Cl$).

POLY(VINYL CHLORIDE) (PVC) PLASTIC

See under PVC *plastic(s)*.

POLY(VINYL FLUORIDE)(PVF)

A polymer of vinyl fluoride[1.1,1.5], semi-crystalline(T_m 195°C to 200°C, depending on method of determination), of density 1.4 g cm^{-3} having a fluorine content of 41% by weight. The service temperature range is quoted as –70°C to 200°C (top temperature for continuous long-term exposure 120°C). PVF resembles PVC in some of its characteristics: e.g. susceptibility to, and mode of, thermal degradation; resistance to hydrolysis; low water absorption. But it is much more resistant to solvents, less permeable to some gasses, and has excellent resistance to weathering. However , it is more flammable than PVC, unlike which it will go on burning after ignition with a small flame. PVF is marketed only as film (*Tedlar; Kel-F* film), and is also used for solution-applied coatings. It finds applications as a facing layer, protective against weathering, soiling and absorption, on wall coverings (including interior panels in passenger aircraft), road signs, greenhouse sheeting, some packaging materials, solar selectors, and building panels.

POLY(VINYL FORMAL) (PVFM[1.1, 1.5, 2.1, 2.2, 2.3, 2.11])

A poly(vinyl acetal) resulting from reaction of poly(vinyl alcohol) with formaldehyde[1.1, 1.5] (HCHO; i.e. R=H in the general reaction scheme under *poly(vinyl acetal) (1* - q.v.).
PVFM is usually produced by the direct process, with poly(vinyl acetate) as the starting material (see under *poly(vinyl acetal) (1)*). The resulting polymer grades typically have low residual hydroxyl group contents, with residual acetate group contents ranging between about 10% and 40%. Commercial PVFM polymers (*Formvar; Mowital; Redux)* have no significant plastics applications. Their main use is as components (jointly with selected phenolic resins) of electrical wire enamels, some surface coatings for metals, and some structural adhesives.

POLY(VINYLIDENE CHLORIDE)(PVDC)

A polymer of vinylidene chloride[1.1,1.5] (1,1-dichloroethylene : $CH_2=CCl_2$). PVDC is characterised by almost complete head-to-tail arrangement of the monomer units in the molecular chains, i.e.

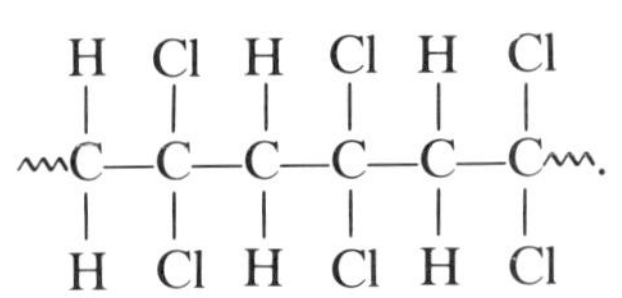

The regularity of chain structure imparted by the symmetrical disposition of alternating $-CH_2-$ and $-CCl_2-$ groups promotes a high degree of crystallinity (T_m about 200°C). The regular structure and high crystallinity are factors instrumental in the polymer's good barrier properties against gases and water vapour, whilst the high chlorine content makes for low flammability (oxygen index about 60); in conjunction with the crystallinity it is also responsible for the high density of PVDC (1.88 g cm^{-3}). Another consequence of the abundance of chlorine atoms, and their alternating with hydogen atoms in the polymer chains, is low thermal stability of PVDC at melt-processing temperatures, which have to be relatively high in view of the high T_m and considerable chain stiffness (also caused by the large numbers and regular disposition of chlorine in the molecular chains). For these reasons, commercial PVDC polymer materials (*Saran*; *Ixan*; *Diofan*) are copolymers of vinylidene chloride with significant amounts (up to about 20%) of vinyl chloride, and in some cases also a few percent of acrylonitrile as a third comonomer: by their presence in the polymer chains, the comonomer units reduce the structural regularity and chain rigidity; this has the dual effect of lessening somewhat the high intrinsic susceptibility of PVDC to thermal decomposition, and enabling melt processing at temperatures lower in some degree than those required for the homopolymer, so that the risk of thermal degradation in processing is generally reduced. At the same time, the modification does not reduce dractically the resistance of PVDC to organic solvents (ketones, chlorinated hydrocarbons, aromatic hydrocabons, esters), which is better than that of PVC. See also *poly(vinylidene chloride) plastic* and *vinylidene chloride plastics.*

POLY(VINYLIDENE CHLORIDE)(PVDC) PLASTIC

Formally defined as a vinylidene chloride plastic (q.v.) based on polymer(s) produced from vinylidene chloride as essentially the sole monomer[1.1,1.5]. The extent to which the qualifying adverb "essentially" contributes to the precision of this standard definition appears questionable, in view of the fact that commercial PVDC plastics are normally based on copolymers of vinylidene chloride which may contain as much as 20% of vinyl chloride, and in some cases also a few percent of acrylonitrile (see poly(vinylidene chloride)). Since the standard definition of vinylidene chloride plastics (q.v.) includes within its scope inter alia plastics based on vinylidene chloride copolymers in which the vinylidene chloride predominates by mass, it would be logical to define a PVDC plastic, without any qualification, as one based on a homopolymer or homopolymers of vinylidene chloride.

NOTE - Alternatively, the significance of the word "essentially" in this context should be made clear, with special reference to the amount of comonomers admissible in what may still be regarded as "essentially" poly(vinylidene chloride). Such a quantifying explanation in given, for example, in the ASTMD standard definition[1.7] of polyethylene plastics (q.v.).

POLY(VINYLIDENE FLUORIDE) (PVDF)

A polymer of vinylidene fluoride[1.1,1.5] (1,1-difluoroethylene : $CH_2{=}CF_2$), with regular head-to-tail molecular chain structure. The polymer is semi-crystalline. Slow cooling from the melt, and/or annealing after moulding, results in a high degree of crystallinity and relatively large crystallite structures, with consequent translucency or opacity and particularly high modulus of the product. Rapid cooling from the melt reduces the degree of crystallinity and crystallite size, so that quenched products having a high specific surface (surface to volume ratio) which promotes rapid cooling (such as films) can be made completely transparent. The crystallite melting temperature is about 170°C. The high fluorine content of PVDF (59% by mass) makes for high density (1.8 g cm^{-3} in ordinary mouldings). The maximum temperature for continuous service is usually given as 150°C (in the absence of stress, agressive environments, or contact with some metals, viz. aluminium, copper or iron, which can promote catalytic degradation). PVDF is a stong, tough polymer, with chemical resistance good enough to qualify it as the material of such elements of some chemical plant as pipes and pipe fittings, valves, seals, and linings for storage tanks. However, it is not resistant to concentrated oxidising acids, to ketones, esters, and such powerful amide solvents as dimethylformamide. It is non-flammable in tests involving ignition with a small flame in air, and has a relatively high oxygen index (typical value : 45). The combination of toughness (maintained at low temperatures) with good electrical insulation properties offered by PVDF is utilized in such applications as cable covering in aircraft wiring but - like PVC and other polar chlorine- and fluorine-containing polymers - it is not suitable for use on high-frequency cables, where dielectric loss is a cardinal consideration. The electrical characteristics of PVDF include two less usual ones, viz. its pyroelectric and piezoelectric properties. The former property is utilised in heat-detecting devices, including security systems, in some of which PVDF sensors can detect - at a distance of 50 metres - the warmth radiated by a human body. The piezoelectric effect, which in PVDF is combined with good response to sound-wave pressure over a wide frequency range, finds important practical use in the application of PVDF films in telephones, microphones (including in particular noise-reduction microphones for military purposes - e.g. for communication by tank crews), and in underwater microphone (hydrophone) devices; the latter also benefit by another property of PVDF, viz. the fact that its acoustic impedance is practically identical with that of water. The weathering resistance of PVDF is generally good, but the polymer is not immune from degradation on prolonged exposure.

Many copolymers of vinylidene fluoride with other fluorinated monomers are either modified versions of PVDF with increased toughness and reduced stiffness, or fluoroelastomers (q.v.).

Commercial versions of PVDF materials include *Dyflor*; *Foraflon*; *Hylar PVDF*; *Kynar*; *Solef*; and *Teflon AF*.

The following standards are relevant to PVDF and to some copolymers of vinylidene fluoride with other fluorine-containing monomers.

ASTM D 3222-91a	Specification for unmodified PVDF moulding, extrusion and coating materials.
ASTM D 5575-94	Specification for copolymers of vinylidene fluoride with other fluorinated monomers (VDF/HFP; VDF/CTFE; VDF/TFE; VDF/TFE/HFP).
ASTM D 4591-93a	Determination of heats of transition of fluorinated polymers (by DSC).

POROMERIC MATERIAL

A polymeric, flexible sheet material resembling leather[1.5], permeable to gases and vapours (in particular to air and water vapour) but typically exhibiting some resistance to penetration by water[1.5], and to abrasion[1.2,1.3]. The term was originally coined for such sheeting developed for use in footwear (cf. e.g. *Corfam*).

POST-CONSUMER MATERIALS

The materials of products which, having served their purpose and become superfluous to the original user, have not been included in, or have been separated from, solid waste, in order to be collected and re-used through recycling, reprocessing, or in other ways (cf. ASTM D 5033-90).

POSTCURE

Completion, by heating, of the cure of a product (a moulding or casting) made from a thermosetting resin[1.1,1.5]. Where, as in some kinds or applications of curable resin systems, the cure needs to be effected in two stages, the second stage - which is actuated by heating - constitutes postcure (cf. also BS 4994:1987). As exemplified by the case of GRP laminates (with which posturing is often practiced), the postcure is typically carried out after demoulding, and normally after the moulding has been allowed to cure for at least 24 hours at room temperature (or 1 hour at 60°C). Once removed from the mould, the moulding should be properly supported during the postcure, to prevent deformation. For a general-purpose resin system, fairly typical postcure conditions would be 3 hours at 80°C.

POSTFORMING

The final shaping of a cured or partly cured thermosetting plastics product (see also *forming*)[1.1,1.5,1.7]

POST-SHRINKAGE

(of a plastics moulding) - see under *moulding shrinkage*.

POWDER BLEND

Term regarded by the three main terminology standards[1.1,1.5,1.7] as synonymous with "dry blend" (q.v.).

POWDER COMPOUND

See under *compound*.

POWDER DENSITY (APPARENT POWDER DENSITY)

Essentially, the bulk density (q.v.) of a powder; but defined by the BS terminology standard[1.1] specifically as the mass in grams of 1 ml of loose moulding powder. See also *apparent density, bulk density*, and *bulk factor*.

POWDER MIX TIME

The time taken by a sample of granular PVC resin to absorb a measured amount of plasticiser under specified conditions of a mixing test in a torque rheometer. This time is indicative of the ease of absorption of the plasticiser by the resin : in a test in which other relevant factors are kept constant it can also provide a quantitative indication of the mutual compatibility of the resin and plasticiser. The powder mix time is determined via the marked drop of motor torque required to maintain constant mixing-rotor speed at the "dry point", when all the plasticiser has been absorbed into the polymer granules and the mix consequently becomes dry and free-flowing (cf. ASTM D 2396-94).

PRECURE

A condition of a thermosetting synthetic resin moulding material in which the thermal hardening of the material has proceeded to a stage where it will not adequately soften and flow under pressure. Precure usually arises as a result of undue delay in the application of pressure after the material has been loaded into the hot mould.

Reproduced, with permission, from BS 1755:Part 1:1982.

PRE-IMPREGNATED FABRIC

A form of prepreg (q.v.) suitable for moulding or laminating under heat and pressure, in which the fibrous or filamentary reinforcement is a fabric, typically of woven construction. Available in sheets or rolls consisting of either single or multiple layers of the fabric prepreg (cf. ISO 8604:1988).

PRE-IMPREGNATED MAT

A form of prepreg (q.v.) suitable for moulding or laminating under heat and pressure, in which the fibrous reinforcement is a mat whose fibres may be disposed randomly or directionally arrayed (cf. ISO 8604:1988).

PRE-IMPREGNATED ROVING

A form of prepreg (q.v.) suitable for moulding or laminating under heat and pressure, in which the filamentary reinforcement consists of continuous roving(s) (cf. ISO 8604:1988).

PREPOLYMER

The product of incomplete polymerisation (which may range from incipient to advanced) of a monomer or monomers[1.1,1.5], capable of attaining a higher, final degree of polymerisation[1.1,1.5] on curing or other appropriate further processing.

PREPREG

A composition, ready for moulding, consisting of a mixture of resin (which may contain fillers), filamentary or fibrous reinforcement, and any requisite additives[1.1,1.5,1.7] (e.g. colorants). The resin component of a prepreg is commonly thermosetting[1.7], but it may also be a thermoplastic polymer. Typical examples of thermosetting-resin prepregs are rovings, mats, and, fabrics impregnated with a thermosetting resin system suitably thickened by curing to "B" stage and/or by evaporation of solvent (see also ISO 8604:1988 for various terms relevant to prepregs).

PRILLS

Granules of thermoplastic feedstocks, and of some additives, specially shaped for free running in conveying and feeding equipment.

PROPORTIONAL LIMIT

The greatest stress that a material is capable of sustaining without any deviation from proportionality of stress to strain (Hooke's law)[1.1,1.3,1.5,1.10]. The actual value of the proportional limit determined for a material in a given type of test can vary substantially with a number of factors, including the sensitivity and accuracy of the test apparatus, excentricity of loading, and sometimes the scale of the stress-strain plot obtained. In quoting the result of a particular determination of the proportional limit, the sensitivity and accuracy of the test equipment, and the test method should be given.[1.10] The proportional limit should not be confused with elastic limit (q.v.).

PROPYLENE PLASTICS

See under *polypropylene plastics.*

"PSEUDO-RANDOM" COPOLYMER

A polymer produced with the aid of special metallocene catalysts, which impose a certain linking order on the comonomer units. For example, in pseudo-random ethylene/styrene copolymers (as patented by Dow Plastics) styrene units can only be linked to one another in a particular orientation.

PULTRUSION

A process for the production of profiles, rods and tubes from reinforced plastics, by pulling continuous strands of resin - impregnated material through a heated die and then through a heating chamber, if necessary, to postcure the resin. The process promotes high longitudinal strength in the products[1.1,1.5]. Standards relevant to pultrusion and pultruded products include ASTM D 4385-84a(1988) - classification of visible defects in pultruded products; ASTM D 5028-90 - test for curing properties of pultrusion resins; ASTM D 5117-90 - test for longitudinal wicking of solid, glass-fibre-reinforced pultruded rod stock; ASTM D 3916-94 - test for tensile properties of pultruded plastic rod; ASTM D 3917-94 - specification for dimensional tolerance of glass-reinforced pultruded products; and ASTM D 3918-80 (1990) - definitions of terms relating to reinforced plastic pultruded products.

PUNKING

Smouldering (q.v.) occurring in some phenol-formaldehyde foams as a result of localised contact with a source of sufficiently intense heat. Typically, the smouldering once started continues (progressing as a spreading front) even after the heat source has been removed. Normally the process proceeds until all the foam has been converted to a char (q.v.). A standard test is available for determining the susceptibility of phenol-formaldehyde foams to punking (BS 5946 : 1980).

PVC PLASTIC(S)

An informal but widely used term for the materials defined by the main plastics terminology standards[1.1,1.5,1.7] as *vinyl chloride plastic(s)* - q.v. Two of the standards[1.1,1.5] also define *poly(vinyl chloride) plastic* as a plastic based on polymers that are essentially homopolymers of vinyl chloride.

PVC RESIN

In popular general usage, reflected also in a number of material specifications and test standards (although not directly recognised by the three main terminology standards[1.1,1.5,1.7]), a PVC resin is any polymer or copolymer of vinyl chloride used as the base polymer of a vinyl chloride plastic (q.v.). See also *resin* (**1**). Standards directly relevant to the technical aspects of PVC resins include the following.

Standard material specifications

ISO 1060	Plastics - Homopolymer and copolymer resins of vinyl chloride. Part 1:1982 - Designation. Part 2:1985 - Determination of properties.
ASTM D 1755-92	Specification for poly(vinyl chloride) resins. *Note*: this standard refers to several of the ASTM test methods listed below.
ASTM D 2474-92	Specification for vinyl chloride copolymer resins.

ASTM D 2873-94	Interior porosity of PVC resins by mercury intrusion porosimetry.

Viscosity and molecular weight

ISO 174-1974	Determination of viscosity number of PVC resins in dilute solution. Standard discontinued in 1995, replaced by ISO 1628-2:1988.
ASTM D 1243-95	Test for dilute-solution viscosity of vinyl chloride polymers.
ASTM D 3591-77 (1991)	Determining logarithmic viscosity number of poly(vinyl chloride) in formulated compounds.
DIN 53726 (1983)	Determination of viscosity number and *K* value of vinyl chloride polymers.

Chlorine content

ISO 1158:1984	Determination of chlorine in vinyl chloride homopolymers and copolymers.
ASTM D 1303-55 (1979)	Test for total chlorine in vinyl chloride polymers and copolymers. Standard discontinued in 1986.

Vinyl acetate content of VC/VAC copolymers

ISO 1159:1978	Determination of VAC in VC/VAC copolymers (chemical method; infra-red spectrophotometry widely used in industry).

Ash and/or sulphated ash content

ISO 1270:1975	PVC resins - Determination of ash and sulphated ash. Standard discontinued in 1994; replaced by ISO 3451-5:1989 (identical with BS 2782:Part 4:Method 470 E:1991).

Volatile matter

ISO 1269:1980 (1985) identical with BS 2782:Part 4 Method 454 D:1978 (1983)	Determination of volatile matter (including water) I vinyl chloride homopolymers and copolymers.
ASTM D 3030-84 (1990)	Test for volatile mater (including water) in vinyl chloride resins.

Impurities and foreign matter

ISO 1265:1979 (1986)	PVC resins - Determination of number of impurities and foreign particles.
ASTM D 2222-94	Test for methanol extract of vinyl chloride resins.

Bulk density

ISO 1068:1975 identical with BS 2782:Part 6:Method 621 D:1978 (1989)	Determination of compacted apparent bulk density of PVC resins.

Particle size

ISO 1624-1978	Vinyl chloride homopolymer and copolymer resins -Sieve analysis in water.
ISO 4610:1977, identical with BS 2782:Part 4:Method 454F:1978 (1983)	Vinyl chloride homopolymer and copolymer resins - Sieve analysis using air-jet sieve apparatus.
ASTM D 1705-92	Particle size analysis of powdered polymers and copolymers of vinyl chloride.

Dissolving procedure

ASTM D 5226-92	Standard practice for dissolving polymer materials (solvent data for PVC in the Annex, Section A1.2.63).

pH of aqueous extract

ISO 1264:1980 (1985) identical with BS 2782:Part 4:Method 454C:1978 (1983)	Homopolymer and copolymer resins of vinyl chloride - Determination of pH of aqueous extract.

Vinyl chloride (VC) monomer content

ISO 6401:1985	Vinyl chloride homopolymers and copolymers - Determination of residual VC monomer - Gas-chromatographic method.
ASTM D 3680-78	Residual VC monomer content by solution injection method (for PVC homopolymers, copolymers and compounds).
ASTM D 3749-93	Determination of residual VC monomer content of vinyl chloride homopolymers by gas chromatography (head-space method).
ASTM D 4443-84 (1989)	Determination of residual VC monomer content in parts per billion in vinyl chloride polymers and copolymers.

Miscellaneous properties relevant to processing

ISO 305:1990	Determination of thermal stability of PVC polymers and copolymers and their compounds, discoloration method.
ISO 4574:1978	PVC resins for general use - Hot plasticiser absorption.
ISO 4608:1984, identical with BS 2782:Part 4:Method 454 E:1992	Vinyl chloride homopolymer and copolymer resins - Plasticiser absorption at room temperature.
ASTM D 2396-94	Powder-mix time of PVC resins in a torque rheometer.
ASTM D 3367-75 (1990)	Test for plasticiser sorption of PVC resins under applied centrifugal force.
ASTM D 3596-92	Determination of gels (fish eyes) in general-purpose PVC resins.
ASTM D 4202-92	Test for thermal stability of PVC resins.

See also PVC *plastic, vinyl chloride plastic*, and *vinyl resin*.

PYROLYSIS

Heat-actuated irreversible chemical breakdown of a material, not involving oxidation[1.3,1.9]. The breakdown is manifested in simultaneous changes of phase and chemical species in the material concerned[1.12]. Pyrolysis of a polymer or its derivative plastic typically results in the formation of decomposition products of lower molecular weight, including simple, volatile, hydrogen-rich carbon compounds, linear or cyclic. Another main product of pyrolytic decomposition is a carbonaceous residue termed char (q.v.).

QUANTITATIVE DIFFERENTIAL THERMAL ANALYSIS; QUANTITATIVE DTA

See under *differential thermal analysis.*

RADIANT FLUX (RADIANT POWER)

The time rate of flow of radiant energy (SI unit : watt). Where the radiant energy is that of visible light, the flux is usually referred to as *luminous flux:* the SI unit for this is the *lumen,* which relates the luminous energy to its visual effect (in that it is the luminous flux emitted from a point source of uniform intensity of one candela into unit solid angle). Thus a lumen is equal to candela x steradian

$$\text{lm} = \text{cd.sr}$$

REACTION INJECTION MOULDING (RIM)

An injection moulding process in which two or more mutually reactive components[1.5] (the joint precursors of the polymer of the ultimate moulding), which may contain additives (e.g. fillers[1.5]), are impingement-mixed under pressure in a mixing chamber[1.5] and the mixture is immediately injected into a closed mould[1.5]. In comparison with ordinary injection moulding RIM offers two main advantages, viz. production of thermoset (notably polyurethane) mouldings by injection from liquid reactants, and particular suitability for producing, relatively easily and at comparatively low pressures, mouldings with heavy wall sections: this makes the process especially suitable for large, thick, flat parts. The relatively low moulding pressures employed permit the use of tools less expensive than those necessary in ordinary injection moulding. See also *reinforced reaction injection moulding.*

REFRACTIVE INDEX

For a given wavelength (frequency) of electromagnetic radiation, the refractive index (usual symbol n) of a material is the ratio of the velocity of the radiation in vacuum to its velocity in the material. In the context of plastics materials, the electromagnetic radiation of practical interest is visible light. It is relevant to note that the value of the refractive index of a given material varies somewhat with the wavelength of light used in its measurement, and with temperature. Unless otherwise indicated, the values of n quoted for plastics materials are normally those measured at 20°C. On the basis of the relevant application of Maxwell's electromagnetic equations to transparent non-conducting materials, it can be shown that for such materials the refractive index is equal to the square root of the dielectric constant (permittivity), i.e. $n = \sqrt{k}$, or $k = n^2$ (cf. e.g. Jenkins and White, 1953). The refractive index of plastics materials is discussed, inter alia, in BS 4618:Section 5.3:1972. Its determination is the subject of ISO 489:1983 (identical with BS 2782:Part 5:Method 531A:1992); Appendix B of BS 3532:1990; and ASTM D 542-95.

REGULAR POLYMER

A polymer the molecules of which can be described by only one species of constitutional unit in a single sequential arrangement.

See note to *constitutional repeating unit.* See also *irregular polymer.*

Reproduced, with permission, from ISO 472:1988.

REGULATOR

In a polymerisation reaction, a minor additive serving to control the relative molecular mass[1.1,1.5] of the polymer being formed.

REINFORCED PLASTIC

A plastic containing a reinforcing filler (q.v.). Most commonly the filler consists of fibres of high strength[1.1,1.5,1.7] and/or high modulus, whose embedment in the polymer matrix (base resin[1.1,1.5,1.7]) imparts to the composition some mechanical[1.7] strength[1.1,1.5] properties superior[1.7] or greatly superior[1.1,1.5] to those of the base resin[1.1,1.5,1.7]. The fibrous reinforcing filler may be in the form of loose fibres (in both thermoplastics and thermosets) or - predominantly in thermosets - fibre rovings, mat, or fabric[1.7] (woven or knitted) in some cases of complex, three-dimensional construction (e.g. a honeycomb). See also *reinforced thermoplastics* and *unsaturated polyester.*

REINFORCED REACTION INJECTION MOULDING (RRIM)

A reaction injection moulding process involving the use of a solid reinforcement for the material. The reinforcement may be fibrous (e.g. glass fibres) or particulate (e.g. mica; talc)[1.7].

REINFORCED THERMOPLASTIC(S) (RTP)

In its most common form, viz. a moulding or extrusion compound and the resulting products, a reinforced thermoplastic is a composite, consisting essentially of a thermoplastic polymer (the base polymer) in which a reinforcing filler has been dispersed. By far the most widely used reinforcing fillers are fibres of high modulus and/or high strength, such as glass fibres or carbon fibres (either relatively short, or the so-called long-fibre type), and also - albeit less commonly - specialist fibres, e.g. aramid fibres, or whiskers. Other additives may be present in an RTP, in relatively minor proportion, for instance particulate fillers, lubricating additives (molydenum disulfide; PTFE powder), antistatic additives, etc. (cf. e.g. Titow and Lanham, 1975). Examples of special versions of RTP include cellular products, the so-called glass-mat thermoplastics (GMT), and film-stacked composites (q.v.). GMT are composites in sheet form, consisting of a polymer matrix - typically polypropylene, a thermoplastic polyester (PBT or PET), polycarbonate, or (polycarbonate + polyester) blend - reinforced with layers of chopped glass fibres, or glass-fibre mats. GMT find their main applications in the production, by thermoforming, of relatively large parts used in motorcars, such as bumber beams, seat panels, and instrument panels (cf. e.g. Nichols, 1994).

REINFORCING FILLER[1.5] (REINFORCING AGENT[1.1])

A filler (fibrous or particulate[1.1]) incorporated in a plastic to improve some mechanical property or properties[1.5], such as strength and stiffness[1.1], of the base polymer of the composition.

RELATIVE DENSITY (SPECIFIC GRAVITY)

Formally defined as the ratio of the mass in air of a unit volume of an impermeable material at a standard temperature (normally 23°C) to the mass in air (of equal density) of an equal volume of gas-free distilled water at the same temperature. Cf. e.g. ASTM D 792-91.

RELATIVE IMPACT STRENGTH

The ratio of the impact strength of notched specimens to that of unnotched but otherwise identical specimens, *or* the ratio of the impact strengths of notched specimens differing only in the type or notch[1.5].

The term "notch sensitivity" is often used in practice as a synonym for relative impact strength in the above two senses, although such usage is deprecated by ISO[1.5].

RELATIVE LUMINOUS REFLECTANCE FACTOR

The ratio of luminous flux (q.v.) reflected from a specimen to the luminous flux reflected from a standard surface under the same test conditions, the angles of incidence and reflection being measured with respect to the normal to the surface (BS 2782:Part 5:Method 520A:1992; ASTM D 523-89). Specular gloss (q.v.) is the relative luminous reflectance factor of a specimen measured at a specular angle (q.v.).

RELATIVE MOLECULAR MASS M_r; MOLECULAR WEIGHT *MW*

The ratio of the average mass per formula unit of a substance to 1/12 of the mass of an atom of nuclide ^{12}C.

NOTE - Relative molecular mass (molecular weight) is a pure number and is not to be associated with any units.

Reproduced, with permission, from ISO 472:1988.

In a simple case in practice, a given amount of linear polymer (i.e. a test specimen, sample, etc.) will consist of chain molecules made up of the same repeated, monomeric units, but differing in size (chain length). A single molecular chain consisting of *i* monomeric units is said to have a degree of polymerisation of *i*. The molecular weight of such an individual chain (neglecting the possible small effects of any end groups) may be designated M_i and the species may be referred to as an "i-mer" or "i^{th} species". Since, at least in the ideal case, M_i is the sum of the individual molecular weights of all the monomeric units in the chain, its value will be different for molecular chains differing in the value of i. The scatter of M_i values (the molecular weight distribution) in a given amount of polymer may be wide or narrow but, as the chains are not all of identical length, a single molecular weight figure quoted for the whole amount of polymer can only be an average value. Depending on the method of determination, this value will in practice normally be a *number average* molecular weight ($\overline{M}_n$), *weight-average* molecular weight ($\overline{M}_w$), or *viscosity - average* molecular weight ($\overline{M}_v$ - see also under *viscosity*). For the same polymer sample these values are numerically in the sequence) $\overline{M}_w > \overline{M}_v > \overline{M}_n$, with $\overline{M}_v$ usually closer to $\overline{M}_w$ than to $\overline{M}_n$. In the ideal theoretical case of all chains being of identical lenght (the same value of i) $\overline{M}_n = \overline{M}_w$.

RELATIVE PERMITTIVITY (RELATIVE DIELECTRIC CONSTANT)

The ratio (ε_r) of the capacitance (C_p) of a capacitor with a given configuration of electrodes, in which the material concerned serves as the sole dielectric, filling completely the space between and around the electrodes, to the capacitance (C_r) of the same configuration of electrodes with vacuum as the dielectric[1.1,1.5] i.e. $\varepsilon_r = C_p / C_r$.

In determinations of relative permittivity in practice, sufficiently close accuracy is achieved by the use of dry air instead of vacuum.

The symbol k' is sometimes used instead of the IUPAP standard symbol ε_r for relative permittivity, to indicate that the latter can be defined as the real part of the *relative complex permittivity*. (For more comprehensive definitions of both terms see e.g. ASTM D 1711-94 and IEC Publication No. 250).

RELATIVE TEMPERATURE INDEX (RTI)

The temperature index (q.v.) of a material relative to that of a reference material, i.e. obtained for the same time period as that used in the determination of the known temperature index of the reference material, and determined - on a comparative basis - by the same test and interpretative procedures. (Cf. ISO 2578 and IEC Publication 216, Parts 1 to 5). See also *temperature index, relative thermal index, and thermal endurance.*

RELATIVE THERMAL INDEX

The temperature (conventionally stated in °C) below which a polymeric material can be expected to retain to an acceptable degree, over a normal service period, the properties relevant to its service performance. Use of the index as a guide in selection of materials for particular applications is subject to the assumption that, apart from elevated temperature, the material will not be experiencing other significant degradative factors (such as e.g. stress or/and aggressive environments). The relative thermal index of a polymeric material can be determined by a procedure practiced by Underwriters Laboraties Inc. In this, the effects of laboratory thermal ageing on the material are evaluated in comparison with those on a relevant reference material whose resistance to thermal ageing in actual service is additionally known. The procedure is the subject of UL Standard 746B (approved also as American National Standard ANSI/UL 746B). The standard contains, inter alia, a list of relative thermal indices for a number of polymeric materials. See also *relative temperature index, temperature index,* and *thermal endurance.*

RELATIVE VOLUME LOSS (in an abrasion test on rubber)

In some test standards for the abrasion resistance of rubber (ISO 4649 : 1985; BS 903:Part A9:1988), the volume - in cubic millimetres - lost from the surface of a rubber when abraded in a prescribed test by a specified abradant which causes the appropriate standard rubber (q.v.) to lose a specific mass under the same test conditions.

REPROCESSED PLASTIC

The standard definitions of this term vary considerably from source to source. The simplest and most general definition (that of the BS terminology standard[1.1]) describes a reprocessed plastic in essence as a plastics material obtained by suitable processing of suitable products. The ISO terminology standard[1.5] defines a reprocessed plastic specifically as a thermoplastic which has been prepared from scrap industrial plastic by a processor other than the processor of the source material. It also adds the comment that, in reprocessing, the original material may be modified by incorporation of additives, such as stabilisers, fillers, plasticisers and the like.

NOTE - In fact, in good industrial reprocessing practice, addition of suitable stabilisers to PVC and polyolefin feed stock is often necessary, and their choice and incorporation levels are matters requiring relevant technical knowledge.

The definition in the ASTM terminology standard[1.7] shares with the ISO definition the restriction of reprocessed plastics to thermoplastics, and the comment about modification by additives. However, it gives the original source material as "usually melt-processed scrap or reject parts, or non-standard or non-uniform virgin material". It also refers to the reprocessor as "a plastics processor" (i.e. not necessarily one different from the processor of the original source material, as in the ISO definition). The ASTM standard also points out that its use of the term "scrap" does not necessarily imply inferiority of this kind of feed stock to the virgin material from which it was initially made. (Cf. also ASTM D 5033-90).

RESIN

(1) In its somewhat informal popular use in the context of plastics materials, the term is virtually synonymous with "polymer" (especially in relation to the base polymer of a plastics material)[1.1,1.5,1.7]. This usage is particularly widespread with thermosets (cf. "epoxy resin", "phenolic resin", "polyester resin"), although quite often encountered also with thermoplastics; e.g. it is common, and almost traditional, to refer to the base polymer of PVC compositions as "PVC resin".

(2) According to a more formal, general definition[1.1,1.5,1.7], a resin is an organic material in a solid, semisolid or quasi-solid state, possessing the following characteristics: (a) molecular weight frequently high, but not expressible as a definite value; (b) a tendency to flow under applied stress; (c) usually softening or melting on heating within an identifiable temperature range; and (d) usually a conchoidal mode of fracture.

(3) A terminology standard for adhesives (ASTM D 907-93) additionally defines a *liquid resin* as "an organic polymeric liquid which, when converted to its final state for use becomes a resin".

RESINOGRAPHY

The science of morphology, structure, and related descriptive characteristics as correlated with composition or condition and with properties or behaviour of resins, polymers, plastics, and their products.

NOTE - Emphasis is on the description of such materials to provide a better understanding of why as well as which properties are obtained and how the material is produced.

RESINOID

A term mentioned, in passing, in reference 1.7, and defined in a terminology standard for adhesives[1.16] as the generic term for any synthetic thermosetting resin in any of its states distinguishable in the course of the curing process. The standard definition refers specifically to two of the stages, viz. the initial state in which the resin is temporarily fusible (cf. also *A-stage; resol; novolak),* and the final infusible state (cf. also *C-stage; resite).* In view of the generic nature of the term it must obviously also include a thermosetting resin in the transition state (cf. *B-stage; resitol)* in which it may be swellable by solvents, and may become rubbery on heating, but is no longer soluble or fusible.

RESISTANCE TO ABRASION

See *abrasion resistance.*

RESITE

A C-stage (q.v.) phenolic (phenol-formaldehyde[1.1,1.5]) resin, i.e. one in the final crosslinked state in which it is infusible and insoluble in the relevant solvents (alcohol and acetone in the case of a phenol-formaldehyde resin[1.1,1.5]).
A B-stage phenolic (phenol-formaldehyde[1.1,1.5]) resin, i.e. one in the transition stage of the polycondensation reaction (curing process[1.1,1.5]), in which crosslinking

has advanced to the point where the material can be softened to a rubbery state by heating and is swellable by the relevant solvents (alcohol and acetone in the case of a phenol-formaldehyde resin[1.1,1.5]), but is no longer soluble or fusible.

RESOL

An A-stage (q.v.) phenolic resin - i.e. one that is still fusible and soluble in some solvents - that contains enough reactive (methylol) groups for its cure to be advanced to the final infusible and insoluble state of the material[1.1,1.5]. Cf. *novolak* (q.v.), which contains no spare methylol groups. See also definition in ISO 10082:1991.

RETARDER

A minor additive incorporated in a reactive chemical system to slow down the rate of reaction[1.1,1.5].

REWORKED MATERIAL[1.1]; REWORKED PLASTIC[1.5,1.7] ("REWORK")

The (thermoplastic[1.5]), material – potentially waste – generated in a manufacturing process (moulding, extrusion, etc.[1.5,1.7]) in the form of trimmings[1.5], reject mouldings[1.5], and the like, and that has been[1.1,1.5], or is to be, reused (after regrinding, pelletising, solvating[1.7], etc, as appropriate) in the same processor's plant[1.5], within the same process[1.1] for the manufacture of material or product of the same kind[1.1].

Many material specifications stipulate that any rework used in a process should be clean and essentially equal in quality to the relevant virgin material[1.5]. A further common requirement is that the amount of rework in a plastics material or product should be limited to a stated percentage.

RIGID PLASTIC

A plastic whose flexural modulus (or tensile modulus if more relevant) measured under standard conditions exceeds 700 MPa[1.1,1.5,1.7]. Typical standard conditions may be 23°C ± 2°C, 50% ± 5% RH, and 86 kPa - 106 kPa (0.86 bar - 1.06 bar) pressure (ISO 291).

RIGISOL

See under *paste* **(1)**.

ROLLER (ROLL)

A cylinder, typically serving as a functionally important rotating component of equipment for various kinds of plastics processing, including compounding machinery (e.g. two-roll mills), coating and spreading machines, equipment for the production of sheeting and film (e.g. calenders) and for their surface decoration.

The rolls of a calender are sometimes referred to as "bowls"[1.1] or, occasionally, "boles"[1.5]. The rollers in plastics-processing equipment are usually of metal (and may be hollow or cored), but solid rubber or plastics rollers, or ones covered with these materials are also used for some purposes. Certain aspects of the latter kind of roller, including the relevant terminology, are covered in BS 7442:Part 1:1991; ISO/TR 8517:1988; and ISO 6123.

RUBBER

A material[1.3], polymeric in nature (more specifically an elastomer[1.2]), that is able to recover vigorously and rapidly from substantial deformations[1.3], and that already is, or can be, modified to a condition of virtual insolubility in (although it remains swellable by) a number of boiling organic solvents, in particular benzene, ethanol/toluene azeotropic mixture, and methyl ethyl ketone[1.2,1.3]. Other typical characteristics of a rubber in its modified condition (but containing no diluents) include the following.

(a) Ability to retract within 1 min to less than 1.5 times its original length after being stretched to twice that length, held so extended for 1 min, and released; the stretching, maintenance of stretch, and retraction all taking place at room temperature (between 18°C and 29°C).[1.2,1.3]

(b) Unsuitability for easy remoulding to a permanent shape by the application of heat and moderate pressure[1.2].

"S4 TEST"

A small-scale, steady-state, dynamic fracture test (ISO 13478:1997) for resistance of pressurised plastics pipes to rapid crack propagation (RCP). The test measures a critical pressure below which RCP can self-arrest and above which it can proceed freely. As pointed out by Leevers et al. (1995) the dynamic fracture resistance (G_D) of the polymeric material of the pipe can be worked out to a good approximation from the data obtained in the S4 test.

SALT FOG TESTS

See *salt spray tests*.

SALT SPRAY (FOG) TESTS

A group of tests whose results are indicative of one aspect of resistance of a material to corrosion in a marine atmosphere. Such tests are more commonly applied to metals and coatings on metals than to plastics materials, as - with the partial exception of some laminates - most of the latter are more susceptible to damage by the UV component of sunlight (to which they would normally be accessible in the course of natural exposure in a marine atmosphere) than to any disruptive action of salt spray or mist. The agents of deterioration of materials in marine atmospheres and their effects on plastics are discussed in BS 4618:Part 4:Section 4.4:1973. Major standards dealing with salt-spray and related tests are: ISO 4611:1987; BS 7479:1991 (equivalent to ISO 9227:1990); ASTM B 117-90;

and ASTM G 85-85 (1990) which includes and acidified sea water test and an SO_2 salt-spray test.

SCORCH (SCORCHING)

Premature vulcanisation (crosslinking) of a rubber compound[1.2,1.3]. Such undesirable crosslinking also sometimes occurs in the processing of other chemically crosslinkable polymers, e.g. polyethylene. *Mooney scorch* is the time to incipient cure of a rubber compound in a test in the Mooney shearing-disc viscometer under specified test conditions[1.3]. In plastics technology, the terms "scorching" or "scorch mark" are also occasionally used synonymously with "burn" or "burn mark" (q.v.).

SECANT MODULUS

The ratio of stress to strain at a given point on the stress-strain curve. In some standards defined - in fully equivalent terms - as the slope of a straight line drawn from the origin to a specified point on the stress-strain curve[1.4] (cf. e.g. ASTM D 907-94).

SECONDARY STRESS

An engineering term, sometimes used to denote a bending stress due to deflection (as distinct from "direct" stresses resulting from extension or compression).

SECOND MOMENT OF AREA

See under *section modulus*.

SECTION MODULUS (MODULUS OF SECTION)

An engineering term for the ratio of the second moment of area to the distance between the neutral axis and the farthest element stressed in bending. The section modulus is a concept employed in calculations of bending stresses in beams and bars: it has the dimensions of length cubed (L^3); its usual units are m^3, mm^3 or cm^3, and the symbols commonly used to denote it are Z and W. Deplorably - though fortunately hardly ever in the plastics context - section modulus is sometimes referred to as "elastic modulus". This loose terminology cannot be too strongly deprecated.

NOTE 1. *Second moment of area.* For a plane area, the second moment of area about an axis XX in the same plane (I_{XX}) is defined by the expression $I_{XX} = \sum ar^2$ where a is an element of the total area, and r is the perpendicular distance from the axis XX. Another relevant relationship is $I_{XX} = I_{GG} + Ah^2$, where I_{GG} is the second moment of area about a parallel axis through the centroid of the area, A is the total area, and h is the perpendicular distance between the axes XX an GG. For a rectangle of width b (parallel to the axis GG) and depth h, $I_{GG} = (bh^3)/12$.

NOTE 2. *Neutral axis*. The engineering term for the line of zero stress in a beam undergoing bending. This line is a transverse section of the longitudinal plane (neutral surface) which passes through the centre of area of the section.

SELF-HEATING

A rise in a material's temperature brought about by an exothermic reaction within the material[1.8,1.9].

SELF-IGNITION

Ignition resulting from self-heating[1.8,1.9] (q.v.). See also *self- ignition by temporary glow.*

SELF-IGNITION BY TEMPORARY GLOW

In the hot-air ignition furnace test of ASTM D 1929-77 (1985) - and, in some cases, in actual service conditions - external heating of a plastics material may result in relatively slow decomposition, causing sporadic temporary glowing at various points in the material, without general ignition taking place. The temperature at which this phenomenon occurs is regarded by ASTM D 1929 as a special case of spontaneous ignition temperature (called "self-ignition temperature" in this standard) termed "self-ignition by temporary glow". See also *spontaneous-ignition temperature.*

SELF-IGNITION TEMPERATURE

As defined in ASTM D 1929-77 (1985) this term is virtually synonymous with spontaneous ignition temperature (q.v.). However, its use in this sense is deprecated by ISO[1.8], because the relevant ISO standard[1.8] (as well as that of the IEC[1.9]) defines self-ignition as *ignition resulting from self-heating* (q.v.).

SEMI-CRYSTALLINE POLYMER

A polymer whose morphology is characterized by the presence of both crystalline and amorphous phases[1.5]. The term is favoured over the non-standard "crystalline polymer" as no common polymer in bulk is entirely crystalline, although some may contain high proportions of the crystalline phase (cf. e.g. up to 80% for regular polyethylene; up to about 90% for PTFE). A non-standard alternative to "semi-crystalline polymer" is "partly crystalline polymer". A standard test method for determination (by DSC) of heats of fusion and crystallization of polymers is given in ASTM D 3417-83 (1988).

SEMI-RIGID PLASTIC

As a matter of general classification[1.7], a semi-rigid plastic is defined by all the three main terminology standards[1.1,1.5,1.7] as one which, under stated conditions, has a modulus of elasticity in flexure or in tension between 70MPa and 700MPa

(10 000 psi and 100 000 psi[1.7]). The ISO[1.5] and BS[1.1] definitions (which are mutually identical) admit the tensile modulus only if the flexural modulus is inapplicable (the ASTM standard[1.7] has no such preference), and indicate that the "stated conditions" would usually include standard temperature and relative humidity as given in ISO 291. The ASTM standard[1.7] specifically cites 23°C and 50% relative humidity, together with ASTM test methods D638, D747, D790 and D882.

SEQUENTIAL ARRANGEMENT (in polymer molecules)

The arrangement of constitutional units in a polymer chain.

NOTE - The constitutional unit 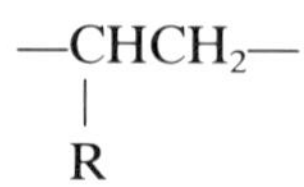 in the polymer molecule represented by the fragment

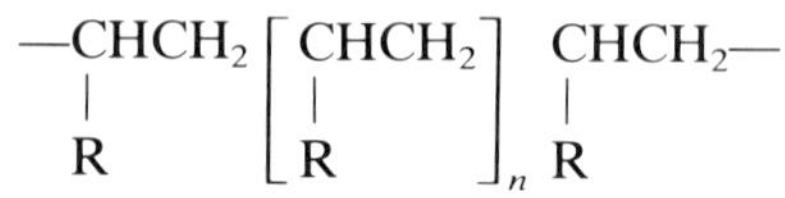

is in one sequential arrangement only (head-to-tail). In the polymer molecule represented by the fragment

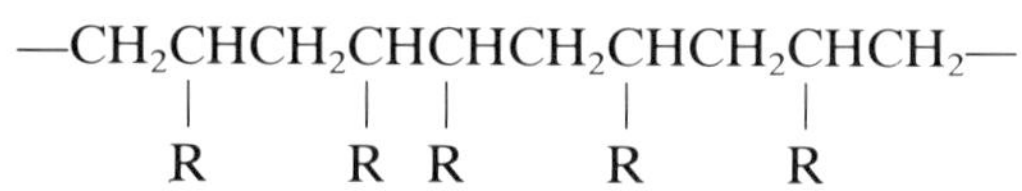

there is more than one sequential arrangement of the constitutional unit.

Reproduced, with permission, from ISO 472:1988.

SHEAR STRESS

The stress, or stress component, tangential to a given plane on which the forces concerned are acting[1.1,1.10]. Recommended standard symbols[1.1,1.5]: τ and σ_{ij}.

SHEET MOULDING COMPOUND (SMC)

An intimate mixture of a resin (typically an unsaturated polyester[1.7]) and chopped or unchopped fibrous reinforcement, suitable for moulding or laminating under heat and pressure. The compound, which is produced in the form of sheet[1.7] typically from 1mm to 25mm thick, may contain such additives as catalysts, fillers, and colorants, and will usually incorporate chemical thickeners to impart the requisite high viscosity (ISO 8604:1988). Although most SMC's in current use are still based on thermosetting resins[1.7], thermoplastic-based versions are also available; among the advantages claimed for them in applications for which they are suitable are virtually indefinite shelf life and recyclability (cf. e.g. Du Pont's "XTC" PET-based sheet for motor-car exterior panels).

SIDING

In the plastics context, cladding elements (such as panels or extruded profiles) applied to an existing exterior wall of a building for weather protection or/and decorative effects. Requirements applicable to standard PVC-U siding are covered by ASTM D 3679-92, and the installation practice by ASTM D 4756-91.

SILICONE (SI)

A polymer with molecular-chain backbone made up of alternating silicon and oxygen atoms, and having side groups or side chains attached to the silicon atoms. Apart from those versions of the silicones which are used as the base polymers of silicone plastics (q.v.), other groups of silicone polymers serve in a variety of applications, viz. as elastomers (silicone rubbers), oils (including constant-viscosity hydraulic oils), greases, water-repellent components in various surface coatings and surface-treatment formulations (including certain paints, fabric coatings, polishes, and damp-proofing treatments for buildings), mould-release agents, and antifoaming agents.

NOTE 1 - Whilst the three main plastics terminology standards define only *silicone plastic(s)*, two[1.1, 1.5] give SI as the letter symbol for *silicone* - thus: "silicone (SI) plastic" (the third standard[1.7] gives no letter symbol). Of the corresponding main standards for letter symbols (abbreviated terms), the ISO standard[2.11] and the BS standard[2.3], also have "SI" for "silicone", whereas the ASTM standard[2.2] gives this symbol for "silicone plastics".

NOTE 2 - In the main standards for letter symbols (abbreviated terms) for rubbers[2.4, 2.5, 2.6], silicone rubbers form the "Q" group: in the symbol for a particular kind of silicone rubber this designatory letter is preceded by letters which indicate substituent groups on the silicon atoms of the polymer chain (in ascending order, from left to right, of the percentage present - i.e. the symbol for the most abundant substituent nearest the terminal "Q"). For example, the letter symbols MQ, FMQ and PVMQ stand, respectively, for silicone rubber containing methyl substituent groups, silicone rubber with methyl and fluorine substituents, and silicone rubber with methyl, vinyl and phenyl substituents.

Some of the non-plastics applications of silicones are governed by the following standards.

Silicone elastomers

For surgical implants: BS 7253 -3 : 1990 (obsolescent).
For electrical insulation: BS EN 60684 - 3-123: 1995 and BS 6893-21:1995 (extruded insulating sleeving).
Use in aircraft: BS EN 2259 and BS EN 2260 to 2262.
SI-elastomer coated fabrics for water-resistant clothing: ISO 80962 - 2: 1989.

Silicone liquids

Specification for SI liquids for electrical purposes: BS 7704: 1993.
Maintenance of SI transformer liquids: BS 7713: 1993.
Silicone-based building sealants: BS 5889:1989 (obsolescent).

SILICONE (SI) PLASTIC

A plastic based on a silicone (q.v.) polymer or polymers[1.1, 1.5, 1.7]. The silicone polymers (widely referred to as silicone resins) used in silicone plastics are thermosetting materials, crosslinkable by catalytic curing agents incorporated in the resins or resin-based moulding compounds. Because of their good electrical properties, maintained over a wide range of temperature and frequency, coupled with low moisture absorption, the principal applications of silicone resins and their plastics are as insulation, encapsulants, and support for electrical components. In such applications they also offer greater heat resistance, higher service temperature, and dimensional stability than most other thermosets.

SI-resin/glass fabric composites are also used as sheets and boards for electrical purposes. Flammability of silicone plastics is limited (a moulded specimen horizontally held will burn in contact with a small test flame, but burning will not normally continue on a removal of the flame). Their chemical resistance is restricted, in that they are vulnerable to a number of solvents, and to concentrated alkalis and acids.

Commercial silicone materials are available in the following ranges: *Baysilone; Claritech; Commax; Duratech* (SI rubber compounds); *Elastofil* (SI rubber compounds); *Recotech* (SI rubber compounds); *Rimplast; Silastic; Silipact; Siltemp; Sycar; Textolite; Thermatech.*

Standards dealing with silicone materials and products include ASTM PS 15-94 (thermosetting moulding compounds for electrical purposes, including SI compounds inter alia); BS 4584 -13:1977 (1991) (composite sheet base materials for printed wiring boards); BS EN 60893-3-6: 1995 (rigid SI resin laminated sheets for electrical purposes).

SINGLE-POINT DATA

The result(s) - expressible as a single value - of determination of a property of a plastics material in a test under one set of specified conditions (cf. ISO 10350:1993). Such result(s) would typically provide one point on a plot of corresponding multi-point data (q.v.).

"SINGLE SITE" CATALYSTS

Polymerization catalysts of the metallocene (q.v.) type, sometimes referred to also as "constrained geometry" catalysts, used in the production of the so-called "pseudo-random" ethylene copolymers (q.v.) and "branched linear" polyolefins (q.v.). Unlike the reactive sites of the conventional Ziegler-Natta catalysts - which can vary widely in their reactivity levels - those of a single-site catalyst are all of identical reactivity. This affords close control over, and reproducibility of, the molecular weight distribution and the kind of branching in a polyolefin produced with the aid of such catalysts. The single-site catalysts used to make the "pseudo-random" copolymers also provide control over the order and orientation in which comonomer units link up to form the polymer chain.

SIOPLAS PROCESS

A two-step process for the manufacture of chemically crosslinkable polyethylene products. In the first step, a suitable grade of polyethylene is melt-compounded with vinyltrimethoxy silane and a peroxide grafting activator, whereby siloxane side-branches are grafted onto the polyethylene chains. The graft is pelletised and used in the second step, in which it is blended with a masterbatch of crosslinking catalyst, and melt-processed in the production equipment - normally an extruder - together with stabiliser(s) and such other additives as may be required. The resulting product, most commonly a cable covering or heat-shrink tubing, is then treated, off-line, in a steam or hot-water bath to crosslink the material. This operation is the same as in the *Monosil* process (q.v.). See also *crosslinked polyethylene, Si-Link, Aquathene,* and *Visico NCPE 4421.*

SMOULDERING

The slow[1.5,1.8,1.9] combustion of a (solid[1.12,1.13]) material[1.5,1.8,1.9,1.13], without flame[1.12,1.13] or light being visible[1.5,1.8,1.9], generally [1.5,1.8,1.9,1.13] (often[1.12]) evidenced by visible[1.12]) smoke[1.5,1.8,1.9,1.12,1.13] and/or increase in temperature[1.5,1.8,1.9], and/or by effluent[1.9].

SOFTENING TEMPERATURE

The temperature at which, in closely defined conditions of a relevant test, a specimen of a plastics material softens sufficiently to become deformed to a specified extent[1.5]. See *Vicat softening temperature, deflection temperature under load,* and - for tests of effects of temperature on the modulus of rigidity of plastics - ISO 6721:Part 2:1994 and ASTM D 1043-92.

SOL FRACTION

See under *gel fraction.*

SOLUBILITY TEMPERATURE

The temperature at which - in test conditions prescribed by DIN 53 408 - particles of a PVC resin dissolve in a plasticizer. Like the *fusion point* (q.v.), the solubility temperature gives a numerical indication of the compatibility of the plasticizer with the resin and its solvent power for it.

SOLVENT BONDING (SOLVENT WELDING)

A method whereby thermoplastic products[1.1,1.5] or parts (e.g. PVC or ABS pipes and their fittings) are bonded together by applying to the surfaces to be joined a solvent that softens the plastics material[1.1,1.5] (or a solution in such solvent of the polymer concerned; in this context the solution is sometimes referred to as "solvent cement", "bodied cement", or "solvent dope"). The solvent-treated surfaces are then pressed together[1.1,1.5] and held for the appropriate time, sometimes at an elevated temperature, to consolidate the bond. According to the standard definition of solvent bonding[1.1,1.5], this final stage involves "removing the solvent by evaporation, absorption, or polymerisation". It has been shown (Titow et al., 1973) that in the bonding of such glassy polymers as bisphenol-A polycarbonate and polysulfone with good (i.e. strongly swelling) solvents (selected chlorinated hydrocarbons) a large proportion of the solvent originally applied can be permanently retained in the material neighbouring the bond plane. For a discussion of the phenomenology and mechanism of solvent bonding of some glassy polymers see Titow et al (1973) and Titow (1978). Standards relevant to solvent and solvent-cement bonding of plastics include the following.

ISO 2044:1974 (PVC solvent-welded pressure-pipe fittings).
ISO 7387-1:1983 (Test methods for solvent adhesives for PVC-U pipes).
BS 4346:Part 1:1969 (PVC fittings for solvent welding); and Part 3:1982 (solvent cement for PVC pipes and fittings).
BS 6209:1982 (Specification for solvent cement for thermoplastic pipe systems).
ASTM D 2672-95 (Solvent-cemented PVC pipe joints).
ASTM D 2855-93 (Standard practice for making solvent-cemented joints for PVC pipes and fittings).
ASTM D 2846-95a, Section 7 (Chlorinated PVC solvent cements and cemented joints for pipes and fittings).
ASTM D 3138-83 (Solvent cement for transition joints between PVC and ABS pipe components).
ASTM F 402-93 (using solvent cements for thermoplastic pipes and fittings).
ASTM F 493-93a (Solvent cements for PVC-C pipes and fittings).
ASTM F 545-93 (PVC and ABS solvent-cemented pipe joints).
ASTM F 656-93 (Primers for use in solvent cementing PVC pipes and fittings.

SPECIFIC GRAVITY

See *relative density*.

SPECIFIC HEAT CAPACITY (SPECIFIC HEAT)

The quantity of heat necessary to produce a unit change of temperature in a unit mass of material, i.e. the quotient obtained when heat capacity (q.v.) is divided by mass. Usual symbol : c. SI units : joule per kilogram kelvin (J/kg.K). For the same material, the value of c determined at constant volume (c_v) can differ from that determined at constant pressure (c_p). A relevant standard method of determination of mean specific heat capacity (with special reference to thermal insulation materials) is given in ASTM C 351-92 b.

SPECULAR

Relating to light (or, more generally, to *radiant flux* (q.v.)) reflected at the specular angle (q.v.) and without diffusion, from a surface. See also *specular reflection*.

SPECULAR ANGLE

The angle of reflection equal and opposite to the angle of incidence (ASTM E 284-89). Observations of *radiant flux* (q.v.) reflected at a specular angle are sometimes said to be made "in the mirror direction" (cf. e.g. ASTM D 2457-90). See also *specular reflection* and *specular gloss*.

SPECULAR GLOSS

In the broadest terms, specular gloss may be thought of as the degree to which a surface approaches that of an ideal mirror in its capacity to reflect incident light.

A beam of light incident upon such a mirror will be reflected completely, i.e. without scattering (diffusion) and with negligible absorption, and the reflection will take place at an angle equal to the angle of incidence of the beam (cf. *specular angle, specular reflection*). In practice, light incident on the surface of a reflective plastics material will normally undergo some scattering - the more the rougher the surface - and also some absorption. Specular gloss as determined in a test is a measure of the degree to which incident light is reflected mirror-fashion by the surface tested: it is expressed on a scale of 0 - 100 in which 100 is the rating of a standard reference surface reflecting at the same specular angle under identical test conditions. Specular gloss expressed in this way may thus be regarded as a special case of the *relative luminous reflectance factor* (q.v.). The reference surface most widely used is the highly polished, plane surface of black glass with a refractive index of 1.567*. The angles of reflection at which the specular gloss of plastics surfaces (as well as certain others) is usually measured, are*: 60° (most widely used, especially for comparing similar materials and products of medium surface gloss or for assessing surface changes in the same material): 20° (typically used for high-gloss materials which give specular gloss values of over 70 in measurements at 60°); and 85° (the angle recommended* for surfaces having 60° specular gloss values below 10 - cf. also ASTM D 523-89).

Measurements of specular gloss may be used to assess in numerical terms the effects on a plastics surface of weathering or ageing, to provide a measure of surface roughness, or to determine haze in reflection (q.v.) - cf. ASTM D 4039-87.

In addition to the standards already cited as directly relevant to the specular gloss of plastics surfaces, the following ones are of some peripheral interest: ISO 2813-1978 and its British equivalent BS 3900:Part D5.1980 (measurement of specular gloss of non-metallic paint films at 20°, 60° and 85°); BS 3962:Part 1:1980 (assessment of low-angle glare of finishes on wooden furniture by measurement of specular gloss at 85°); ASTM D 1834-80 (1985) (standard test method for 20° specular gloss of waxed paper).

*cf. e.g. BS 2782:Part 5:Method 520A:1992 and ASTM D 2457-90

SPECULAR REFLECTION

Reflection of light (or, in the widest sense, of *radiant flux* - q.v.) as by a mirror (cf. speculum - the Latin for "mirror") i.e. without diffusion (scattering), and in conformity with the laws of regular optical reflection (in particular with respect to equality of the angle of reflection to the angle of incidence - cf. *specular angle*).

SPECULAR TRANSMITTANCE

The value of *transmittance* (q.v.) when the transmitted radiant flux is only that measured in the same direction as that of the incident radiant flux. (cf. ASTM D 1746–96)

SPIN-WELDING

A friction welding (q.v.) technique or operation suitable for joining circular plastics parts of appropriate design. Normally, one of the two parts is held stationary and the other is rotated in pressure contact with it. Typically illustrative conditions would be: relative joint velocity 3 m s^{-1} - 12 m s^{-1}; joint pressure 3.5 MPa - 5.0 MPa; weld time 0.25 s - 0.5 s followed by a holding time of 0.5 s - 1.0 s. With normal, simple equipment a fairly representative upper limit for the practicable joint area would be about 60 cm^2.

SPONTANEOUS-IGNITION TEMPERATURE

See *temperature of spontaneous ignition*.

STAINING

See under *migration*.

STANDARD RUBBER(S)

Rubber(s) formulated and compounded in accordance with the directions given in ISO 4649-1985 (Annex B) or in BS 903:Part A9:1988, or other rubber compounds complying with all the relevant requirements of these standards. The main purpose of the standard rubbers specified in the two standards is their use as reference materials in abrasion-resistance tests involving determinations of relative volume loss or/and abrasion resistance index of rubbers. (See also the calibration compounds in ASTM D 2228-88).

STANDARD TEMPERATURE AND PRESSURE (STP)

Arbitrary standard reference conditions, often regarded as identical with *normal temperature and pressure* (q.v.), but occasionally also defined as the temperature of 60°F (15.55°C) and pressure of 60 in Hg (762 mm Hg).

STATIC ELECTRICITY CHARGE; "STATIC CHARGE"; "STATIC"

Charges of static electricity thousands of volts in magnitude can build up, especially in dry atmospheres, on the surfaces of plastics materials and products. In general, the higher the surface and volume resistivity of he material the easier the charge build-up and the greater its magnitude. The following volume resistivity values (at room temperature and 60% relative humidity) provide a rough guide to the condition of a plastics material in regard to static electricity charge accumulation:

Below 10^5 ohm.cm	:	conductivity (increasing with decreasing resistivity).
10^5 ohm.cm to 10^8 ohm.cm	:	"antistatic" condition (decreasing with increasing resistivity).
Above 10^8 ohm.cm	:	susceptibility to static charge accumulation (increasing with increasing resistivity).

Static electricity charges can constitute hazards in environments where their discharge in the form of a spark can ignite solvent vapours, or cause dust explosions, or damage electronic circuits. They can also give rise to electromagnetic interference (q.v.), and attract contaminants (dust, dirt) to plastics surfaces. Charge accumulation can be prevented, and charges already present can be dissipated, by ionising the air near the plastics surface. In industrial operations this is done with the aid of a radioactive source, or high-voltage electrical discharge equipment. This kind of treatment is most appropriate for continuous sheet materials, which can be passed under the ionising device; but the effect is of limited duration. For more permanent effects use is made of antistatic agents (q.v.).

Whilst a plastic's tendency to accumulate static electricity charge is indicated by its resistivity, the build-up, magnitude, and decay of charges can be measured directly on plastics surfaces with the aid of field meters and electrometers of various kinds. Some of these instruments are used qualitatively, to detect the presence, or areas of greatest concentration (density), of electrostatic charge.

NOTE - Microprocessor-controlled field scanners are available for the measurement of static charges in the range 1 V to 10^5 V with a claimed accuracy of about 0.1%.

BS 5958:Parts 1 and 2:1991 (a code of practice, in two parts, for control of undesirable static electricity) provides useful basic information on the generation of static electricity, its measurement and control. Measurement of the rate of charge decay in laboratory or factory conditions can provide an indication of the effectiveness of antistatic agents (q.v.). The results of such determinations are often quoted in terms of "half time" of charge decay, i.e. the time required for a charge to be reduced by a factor of 2. A US Federal Standard (Test Method Standard 101 B, Method 4046) gives a method for the measurement of static charge decay rates. Other US standards concerned with, or referring to, this subject are MIL-B-81705 A and B, and NFPA (National Fire Protection Association) Code 99 (replacing the former Code 56 A). In a qualitative test, sometimes referred to as the "soot (or smoke) chamber test", the static charge on the surface of a polyethylene bottle is assessed in terms of the attraction to the surface of smoke from burning toluene (cf. ASTM D 2741-95).

The "antistatic" requirements to be met by plastics products are typically given in relevant standards in terms of maximum allowable electrical surface or volume resistivity as determined in specified tests. For example, the maximum allowable resistivity of PVC conveyor belting for underground use in coal mines is laid down as $3\times10^8\Omega$ by both BS 3289:1990 (1995) and the relevant National Coal Board specification (NCB 158/1989); the upper limit set by BS 2050:1978 (1996) for the surface resistivity of many products used in hospitals (e.g. anaesthetic tubing, furniture buffers, mattresses and pads) is $10^6\Omega$, for flooring for explosive-handling areas it is $5\times10^4\Omega$, and for conducting footwear (soles and heels) for use in such areas $1.5\times10^5\Omega$; the limit for small "antistatic" mouldings is $10^6\Omega$.

Determination of electrical resistivity is discussed in sections 2.3 (volume resistivity) and 2.4 (surface resistivity) of BS 4618. Relevant test methods are given in ISO 3915:1981; BS 3187:1978 (1996); BS 2782:Part2:Method 230 A:1982 (volume resistivity) and 231 A:1991 (surface resistivity); IEC Publication 93:1980; ASTM D 257-93; and BS 7506:Part 2:1996, which replaces BS 2044:1984 (now withdrawn) and Methods 250 A to C:1976(1983) of BS 2782 (determination of antistatic behaviour of plastics films).

STAUDINGER INDEX

Term sometimes used for *instrinsic viscosity* (*limiting viscosity number*); see under *viscosity.*

STAUDINGER-KÜHN EQUATION

Also known as *Mark-Houwink equation, Mark-Houwink-Sakurada equation,* or *M-H-S equation*. See under *viscosity.*

STEREOLOGY

A mathematical computer-based procedure, whereby two-dimensional images (e.g. photomicrographs, electron micrographs) can be analysed to obtain a three-dimensional representation of the structure of the object concerned. The use of stereology for quantitative characterization of the structure of plastics foams in discussed in a paper by Rhodes and Khaykin (1991).

STEREOREGULAR POLYMER

A regular polymer, the molecules of which can be described in terms of only one species of stereorepeating unit in a single sequential arrangement.

NOTE - A stereoregular polymer is always a tactic polymer, but a tactic polymer is not always stereoregular since a tactic polymer need not have every site of stereoisomerism defined.

See notes to *stereorepeating unit, isotactic polymer, syndiotactic polymer.*

Reproduced, with permission, from ISO 472:1988.

STEREOREPEATING UNIT

A configurational repeating unit having defined configuration at all sites of stereoisomerism in the main chain of a polymer molecule.

NOTE - In a stereoregular polypropylene, the three simplest possible stereo-repeating units are:

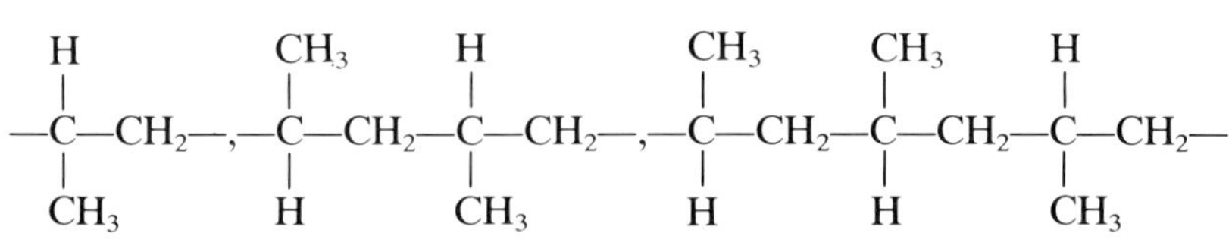

The corresponding stereoregular polymers are:

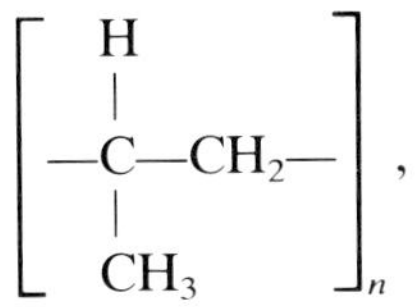

(an isotactic polymer)

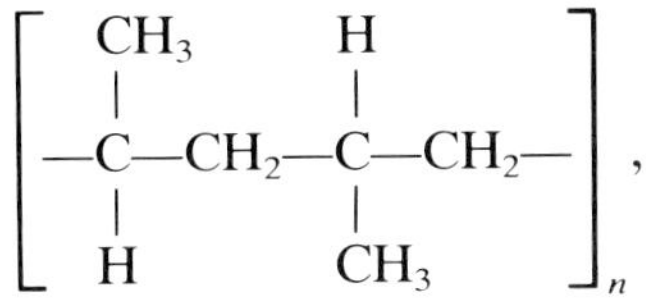

(a syndiotactic polymer)

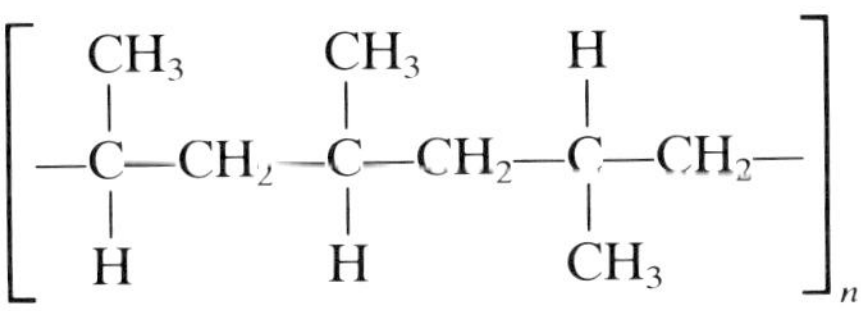

Reproduced, with permission, from ISO 472:1988.

STORAGE MODULUS M' (Pa); ELASTIC MODULUS (deprecated); [and STORAGE COMPLIANCE C' (Pa^{-1})]

The real part of the complex modulus [compliance].

NOTE - It is a measure of the energy stored and regained during a loading cycle.

See also *complex modulus* and *complex compliance*.

Reproduced, with permission, from ISO 472:1988.

STRAIN

1. In general, the unit change - caused by an applied force - in the size or shape of a body, referred to the body's original size or shape[1.1,1.3,1.5,1.10]. The term "engineering strain" is sometimes also used for strain defined in this way[1.10], to distinguish it from "true strain" (q.v.). The usual symbol for strain is e or ε (the latter is sometimes preferred for true strain). Being a ratio, strain is non-dimensional, but in practical contexts it is sometimes quoted in the appropriate units per the same unit (e.g. as millimetres per millimetre for linear strain), or as a percentage. In formal mathematical terms, strain at a point is defined by six components of strain: three normal (linear) components and three shear components, referred to a set of coordinate axes[1.5,1.10].

2. In the context of common tests for "short-term" mechanical properties of plastics, viz tensile, compression, and shear tests, only one component of strain is normally considered, being measured and reported simply as strain. When measured in this way in the above three deformation modes, the strains are defined by the following expressions (ISO 31-3:1992).

Linear strain (relative elongation):

$\varepsilon = \frac{\Delta \ell}{\ell_0}$, where ℓ_0 is the original, specified length of the test specimen (commonly the length of its central portion between gauge marks), and $\Delta \ell$ is the increase in length.

Volume strain (bulk strain):

$\vartheta = \frac{\Delta V}{V_0}$, where V_0 is the original (reference) volume of the test specimen, and ΔV is the increase in volume.

Shear strain

$\gamma = \frac{\Delta x}{[d]}$, where Δx is the parallel displacement of the upper surface with respect to the lower surface of a layer of thickness d.

The shear strain may also be defined[1.10] as the tangent of the angular displacement, caused by a force, between two lines originally perpendicular to each other through a point in a body.

STRESS

The intensity at a point in a body of the force(s) or components of force that act on the unit area of a given plane through the point[1.1-1.3,1.5,1.10]. In formal mathematical terms, stress at a point is defined by six components of stress: three normal (linear) components and three shear components referred to a set of coordinate axes[1.1,1.5].

In tension, compression, or shear tests prescribed in product or test-method specifications, stress is calculated on the basis of the original dimensions of the appropriate cross-section of the specimen[1.1-1.3,1.5,1.10]. (Typically, the force applied to the specimen at the relevant time(s) in the test is divided by the relevant original cross-sectional area). The stress value so calculated is occasionally termed "engineering stress",[1.10] in contradistinction to true stress (q.v.). Stress is expressed in units of force per unit area (for plastics usually MPa in the SI system, or lbf in^{-2}, or kgf m^{-2}). δ and τ are the symbols put forward by ISO as standard respectively for normal stress and shear stress (ISO 31- 3:1992).

STRESS CRACKING

The formation of cracks at the surface or/and in the interior of a polymer or plastic under stress(es) which may be lower than the short-term mechanical strength of the material. The stress(es) responsible for the cracking may be internal (e.g. cooling, packing, or orientation stress(es) in mouldings), or externally imposed, or both. See also *environmental stress cracking* and *crazing.*

NOTE - The principal terminology standards[1.1,1.5,1.7] all define *stress crack* (not *cracking*), and one[1.7] makes its reference to stress specific to tensile stress. However, the definition here brings out all the elements relevant to both the noun and the verb.

STRESS RELAXATION

The reduction in stress, occurring in a given time[1.3] (i.e. time-dependent[1.1, 1.5]) in a material under strain[1.2, 1.3].

STYRENE/α-METHYLSTYRENE (S/MS[1.1,1.5,2.1,2.3]; SMS[2.2,2.11]) PLASTICS

Plastics whose base polymers are copolymers of styrene with α-methylstyrene[1.1,1.5]. Except for their higher softening temperature (T_g up to about 113°C, depending on α-methylstyrene content) the material properties of the copolymers are generally similar to those of polystyrene, although their heat stability in processing and service is somewhat inferior. The copolymers are available inter alia in the *Amoco R* series. The main material standards relevant to S/MS plastics (inter alia) are BS ISO 1622 and ASTM D 4549 (for details see under *polystyrene plastics*).

STYRENE/ACRYLONITRILE (SAN) PLASTICS

Plastics whose base polymers are copolymers of styrene with acrylonitrile[1.1, 1.5]. Commercial copolymer grades are available with different styrene:acrylonitrile ratios, 76:24 being fairly common in SAN moulding materials. In general SAN is comparable with ordinary polystyrene in clarity (a light bluish tint in typically present in clear SAN, incorporated to counteract the slight yellow tinge natural to the material); it is superior to polystyrene in impact resistance, and also somewhat more resistant to a few chemicals (some oils and fats; some aliphatic hydrocarbons), as well as to abrasion, to environmental stress cracking, and thermal stress cracking under sharp temperature changes.

It is inferior to polystyrene in its tendency to yellowing on heating and exposure to UV radiation, and in electrical properties; its water absorption is substantially higher than that of polystyrene.

SAN polymers and/or compounds (including glass-fibre reinforced grades) are available in the following commercial material ranges (See also Section 1 of Part 3 of this book); *Cevian N; Denisan; Dow San; Kibisan; Kostil; Lastil; Lupan; Luran; Lustran SAN; Multi-SAN; Polidux; Restil; San; Tyril.*

Standards relevant to SAN plastics include

BS ISO 4894	Specification for SAN extrusion and moulding materials.
Part 1: 1990(1997)	Designation.
Part 2: 1995	Preparation of test specimens and determination of properties.
ASTM D 5508-94	Determination of residual acrylonitrile monomer in SAN copolymers and nitrile/butadiene rubber.

STYRENE/MALEIC ANHYDRIDE (SMA[2.1, 2.3]; S/MA[2.2]; SMAH[2.11])

A copolymer of styrene and maleic anhydride 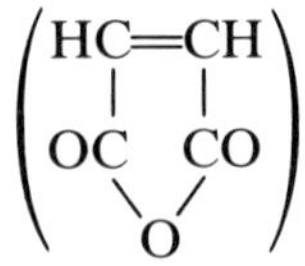 , with the monomer units alternating in the polymer chain. Some commercial versions contain a third, rubbery comonomer (butadiene in the *Elix 300* series). Two characteristics which distinguish SMA from many other copolymers of styrene are transparency and relatively high T_g, which is responsible for the similarly high softening point (both about 120°C to 125°C). Commercial materials are available in the following ranges: *Cadon; Dylark; Elix; Stapron; Styrafil* (glass-filled compounds). Applications include polystyrene-type mouldings and extrusions. The *Elix 300* series terpolymers are used as modifiers in PVC.

ASTM D 4634 - 86 (1993) provides a standard specification for SMA moulding and extrusion compounds.

STYRENE PLASTIC

A plastic based on polymers of styrene or copolymers of styrene with other monomers, the styrene being in the greatest amount by mass.

Reproduced, with permission, from ISO 472:1988.

The other two main terminology standards[1.1,1.7] provide essentially the same definition.

Plastics falling within the scope of the formal definition thus include the following industrial materials: polystyrene (PS) plastics; acrylonitrile/butadiene/styrene (ABS) plastics; styrene/acrylonitrile (SAN) plastics; acrylonitrile/styrene/acrylate (ASA) plastics; acrylonitrile/ethylene/styrene (AES) plastics; styrene/α–methylstyrene (SMS) plastics; styrene/maleic anhydride (S/MA or SMAH) plastics; and elastomeric styrene block-copolymer plastics (see under *thermoplastic elastomer*).

On a broad interpretation, the definition should extend also to toughened polystyrene (q.v.) as well as to plastics based on polystyrene blended with various elastomeric impact modifiers (such as e.g. polybutadiene, or copolymers of butadiene with styrene (S/B) or with fumaric acid esters).

The following standards are relevant to styrene plastics. Others, relating more specifically to particular materials, are mentioned, together with some trade names, under the definitions of some of the individual styrene plastics.

BS ISO 6402	Impact resistant moulding and extrusion materials (ACS; AES; and ASA plastics).
Part 1: 1990(1997)	Designation.
Part 2: 1994	Preparation of test specimens and determination of properties.
ASTM D 4474-91	Specification for injection and extrusion materials of styrenic thermoplastic elastomer (TES).
ASTM D 4703-93	Compression moulding of plaques (sheets) of thermoplastic materials for test specimens.

STYRENE-RUBBER PLASTIC

A styrene plastic (q.v.) combining a styrene polymer or polymers with a rubber or rubbers, and in which the proportion of the styrene polymer(s) is predominant by mass[1.1,1.5,1.7]. See also *toughened polystyrene* and *polystyrene plastics.*

"SUPERMICROCELLULAR" FOAM

A version of microcellular plastic (q.v.) with extra-small cell size (typically between 0.1 μm and 1 μm) and very high cell population density (10^{12} to 10^{15} cells per cm^3), produced by blowing with CO_2 introduced into the plastic in the supercritical liquid state. Foams of this kind have been made in PVC (flexible and rigid), polyethylene (LD and HD), and polycarbonate. A further development, "ultra-microcellular" foams is under way (in 1993). Amongst the reported characteristics of these materials are cell sizes down to 0.01 μm and cell population densities of up to 10^{18} per cm^3. Transparent versions are possible, as the cells would be too small in relation to the wavelength of visible light to cause scattering.

SURFACE CONDUCTANCE

The reciprocal of surface resistance (q.v.).

SURFACE CONDUCTIVITY

The reciprocal of surface resistivity (q.v.).

SURFACE RESISTANCE

The resistance between two electrodes surface-mounted on an appropriate test specimen, as given by the quotient resulting when the direct voltage applied to the electrodes is divided by that part of the current which flows between them through a thin layer on the surface of the specimen[1.1,1.5]; the layer may be e.g. one of moisture or other poorly conducting[1.5] (semi-conducting[1.1]) material. In practical tests it is difficult to avoid including some volume resistance in the surface resistance being determined (cf. e.g. ASTM D 1711-94). The unit of surface resistance is the ohm. For methods of determination see under *surface resistivity.*

SURFACE RESISTIVITY

In relation to its measurement in practice by standard methods, the surface resistivity of a plastics material may be defined as the surface resistance (q.v.) between two surface-mounted electrodes forming the opposite sides of a square of a convenient size (the actual size is not significant)[1.1,1.5]. Other definitions, in terms of potential gradient in relation to current per unit width of surface, can also be formulated (cf. Refs. 1.1 and 1.5; BS 4618; Section 2.4:1975; ASTM D 257-93; IEC Publication 93-1980; and ASTM D 1711-94). The unit of surface resistivity is the ohm. Standard methods of determination include those given in IEC Publication 93; BS 2782:Method 231A:1991; and ASTM D 257-93.

SWELL RATIO

In a solvent-extraction test essentially as carried out to determine the gel content (q.v.) of a crosslinked polymer, the ratio of the volume of the solvent-swollen gel to its volume after removal of the solvent (cf. ASTM D 2765-90).

SYNDIOTACTIC POLYMER

A regular polymer the molecules of which can be described in terms of alternation of configurational base units that are enantiomeric.

NOTE - In a syndiotactic polymer, the configurational repeating unit consists of two configurational base units that are enantiomeric. An example is the configurational repeating unit:

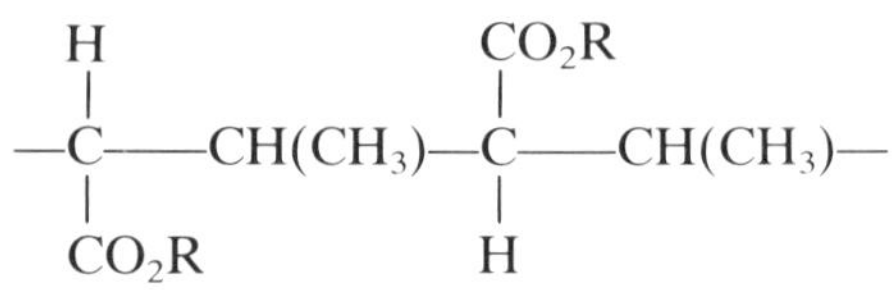

and the corresponding polymer is:

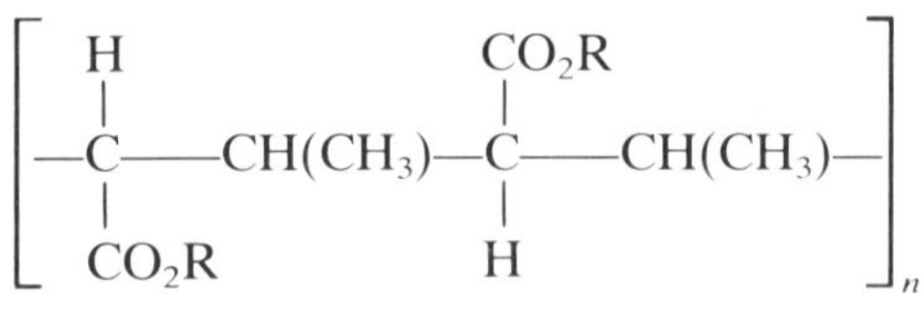

It is not a stereoregular polymer because the configuration at the stereoisomeric centre $-CH(CH_3)-$ is not defined. Similarly

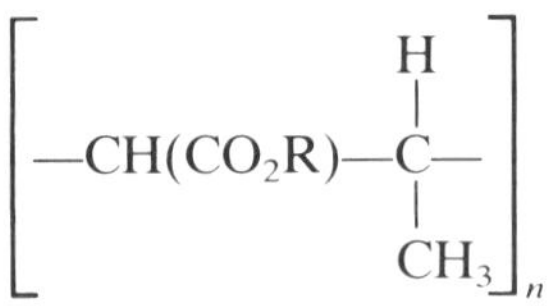

and

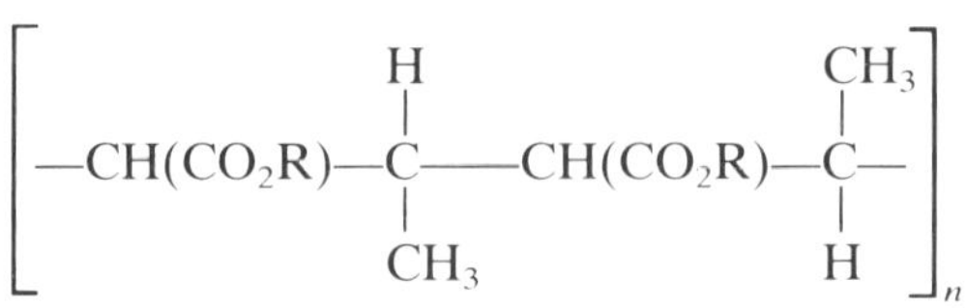

are not stereoregular polymers because of the lack of knowledge about the configuration of the stereoisomeric centre $-CH(CO_2R)-$. The following disyndiotactic polymer is stereoregular:

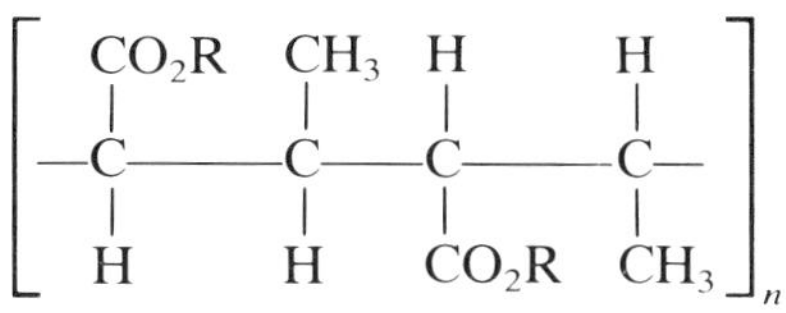

See note to *stereorepeating unit.*

Reproduced, with permission, from ISO 472:1988.

SYNDIOTACTIC POLYSTYRENE

Semi-crystalline polystyrene produced with the aid of metallocene "single-site" catalysts. Claimed to have a crystalline melting point of 270°C (with a T_g of 100°C, i.e. the same as that of the standard commercial atactic polymer), and to be comparable in some properties with such engineering polymers as PPS and PA 66. Developed jointly by Dow and Idemitsu Petrochemical Co., Japan.

SYNTACTIC CELLULAR PLASTIC (SYNTACTIC FOAM)

A form of cellular plastic (q.v.) in which hollow microspheres (q.v.), incorporated as a filler, constitute the low-density element[1.1,1.5,1.7].

TABER ABRASER

Test apparatus widely used for determinations of abrasion resistance of plastics materials. The following standards deal with the use of this equipment: ISO 9352: 1989, identical with Method 370:1990 of BS 2782: Part3 ; ASTM D 1044-94; and ASTM D 4060-95.

A paper by Hill and Nick (1966) presents the results of an evaluation of determination of abrasion resistance of polymeric materials (principally surface coatings) by the Taber abraser equipment and method.

TACTIC POLYMER

A regular polymer the molecules of which can be described in terms of only one species of configurational repeating unit in a single sequential arrangement.

See note to *stereoregular polymer.*

Reproduced, with permission, from ISO 472:1988.

TEAR-PROPAGATION FORCE

The force necessary to maintain the propagation of a tear initiated in a plastics film or thin sheet[1.5].

TEAR-PROPAGATION RESISTANCE

Tear-propagation force (q.v.) per unit thickness of the material (or the specimen) concerned[1.5].

TEAR RESISTANCE; TEAR STRENGTH

These terms are used interchangeably. The concept they denote has been formally defined as the force required to tear a test specimen of a thin material[1.5]. Implicit in this general definition are the assumptions that the "thin material" is a sheet or film which can be torn without brittle fracture, and that the shape and size of the specimen are specified, as are the appropriate test procedure and conditions. The force is usually expressed in newtons, but other units may be used (e.g. kgf).

In most standard tear tests on plastics films and sheets, the force measured is that required to propagate a tear already incipient in a cut or notch made in the specimen (as in the Elmendorf and "trouser tear" test methods - see below).

In the rubber context, the standard definitions of tear resistance[1.2,1.3] refer to the *maximum* force required to tear a test specimen, when the force is acting substantially parallel to the major axis of the specimen (in some cases also in the plane of the tear-initiating cut - cf. ISO 34-1:1994).

Some ISO test-method standards (cf. ISO 34-1:1994 and ISO 6383-1:1983) confuse the issue somewhat by offering their own definitions of tear resistance or tear strength which tally with the ISO vocabulary standard's[1.5] definition of *tear propagation resistance* (q.v.) in that they are based on tearing force *per unit specimen thickness* (so that the units are N/mm).

The following standards are concerned with the determination of tear resistance of plastics sheeting and films, or - in some cases - of flexible cellular materials.

ISO 6133:1981	(under revision in 1996) - Gives method of analysis of multi-peak traces obtained in tear-resistance and adhesion-strength tests.
BS 1763:1975	(under revision in 1996) - Deals with thin PVC sheeting (including tear resistance, inter alia)
BS 2739:1975	(under revision in 1996) - Deals with thick PVC sheeting (including tear resistance, inter alia)
BS 2782:Part 3:Method 360 A:1991 *ISO 6383-2:1983*	Joint BS-ISO standard dealing with the Elmendorf test method for plastics films and sheets
BS 2782:Part 3:Method 360 B: 1991 *ISO 6383-1:1983*	Joint BS-ISO standard dealing with the trouser-tear test method for plastics films and sheets
BS 2782:Part 3:Method 360 C:1991	Gives the initiation test method for tear resistance of plastics films and sheets
BS 4443:Part 6:1991:Method 15	Test method for tear strength of flexible cellular materials
BS 4443:Part 7:1992:Method 17	Test method for tear strength of flexible cellular materials with integral skins
BS EN ISO 8067:1995	Gives test method for tear strength of flexible cellular polymeric materials
ASTM D 624-91	Deals with determination of tear strength of conventional vulcanized rubber and thermoplastic elastomers

ASTM D 1004-94a	Tear test at very low rates of loading, primarily concerned with measuring tear-initiation force
ASTM D 1922-94a	The Elmendorf test method
ASTM D 1938-94	The trouser-tear test method
ASTM D 2582-93	Test determining resistance to puncture and propagation of tear initiated by puncture (snagging resistance)

TELOMER

A polymer whose molecular chains are end-stopped with groups that cannot react with fresh monomer(s) in the same polymerisation conditions, so that the polymer chains cannot be extended to form larger molecules of the same kind (chemical type[1.1,1.5,1.7]). This is the essence of the IUPAC definition, adopted by all three major terminology standards[1.1,1.5,1.7.]

TEMPERATURE-GRADIENT BAR

Simple apparatus for following the effects of heating time and temperature upon the progress of transformation of a PVC paste into plasticised PVC material (through the various stages from initial gelation to final complete fusion - see under *paste (1)*). Some versions of the apparatus are also known as "hot bench" and "gel block" (cf. e.g. Titow, 1990). An essential feature, common to all forms of the apparatus is a metal element (bar, block, or the like), which is differentially heated to produce along its length several zones of progressively higher, closely controlled temperature. A thin layer of PVC paste is cast on the surface of the element, so that it spans all the heating zones. After an appropriate time (which may itself be one of the variables investigated) heating is discontinued, and the cooled PVC layer - now consolidated in varying degrees in sections corresponding to the heating zones - is removed. The extent of transformation of the material in each of the sections can now be examined by the relevant methods.

TEMPERATURE INDEX (TI)

(1) A numerical expression of the thermal endurance (q.v.) of a material, as determined in a relevant test. The number is based on the temperature, in degrees Celsius, obtained by extrapolation - to a specified time - of an Arrhenius plot of satisfactory-life time versus temperature. The TI is thus indicative of the maximum temperature at which the material concerned may be expected to last satisfactorily (in the sense of maintaining an acceptable level of a relevant property or properties) for a specified time. The time normally considered in the standard tests is 20 000 h. The major standards dealing directly with determination of TI are ISO 2578:1993 and IEC Publication 216, Parts 1 to 5; the former is directed to plastics materials generally, whilst the latter focuses on electrical insulation materials. ASTM D 1711-94 (an electrical-insulation terminology standard) contains, inter alia, a definition of TI, and a standard practice for heat ageing of plastics without load is provided in ASTM D 3045-92. See also *relative temperature index, thermal endurance,* and *thermal stability*.

(2) The temperature at which the oxygen index of a material (or of an individual test specimen) equals the normal percentage concentration of oxygen in the atmosphere (21% to the nearest unit) has also been referred to as the temperature index. This is schematically shown in the figure, which also illustrates the fact that the flammability of a plastics material tends to increase (the oxygen index decreases) with increasing temperature.

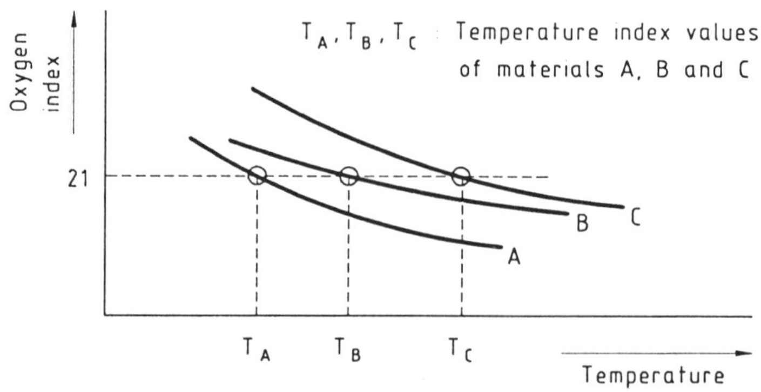

TEMPERATURE OF DEFLECTION UNDER LOAD

See *deflection temperature under load.*

TEMPERATURE OF SPONTANEOUS IGNITION (SPONTANEOUS-IGNITION TEMPERATURE)

Sometimes also called auto-ignition temperature[1.8], or self-ignition temperature (q.v.), although the term "self-ignition" (q.v.) has a special connotation in some terminology standards[1.8,1.9] and hence its use in the context of spontaneous ignition brought about by external heating (see below) is deprecated by ISO[1.8].

When a plastics material is heated sufficiently strongly, it may generate flammable vapors. Under suitable conditions, the concentration of the vapors at the material's surface may reach the critical value for their spontaneous ignition at the prevailing high temperature (cf. e.g. "ignition process" in ISO 10093). Thus the spontaneous-ignition temperature is the lowest temperature at which ignition is brought about by heating a plastics material under specified conditions in the absence of any extraneous ignition source[1.8,1.9]. In the context of the hot-air ignition furnace test of ASTM D 1929-91, the spontaneous-ignition temperature (which this standard terms "self-ignition temperature") is the temperature of heated air passing round the specimen when ignition occurs. Unlike many other small-scale laboratory tests for flammability, the test of ASTM D 1929 is claimed to rank plastics materials in the order of their susceptibility to ignition under actual-use conditions. The accuracy of determination of the spontaneous-ignition temperature (and the flash-ignition temperature - q.v.) by the test method of this standard is indicated as within 5°C to 20°C at the 250°C - 350°C level, and 20°C to 40°C at the 350°C - 500°C level.

TENSILE STRENGTH

(1) In the most general sense, a material's ability to withstand tensile stress[1.1].

(2) In the context of mechanical properties of plastics materials, the stress at which a material fails in tension[1.1]: failure may be by yielding or breaking, hence the expressions tensile strength at yield, or at break. Some versions of the standard definition[1.5,1.10] identify tensile strength as the *maximum* tensile stress sustained before failure (cf. also ASTM D 638M-93). In tests on standard specimens, the tensile strength (σ_t) at yield and/or break is calculated, for each specimen, from the load (force applied) at - or the maximum load before - yield or rupture and the original cross-sectional area of the specimen, i.e. $\sigma_t = F/(bh)$, where F is the load, b is the mean initial width and h the mean initial thickness of the test specimen. The arithmetic mean of the individual σ_t values is normally reported as the tensile strength of the material. The unit of tensile strength in widest use is the Pascal ($1\,Pa = 1\,Nm^{-2}$); pound-force per square inch (lbf in^{-2}, occasionally still designated psi) is also used (especially in the USA); and, sometimes, kilogram-force per square centimetre (kgf cm^{-2}). The mutual conversion factors of these units are:

Pa	lbf in^{-2}	kgf cm^{-2}
1	1.450×10^{-4}	1.020×10^{-5}
6.895×10^{3}	1	7.031×10^{-2}
9.807×10^{4}	14.22	1

In practice, the actual tensile strengths of many common plastics materials run into tens of megapascals (MPa), i.e. thousands of lbf in^{-2}, and hundreds or thousands of kgf cm^{-2}.

Some material or product standards specify tensile strength values or value ranges. A few examples of such standards are listed in the table.

Type of material or product	Polymeric material	Standard containing numerical tensile-property requirements
Extrusion and moulding compounds	PPS (reinforced and filled),	ASTM D 4067-93
	PTFE (filled)	ASTM D 4745-91
	Polyethylene	ASTM D 1248-84 (1989)
	Cellulose propionate	ASTM D 1562-82 (1989)
	Unplasticised PVC and chlorinated PVC	ASTM D 1784-92
	Plasticised PVC	ASTM D 2287-92
	Polystyrene	ASTM D 4549-85 (1994)
	Styrene/maleic anhydride materials	ASTM D 4634-86 (1993)
	Thermoplastic elastomer (ether-ester)	ASTM D 4550-92 (1993)
	ABS	ASTM D 4673-87 (1993) and D 3965-94
	Polyamide (nylons)	ASTM D 4066-94
Sheeting - miscellaneous	Polyethylene	BS 3012: 1970 (1982)*
	Polyethylene (LD and MD)	ASTM D 4635-91
	Rigid PVC	BS 3757:1978 (1990)
	Flexible PVC	BS 1763:1975
		BS 2739:1975
		BS 3878:1982 (1990)
		ASTM D 1593-92
	PCTFE	ASTM D 3595-91
	Poly(methyl methacrylate)	ASTM D 4802-94
Sheeting - membranes for water containment	Chlorinated polyethylene	ASTM D 4068-91
	Flexible PVC	ASTM D 4551-91
GRP laminates		BS 3691:1990 (1995)
		BS 4994:1987
		ASTM D 4357-85 (1990)
* Obsolescent in 1995		

Numerous standard test methods for the tensile properties of plastics materials are available, applicable according to the nature and/or form of the material concerned. The table below lists some of the main English-language standards of this kind.

Material: Nature and/or form	Standard specification for (or including) method(s) of determination of tensile properties	Remarks
Plastics - general	ISO 527-5 parts (1993 -)* BS 2782:Part 3: Methods 320 A-F:1986**	Principal general tensile-testing specifications
	ASTM D 638-93 ISO 6239-1986 BS 2782:Part 3: Method 327 A: 1993	For small specimens Tensile properties of PTFE products
	ASTM D 1708-93 ASTM D 759 (withdrawn in 1983)	For "microtensile" specimens Determination of tensile properties at low and elevated temperatures
Thin sheeting and/or film	ISO/R 1184-1983 BS 2782:Part 3: Methods 326 A-C: 1983 ASTM D 882-91	Withdrawn in 1995*
Cellular plastics	ISO 1926-1979 ISO 7214-1985 (under revision in 1990)	For rigid cellular plastics Test methods (including tensile tests) for flexible and semi-rigid cellular polyethylene
	BS 4370:Part 2: Method 9:1993 ASTM D 1623-78 (1995)	For rigid cellular plastics For rigid cellular plastics
Reinforced plastics	ISO 3268-1978	For glass-reinforced plastics
	European standard EN 61, adopted as BS 2782:Part 10: Method 1003:1977	For glass-reinforced plastics
	ASTM D 3039/D 3039M-95	For fibre-resin composites reinforced by oriented high-modulus fibers
	ASTM D 3916-94	For pultruded rod
Moulded electrical insulation material	ASTM D 651-84	Withdrawn in 1991

* Some parts of ISO 527 still in preparation in 1996

** Method 321:1994 = ISO 527-1:1993
Method 322:1994 = ISO 527-2:1993

* Replaced by ISO 527:Part 3:1995

TENSILE STRESS

See under *normal stress*.

TENSION FATIGUE

The initiation and growth of cracks, with eventual fracture, in a material or test specimen undergoing repeated straining in the tensile mode[1.3].

TETRAMER

An oligomer (q.v.) molecule consisting of four monomeric units, or a substance composed of such molecules[1.1].

NOTE - The ISO and ASTM terminology standards[1.5,1.7] do not define this term. It is noteworthy that the BS definition[1.1] is not restricted to the case where the monomer units are of the same species (cf. the definition of *trimer*).

THERMAL ANALYSIS (TA)

A general term for an instrumental technique (or, collectively, all such techniques) used to measure the functional dependance of a physical property of a material upon temperature[1.1] when the material is subjected to an appropriately controlled temperature programme[1.5,1.15]. Standards relevant to TA include ASTM E 472-86 (1991) (reporting of thermoanalytical data); ASTM E 1142-93b (standard terminology of thermophysical properties); ASTM D 3417-83 (1988) (determination of heats of fusion and crystallisation of polymers by TA); ASTM D 3418-82 (1988) (determination of transition temperatures of polymers by TA); ASTM D 5028-90 (determination of curing properties of pultrusion resins by TA).

THERMAL CONDUCTIVITY

The quantity of heat which passes in unit time (i.e. the rate of heat flow[1.1,1.5], the heat flux), in steady-state conditions, through a unit area of a material, under the influence of a unit temperature gradient in the direction perpendicular to that unit area[1.1,1.5]. Implicit in this definition are the assumptions that the material is homogeneous, and that the steady-state heat flow through it is not affected by its thickness[1.1,1.5]. The usual symbols for thermal conductivity are λ (IUPAP symbol) and k, and the SI units are $(\mathrm{W.m^{-2}})/(\mathrm{K.m^{-1}})$ i.e. W/(m.K). In typical practical determinations of thermal conductivity, the material is in the form of either a flat parallel-sided slab (with the mutually parallel heat-flow lines perpendicular to its sides), or a cylindrical shell (with radial heat-flow lines). The expressions for thermal conductivity for these two cases are.

Flat slab: $\lambda = (Q / A)(L / \Delta T)$

where Q is the heat flow rate, A is the total area concerned, L is the slab thickness, and ΔT is the temperature difference between the two sides of the slab (so that $L / \Delta T$ is the approximate value of the temperature gradient).

Cylindrical shell: $\lambda = \left[Q /(2\pi / \Delta T)\right]\left[\ln(r_2 / r_1)\right]$

where l is the shell length, and r_1 and r_2 are, respectively, the inner and outer radii.

The concept of thermal conductivity is discussed in BS 4618:Section 3.3:1973, and in the annex to ISO 7345:1987 (in connection with thermal insulation).

Some methods of determination are given in BS 874:Part 2:Sections 2.1:1986 and 2.2:1988, and in ASTM C 177-85 (1993).

THERMAL CURVE

Plot of the results of a determination by one of the techniques of thermal analysis (q.v.), such as differential scanning calorimetry (DSC), differential thermal analysis (DTA), or thermogravimetric analysis (TGA). Depending on the technique concerned, the ordinate values on the plot may be those of heat flow (as in DSC), or temperature difference (as in DTA), or mass (as in TGA), and the abscissae are the corresponding values of either temperature or time. See also ASTM E 1142-93b.

THERMAL DEGRADATION (HEAT DEGRADATION)

Degradation (q.v.) caused by exposure to heat[1.3] at an elevated temperature[1.5]. See also *thermal stability*.

THERMAL ENDURANCE

Degree of resistance of a material (in the context of the principal relevant electrical (IEG) standards, specifically an insulating material) to irreversible degradation resulting from prolonged exposure to elevated temperature in an appropriate test. (Cf. IEC Publication 216-1, and ASTM D 1711-91). See also *temperature index (1), thermal stability* and *heat ageing*.

THERMAL INDEX

See *relative thermal index*.

THERMAL INSTABILITY INDEX (of PTFE)

A value representing the reduction in molecular weight of PTFE material (as reflected in a corresponding change in specific gravity) following prolonged heating under specified conditions (see ASTM D 1457-92). For most of the types, grades and classes of PTFE moulding and extrusion compounds covered by ASTM D 1457, the maximum permitted value of the index is 50; in a few cases the stipulated maximum is 15. For the thermal instability index of PTFE resins produced from dispersion see section 10.6 of ASTM D 4895-94.

THERMAL OXIDATIVE STABILITY

A term sometimes applied to the resistance of polyolefin plastics to degradative heat ageing at an elevated temperature in the presence of air or oxygen. The resistance is normally measured and expressed in terms of time elapsing, in specified test conditions, before the onset of significant degradation on heating. Standard test methods have been published for the determination of thermal oxidative stability of propylene plastics at 150°C in air (cf. the oven-heating test of ISO 4577:1983, and the analogous test of ASTM D 3012-90) and in oxygen (cf. ASTM D 2445, discontinued in 1986) as well as of polyethylene pipe material in oxygen (cf. ISO/TR 10837:1991). A requirement (based on the ISO test method) for the thermal stability of polyethylene water pipes is featured in a draft European standard initiated in 1992 (155 N 697E). See also *oxidative induction time*, and *temperature index*.

THERMAL STABILITY (HEAT STABILITY)

A material's ability to withstand heating without suffering significant degradation[1.5] (q.v.). Thermal stability is not an inherent or invariable property of a plastics material: whilst it is to some extent influenced by the material's chemical nature, it is cardinally dependent on such external factors as the temperature, rate, duration and mode of heating, and on the presence of oxygen. In standard tests of thermal stability of plastics, these factors, as well as other relevant conditions, are specified. In such tests, the thermal stability of a plastics material is normally determined, and subsequently reported, in terms of time to the onset of relevant manifestations of significant degradation, such as substantial drop in a designated mechanical property (e.g. tensile strength and/or elongation), evolution of volatile decomposition products (e.g. HCl from PVC), onset of substantial loss of mass in thermogravimetric determinations, darkening, etc. Several relevant standards are listed in the table. See also *heat ageing, oxidative induction time, relative thermal index, temperature index, thermal endurance, thermal instability index,* and *oxidative stability*.

THERMODILATOMETRY

A technique in which a dimension of a substance under negligible load is measured as a function of temperature while the substance is subjected to a controlled temperature programme.

NOTES

1 The record is the thermodilatometric curve: the dimension should be plotted on the ordinate increasing upwards and temperature or time on the abscissa increasing from left to right.

2 *Linear thermodilatometry* and *volume thermodilatometry* are distinguished on the basis of the dimensions measured.

Reproduced, with permission, from ISO 472:1988.

The term dilatometry is also used for the same technique[1.1].
The following two standards are relevant in the plastics context.

ASTM D 696-91	Test method for determination - with a quartz dilatometer - of the coefficient of linear thermal expansion of plastics between –30 °C and 30 °C.
ASTM D 864-52(1978)	Determination of volume expansion with a liquid-in-glass dilatometer.

See also the paper by Bekkedahl (1949).

THERMOGRAVIMETRY (TG)

A method (technique[1.1,1.5]) for measuring - with the aid of apparatus incorporating a sensitive microbalance - changes in the mass of a small specimen of a material as it is being heated or cooled at a rate controlled by an appropriate temperature programme[1.1, 1.5]. The results of the measurements are recorded in terms of mass as a function of temperature[1.1, 1.5] or time[1.1]. The two major terminology standards which provide a definition of thermogravimetry[1.1, 1.5], additionally make the following points.

a) That thermogravimetric measurements are recorded in the form of the thermogravimetric curve (TG curve), in which the mass is plotted on the ordinate (decreasing downwards) and the temperature (T) or time (t) on the abscissa (increasing from left to right)[1.1, 1.5].

b) That the terms "thermogravimetric analysis", and "dynamic thermogravimetric analysis", sometimes used for thermogravimetry, are rejected by the International Confederation for Thermal Analysis (ICTA)[1.1]; and
c) that the occasionally encountered use of the term "thermogram" for thermogravimetric curve, is also rejected by ICTA, on the grounds that the term is already employed in medicine for the results of the diagnostic technique of thermography[1.1].

Thermogravimetry of polymers and plastics is the subject of ISO 7111: 1987 and (in relation to thermal degradation characteristics of electrical insulating materials) ASTM D 3850-84.

THERMOGRAM

Term sometimes used for *thermal curve* (q.v.). Such usage is discouraged because "thermogram" has a different, specific meaning in the medical technique of thermography (see point (c) under *thermogravimetry*). Cf. also ASTM E 1142 -93b.

THERMOMECHANICAL ANALYSIS (TMA)

This term, although not uncommon in general usage, is not listed in any of the three major English-language terminology standards[1.1,1.5,1.7], and is deprecated by another relevant standard[1.15], in favour of *thermomechanical measurement* (q.v.).

THERMOMECHANICAL MEASUREMENT

A technique whereby a material is subjected to a non-oscillatory load (in an appropriate mode, viz. tension[1.5], compression[1.5], flexure[1.5], torsion[1.5], or penetration) under a suitably controlled temperature programme, and the resulting deformation is measured to determine its functional relationship with the temperature[1.5,1.15].

THERMOPLASTIC

(1) As noun: a plastic that can be made repeatedly to undergo the essentially physical changes of softening on heating and re-hardening on cooling within a temperature range characteristic for the material, and that - when softened by heat - can be caused to flow in such processes as moulding, extrusion and thermoforming whereby it is converted, on final cooling, into manufactured articles.

(2) As adjective: of the nature or kind of a thermoplastic. One of the three main terminology standards[1.7] includes the characteristics of a thermoplastic (see 1. above), in virtually identical terms, in the definitions of both the adjective and the noun. The other two standards[1.1,1.5] give these characteristics only in their (mutually identical) definitions of the adjective, defining the noun simply by reference to the latter.

STANDARDS RELEVANT TO THERMAL STABILITY OF PLASTICS

MATERIAL(S)	STANDARD(S)		REMARKS
	DESIGNATION	SUBJECT	
PLASTICS	BS 4618 : Section 4.6 : 1974	Discussion of thermal endurance and relevant test methods	
PLASTICS	ISO 2578 : 1993* (BS 2782 : Method 135 : 1993*)	Determination of time-temperature limits in prolonged exposure to heat	Method useful inter alia, for assessing likely service-life span, at a given temperature
	ASTM D 3045 - 92	Practice for heat-ageing of plastics (in the absence of mechanical stress)	Very closely similar to ISO 2578
	IEC Publication 216 : 1990 Parts 1 to 5	Determination of thermal endurance of electrical insulation materials	Subject and methods similar to those of ISO 2578
	ANSI/UL 746 B - 1992	Polymeric materials : long-term property evaluation (with the aid of heating)	Joint American National and Underwriters Laboratories standard - methods similar to that of ISO 2578
PLASTICS	ASTM D 794 - 93	Practice for determination of permanent effects of heat on plastics	
PLASTICS	ASTM D 1870 - 91	Practice for heat-ageing of plastics in a tubular oven	Closely similar to ISO 1137 (withdrawn in 1990)
THERMOSETTING MOULDING MATERIALS	ISO 11248 : 1993* (BS 2782 : Method 136 : 1994*)	Determination of short-term performance of thermosetting moulding materials at elevated temperatures	
THERMOSETTING MOULDING MATERIALS	BS 2782 : Method 131 C : 1978	Determination of resistance to heat-ageing	Evaluation of crushing strength after heat-ageing
THERMOSETTING MOULDINGS OR LAMINATED SHEET	BS 2782 : Method 131 D : 1978	Determination of resistance to heat ageing	Evaluation of crushing strength after heat-ageing
THERMOSETTING MATERIALS	BS 2782 : Method 132 B : 1976	Determination of the blistering temperature	
PVC COMPOUNDS AND PRODUCTS	ISO 182-1 : 1990* (BS 2782 : Method 130 A : 1991*)	Determination of thermal stability (Congo Red method)	
PVC COMPOUNDS AND PRODUCTS	ISO 182-2 : 1990* (BS 2782 : Method 130 B : 1991*)	Determination of thermal stability (pH method)	

STANDARDS RELEVANT TO THERMAL STABILITY OF PLASTICS

MATERIAL(S)	STANDARD(S)		REMARKS
	DESIGNATION	SUBJECT	
PVC COMPOUNDS AND PRODUCTS	ISO 182-3 : 1993* (BS 2782 : Method 130 C : 1993*)	Determination of thermal stability (conductometric method)	
PVC COMPOUNDS AND PRODUCTS	ISO 182-4 : 1993* (BS 2782 : Method 130 D : 1993*)	Determination of thermal stability (potentiometric method)	
PVC COMPOSITIONS	ASTM D 2115-92	Determination of oven-heat stability	
PVC RESINS	ASTM D 4202-92	Determination of thermal stability (Congo red method)	
PVC COMPOUNDS	ASTM D 2538-94 (section 10)	Thermal stability test (heating and agitation in a torque rheometer)	
PVC SHEET	BS 2782 : Method 131 B : 1983	Determination of extensibility after heat-ageing	
PTFE RESIN	ASTM D 4895-94 (section 10.6)	Determination of thermal instability index	See also *thermal instability index*
PTFE MOULDING AND EXTRUSION MATERIALS	ASTM D 1457-92 (section 13.8)	Determination of thermal instability index	
PMMA SHEET	ASTM D 4804-94 (section 8.1.8)	Determination of thermal stability	
CELLULOSE ACETATE	ISO 1599 : 1990	Determination of viscosity loss on moulding	Providing an indication of thermal stability in processing

* ISO standard adopted as the corresponding BS standard.

THERMOPLASTIC ELASTOMER (TPE)

A polymeric material which is rubber-like[1.3,1.7] - i.e. which exhibits elastomeric properties[1.5] - within its normal service temperature range, but which behaves like a thermoplastic in processing[1.3,1.5], in that it can be melt-processed and - if necessary - reworked by methods applicable to thermoplastics[1.3,1.7], and will remain thermoplastic on repeated heating and cooling in the temperature range relevant to such processing[1.5]. See also *elastomer, thermoplastic,* and *thermoplastic rubber.* The standard abbreviated term (letter symbol) for "thermoplastic elastomer" is TPE[1.3,1.7,2.2]. The new, revised edition of ISO 1043-1:1987 (cf. Refs. 2.1 and 2.11) provides for the use of additional letter symbols for "special characteristics". One such would be "TE", to be added - after a hyphen - to the abbreviated terms for particular thermoplastic elastomers (e.g. "PEBA-TE") - cf. TE in Section 1 of Part 2.

The morphology of a typical TPE is characterised by domains of relatively rigid, "hard" thermoplastic material interspersed with softer, elastic regions. It is those latter regions that - at ordinary temperatures at which the TPE is a solid - impart to the material its elastomeric properties, whilst the more rigid domains act as quasi-crosslinks, functionally somewhat similar to the chemical crosslinks in vulcanised rubber. At the elevated temperatures of thermal processing, these domains fuse in the manner of a thermoplastic, enabling the entire TPE to form a melt. When the melt is cooled, the rigid domains reform, and the material regains its pseudo-crosslinked, rubbery condition. The special morphology of a TPE may be achieved in one of two general ways : by intimately blending together suitable separate "hard" (thermoplastic) and elastic components, or by the design of polymers in whose molecular chains elastomeric segments or blocks alternate with, or are flanked by, more rigid segments or blocks of a thermoplastic nature. In the bulk TPE material of the latter kind, these rigid segments or blocks group together to form the "harder" thermoplastic domains. Among commercial TPE's the blends are represented principally by various olefinic compositions (also by some PVC/elastomer blends), and the other main group by a number of copolymers, as illustrated by the examples below.

Olefinic Blends

1. *Blends of polypropylene or a propylene copolymer with EPDM or with ethylene/propylene rubber (EPM).* In blends of this general kind (in which part of the polyolefin content may be a metallocene-produced polyolefin) the rubbery component may be partly crosslinked in the course of mechanical blending ('dynamic vulcanisation'), or it may be fully vulcanised (which makes the blend a "thermoplastic vulcanisate" (TPV) - cf. e.g. *Santoprene*).

2. *Polypropylene/nitrile rubber blends, in which the rubber is dynamically vulcanised (cf. e.g. Geolast).*

The two components in such blends are not readily compatible: compatibilisers (q.v.) are used to promote full, intimate interdispersion.

Commercial versions of olefinic TPE's include materials in the following ranges : *Alcryn, Chemigum, Dexflex, Ektar, Engage, Evoprene, Exxtral, HiFax, Kelburon, Milastomer, Multi-Flex, Novalast, Novalene, Perriprene, Polytrope, Prolastic, Sarlink, Telcar, Vamac, Vistaflex, Vyram.*

PVC/Elastomer Blends

These include blends of PVC resin with nitrile rubber (cf. e.g. *Crylac, Europrene N OZO*), and with polyurethane elastomers (cf. e.g. *Duralex)*

Block or Segment Copolymers

1. *Styrenic tri-block copolymers of the type* 

The styrenic block-copolymer TPE's are styrene/butadiene/styrene (SBS), styrene/isoprene/styrene (SIS), styrene/ethylene/butadiene/styrene (SEBS), and styrene/ethylene/propylene/styrene (SEPS). The relevant commercial product ranges are *Cariflex, Dynaflex, Estalloy, Evoprene, Finaclear, Finaprene, Hercupene, J-Plast, Kraton, Lacbloc, Multi-Flex, Plenex, Prolastic, Sarlink, Terlac, Thermoflex.*

2. *Polyamide block copolymers*

2.1 Poly(ether block amide) (PEBA), c.f. e.g. the commercial products *Peebax* and *Vestamid.*
2.2 Poly(ester block amide) : cf. e.g. *Estamid, Grilamid EL, Grilon EL.*

3. *Poly(ether/ester) copolymers ("copolyesters"),* represented by the commercial products *Arnitel, Bexloy, Ecdel, Elastotec, Elastuf, Gaflex, Hytrel, Lomod, Pelprene S, Riteflex.*

4. *Polyurethanes (polyester or polyether types)*

The relevant commercial products are *Desmopan, Elastollan, Estaloc, Estane, Lastane, Morthane, Oldopren, Pellethane, Polypur, Rimplast, Tecoflex, Tecophilic, Tecoplast, Texin, Urepan.*

Miscellaneous

1. Fluoropolymers : eg. *Dai-El, Fluorel.*
2. Amorphous, highly chlorinated polyolefin : e.g. *Alcryn.*
3. Syndiotactic polybutadiene (SBD - q.v.)

The following standards are concerned with the technical aspects of TPE's (see also under *thermoplastic rubber*).

ASTM D 4474-91 (styrenic TPE moulding and extrusion materials);
ASTM D 4550-92 (poly(ether/ester) TPE materials);
ASTM D 5021-89 (alloys of chlorinated ethylene/TPE);
ASTM D 5046-92 (fully crosslinked elastomeric alloy);
ASTM D 5538-94 (thermoplastic elastomers - terminology and abbreviations);
ASTM D 5593-94 (olefinic TPE's - standard classification);
ASTM F 913-94 (TPE gaskets).

THERMOPLASTIC POLYESTER PLASTICS

Plastics based on those polyesters (q.v.) which are thermoplastic polymers, or on blends in which the thermoplastic polyester components are in the predominant proportion by mass. The polymers most widely used as bases of thermoplastic polyester plastics are poly(ethylene terephthalate) (PET) and poly(butylene terephthalate) (PBT), with the respective repeated structural units

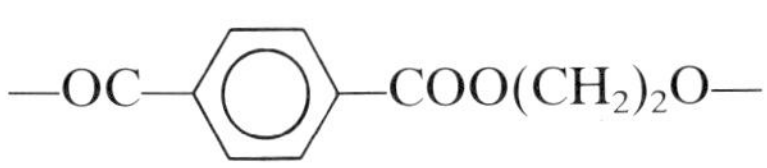

and

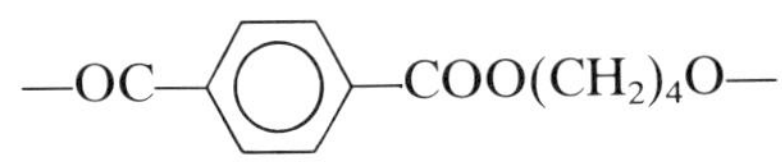

With the exception of the relatively low DTL in the case of PBT, these thermoplastic polyesters are, in general, good engineering plastics, and become particularly good when reinforced with glass fibres. Whilst their tensile strengths are somewhat lower than that of PA 66, they possess good toughness and fatigue endurance. Their dimensional stability and property retention in the presence of moisture are much superior to those of the aliphatic polyamides, because of the polyesters' considerably lower water absorption. Their processability is very good, especially in the case of PBT, making possible fast cycle times in moulding. Among their disadvantageous characteristics are notch sensitivity under impact, susceptibility to hydrolysis on prolonged contact with hot water or steam, low resistance to strong acids and alkalis, to phenols and cresols, susceptibility to swelling by some chlorinated hydrocarbons and, in the case of PBT especially, by some ketones and esters. The normal commercial grades are flammable, but flame-retarded versions are available.

Other thermoplastic polyesters in industrial use include e.g. poly(cyclohexylenedimethylene terephthalate)(PCT) and various copolymers. The following standards are relevant to the general group of thermoplastic polyester plastics (see also under poly(ethylene terephthalate) and poly(butylene terephthalate)).

ASTM D 4507-93a	Standard specification for thermoplastic polyester (TPES) materials (including a number of blends).
ASTM PS 2-94	Provisional standard for moulding thermoplastics (including PBT, PET, PCT, and PCTG) for electronic and electrical parts.

THERMOPLASTIC RUBBER

In general usage, a virtual synonym for *thermoplastic elastomer* (q.v.), despite the distinction between "rubber" and "elastomer" in standard terminology. The ISO rubber vocabulary standard[1.2] (Addendum 16) also regards "thermoplastic elastomer" as simply the common commercial term for thermoplastic rubber. This standard is the only one among the major relevant vocabulary standards to offer a definition of thermoplastic rubber, according to which the term denotes a polymer (or a blend of polymers) that, whilst not requiring vulcanization or cross-linking in processing, possesses - within its normal service-temperature range - properties similar to those of vulcanized rubber.

The following test standards are applicable to both conventional vulcanized rubber and thermoplastic rubbers:

ISO 34-1 and	:	tear strength tests (the title of the ASTM standard refers to
ASTM D 624-91	:	"conventional vulcanized rubber and thermoplastic elastomers").
ISO 36	:	determination of adhesion to textile fabrics
ISO 48	:	determination of hardness
ISO 815	:	determination of compression set
ISO 1431 - 1 & 2	:	determination of resistance to ozone cracking
ISO 1432	:	determination of low-temperature stiffening (Gehman test)
ISO 1827	:	determination of modulus in shear
ISO 2285	:	determination of tension set
ISO 3384	:	test for stress relaxation in compression
ISO 4648	:	determination of the dimensions of test pieces and products for test purposes
ISO 4661	:	preparation of samples and test pieces
ISO 7743	:	determination of stress-strain properties

THERMOSET

(1) As a noun, *either* a synonym for thermoset plastic[1.7] (q.v.), *or* - according to the definition in the ISO and BS terminology standards[1.1,1.5] and to fairly common informal usage - a general term for both thermosetting and thermoset plastics, indicating that the material concerned becomes essentially insoluble and infusible on suitable treatment (cure) by heating or in other ways. See also *thermosetting plastic*.

(2) As an adjective, a description of the infusible[1.7] and insoluble condition of a thermoset plastic (q.v.).

THERMOSET PLASTIC

A thermosetting plastic (q.v.) after its conversion to an essentially infusible and insoluble condition by an appropriate treatment (cure[1.1,1.5,1.7]) such as heating, irradiation, catalysis, etc.

THERMOSETTING PLASTIC

A plastic possessing thermosetting characteristics[1.1,1.5], i.e. one that can be converted to an essentially insoluble and infusible state when suitably treated (cured[1.1,1.5,1.7]) by heating, or in other ways[1.1,1.5,1.7] (e.g. by irradiation, catalysis, etc.[1.1,1.5]). Thermosetting plastics are often referred to simply as "thermosets", and the term "thermosetting resin" is applied to many thermosetting materials, e.g. epoxy resins, unsaturated polyesters, phenolic resins, and others. The groups of thermosetting plastics in wide industrial use are those based on alkyd resins, allyl resins, amino resins, epoxy resins, phenolic resins, polyurethanes, silicones, unsaturated polyesters, and vinyl esters. Standards relevant to thermosetting plastics generally include the following (see also under the individual thermosetting-plastic names).

Properties of thermosetting moulding materials

ISO 2577:1984 (determination of shrinkage). ISO 10724:1994 (evaluation of short-term performance at elevated temperatures). BS 2782:Part 1:Method 131C:1978 (1992) (heat resistance : crushing strength after heating). ASTM PS 15-94 (provisional specification for thermosetting moulding compounds).

Preparation of test specimens from thermosetting moulding materials

By compression moulding : ISO 295:1991 (identical with BS 2782:Part 9:Method 902A:1992); ASTM D 5224-93.
By injection moulding : ISO 11248:1993 (multi-purpose test specimens); ASTM D 3419-93 (specimens for impact tests).
By transfer moulding : ASTM D 1896-93.

Processing characteristics of thermosetting resins and moulding materials

Determination of gelation time of resins : BS 2782:Part 8:Method 835 D:1980(1986) (hot-plate method); ASTM D 2471-94 (determination of gel time and peak temperature).
Cure behaviour of resins : ASTM D 4473-94 (by dynamic mechanical measurements); ASTM D 3795-93 (in torque rheometer).
Determination of thermal flow : ISO 7808:1992 (transfer flow); ASTM D 3123-94 (spiral-flow test for low-pressure moulding compounds).
Determination of the moulding index of moulding powder : ASTM D 731-95.

Properties of products made from thermosetting plastics materials

ISO 75-3:1993 (determination of temperature of deflection under load of high-strength laminates and long-fibre-reinforced plastics); BS 2782:Part 1:Method 131 D:1978 (1992) (determination of heat resistance of thermosetting laminated sheet or mouldings), and Method 132 B:1976 (1993) (determination of the blister temperature of thermosetting materials); ASTM D 5870-96 (property-retention index of thermoplastic/thermoset plastics); ASTM D 2562-94 (classification of defects in products compression-moulded or transfer-moulded from thermosetting plastics materials).

THERMOSETTING RESIN

A polymeric material capable of crosslinking under the influence of heat, pressure, radiation, ultraviolet light, or chemical agents to form a thermoset. See also *A-stage, B-stage,* and *C-stage.*

DISCUSSION - Many important properties of thermosetting resins, including their infusibility, insolubility, resistance to swelling, and most mechanical properties, increase with the degree of crosslinking.

THICK MOULDING COMPOUND (TMC)

A sheet moulding compound (q.v.) of thickness greater than the normal upper thickness limit for such compounds, i.e. above 25 mm (see ISO 8604:1988). In TMC the requisite high viscosity is achieved by the use of chemical thickeners.

THICKENER (THICKENING AGENT)

An additive incorporated in a liquid polymeric system to increase viscosity[1.5]. With most thickeners the level of addition is low (from fractional percentage to a few percent). In the rubber context, the standard definitions are concerned with rubber latex[1.2,1.3] or a rubber latex mix[1.2] as the liquid system; one[1.3] also makes the additional point that the use of a thickener should not involve a major change in the total solids content of the latex.

Some thickeners impart thixotropic properties to the liquid system (see *thixotropy*), thus combining simple thickening action with modification of the system's rheological behavior.

The following are a few examples of thickeners and the systems in which they are used:

SYSTEM		THICKENER
Latex		
	Polyacrylate	Cellulose derivatives (e.g. sodium carboxymethyl cellulose, hydroxy-propyl methyl cellulose, methyl cellulose); sodium alginate; casein.
	PVC (polymer and copolymer)	Cellulose derivatives; polyvinyl alcohol; bentonite clay.
PVC pastes		Aluminum stearate; fine-particle silica (fumed or precipitated); bentonite clay; some proprietary calcium compounds (e.g. *Ircogel 900* - Lubrizol Corp., USA).
Unsaturated polyester resins		Magnesium oxide; fine-particle silica; bentonite clay.

THIOUREA-FORMALDEHYDE RESIN

The kind of amino resin (q.v.) that results from polycondensation of thiourea $(H_2N-\underset{\underset{S}{\|}}{C}-NH_2)$ with formaldehyde[1.1,1.5]. In practice it was more usual to find not pure thiourea-formaldehyde resins, but rather urea-formaldehyde resins (q.v.) in which part of the urea was replaced by thiourea. However, such resins are not widely available, either alone or in moulding compounds, inter alia because of regulatory restrictions in several developed countries on thiourea derivatives coming into contact with foodstuffs.

THIXOTROPY

The property of a liquid or semi-liquid material, whereby its apparent viscosity decreases with time under a constant applied shear rate at constant temperature, and reverts to the initial (higher) value after the shearing has been discontinued. That is, a thixotropic material "thins-out" on stirring or other uniform, isothermal agitation, and thickens back to its original state after the agitation has stopped (cf. e.g. ASTM C 904-94). An example of thixotropic behaviour is provided by "non-drip" paint, which will flow or spread on stirring or brushing although it is gel-like when not worked. Certain kinds of PVC paste behave in the same way.

"TIME TO BLACK"

The length of time elapsing before a PVC material undergoing a test for thermal stability develops a black colour indicative of extensive degradation. The blackening is usually preceded by gradual development of progressively darker colour (fairly typically light yellow to yellow, to light brown, to dark brown). Collins et al. (1976) cite evidence for close correspondence between the time to black of a PVC compound being worked and heated in a torque rheometer in the course of a dynamic-stability test, and the time of first onset of a steep increase in the compound's melt viscosity. See also *time to colour* .

"TIME TO COLOUR"

The length of time elapsing before the first appearance of colour (usually yellow or light brown) in a PVC material undergoing a test for thermal stability, in which colour development serves as an index of significant degradation. See also *time to black*.

TORSIONAL STRESS

The shear stress or stress component on the appropriate transverse cross-section, resulting from a twisting action[1.5,1.10].

TOTAL SHRINKAGE (of a plastics moulding)

See under *moulding shrinkage.*

TOUGHENED POLYSTYRENE

Polystyrene whose impact resistance (toughness) has been improved by incorporation of rubber in a minor proportion[1.1]. See also *styrene-rubber plastic* and *polystyrene plastics.*

TOUGHNESS

The capacity of a material for absorbing a relatively high amount of mechanical energy without failure. In practical contexts, toughness is often equated with ability to withstand impacts without brittle failure, and determined in terms of impact resistance (q.v.). In this connection it is noteworthy that one of the terminology standards[1.1] defines a *toughening agent* (a term it treats as the equivalent of "impact modifier") as a "polymeric material for blending in rigid thermoplastics to increase impact strength". In the above sense toughness may thus be regarded as the opposite of brittleness[1.1,1.5]. In an appropriate mechanical test, a tough plastics material is characterised by a high strain at break and a large area under the stress-strain curve[1.1,1.5], as well as by low notch-sensitivity (q.v.). Cf. also brief definitions in ASTM C 168-90 and ASTM D 123-96.

TRANSITIONAL PORE (in activated carbon)

Conventionally defined as a pore having a diameter in the range 4nm - 500nm (cf. ASTM D 2652-76 (1987)).

TRANSLUCENCY

According to the ISO plastics vocabulary standard[1.5], this is defined in essence as the property of a material whereby the majority of visible light transmitted through it is scattered, so that it is difficult or even impossible to distinguish objects viewed through a translucent material. However, the rather restrictive nature of such a definition would not appear to be entirely justified: it is a fact that there are varying degrees of translucency, and the scattering (diffusion) of light occurring in only slightly translucent materials, or in thin layers of relatively more highly translucent ones, may be low enough to enable objects viewed through such materials to be seen reasonably clearly, especially if they are in close surface contact with the material concerned. This fact is properly accommodated by the wider ambit of another standard definition - that of ASTM E 284:89: according to the tenor of this, translucency is the property of a material ("specimen" is the word actually used in the standard) whereby the light passing through it is being "transmitted diffusely" i.e. with some scattering, to the extent that an object behind the material and not in contact with it is not seen clearly. See also *haze in transmission, transmittance,* and *clarity*.

TRANSMITTANCE

(1) Thermal transmittance. The transmission of heat in unit time through a unit area of a material or object (including any air films at the relevant boundaries) actuated by a unit difference between the temperatures at the two sides concerned (cf. ASTM C 168-90).

(2) Light transmittance (luminous transmittance; total transmission factor). In general, the fraction of incident light which passes through a material or object without being reflected or absorbed (cf. e.g. ASTM C 859-92b). In more formal terms, light transmittance may be defined by the expression

$$\tau = \phi_r / \phi$$

where τ is the light transmittance;

ϕ_r is the total luminous flux transmitted through a test specimen in defined conditions;

ϕ is the incident luminous flux, concentrated in a parallel beam perpendicular to the test specimen.

See e.g. BS 4618:Section 5.3:1972, and Ref 1.14. Luminous flux is defined as the rate of flow of light energy[1.14]. ASTM D 1003-92 provides a method for determining luminous transmittance of transparent plastics.

TRANSPARENCY

In simple terms, transparency may be defined as the ability of a material to transmit light readily with negligible scattering[1.5], so that objects viewed through the material (in a geometrically suitable form - e.g. a thin sheet) can be seen clearly and distinctly[1.5] (cf. also ASTM E 284, ASTM E 349). A test method for the determination of transparency of plastics sheeting (in terms of specular transmittance, q.v.) is given in ASTM D 1746. In the more rigorous, scientific sense, transparency is a composite concept combining the three properties of good light transmission (high regular transmittance - q.v., low haze - q.v. and high clarity - q.v.). Cf. BS 4618:Section 5.3:1972.

TRANSVERSE STRENGTH

See *flexural strength*.

TRIMER

An oligomer (q.v.) molecule consisting of three monomeric units of the same species[1.1,1.5], or the substance composed of such molecules. Both the ISO and BS terminology standards[1.1,1.5] point out that a trimer may result either from appropriate oligomerisation (i.e. trimerisation), or from scission of a longer molecular chain.

TRIPLE-SKIN SHEET

The term favoured by ISO (cf. e.g. ISO 12017) for the product sometimes also referred to a triple-walled sheet or triple-web panel (German "Stegdreifach-platte"). This kind of extruded plastics sheet consists of three mutually parallel layers (two external, and an internal skin), connected and spaced by internal ribs in a manner similar to that found in double-skin sheets (q.v.). Triple-skin sheets are commonly produced from the same thermoplastics as the double-skin sheets.

TRUE STRAIN

In a body subjected to axial force, the natural logarithm of the ratio of the gauge length of the moment of observation to the original gauge length[1.10].

TRUE STRESS

Stress calculated on the basis of the actual cross-sectional area (of a specimen or other object) at the moment of measurement (the supporting area[1.1,1.5]) rather than the original cross-sectional area[1.1,1.5,1.10]. In a tension or compression test, true stress is the axial stress so calculated[1.10].

TUBE

In the most common general senses of the term, a tube is a relatively long cylindrical shell (hollow cylinder[1.5]), or a collapsible cylindrical container[1.5] (squeeze tube) closed by a cap, used for toothpaste, creams, gels, ointments, and the like; many kinds of such containers are made of plastics materials. The BS terminology standard[1.1] also defines *rolled laminated tube* as a tube produced by winding onto a former - under tension, heat and pressure - paper or fabric coated or impregnated with a synthetic resin.

TUBING

In plastics terminology, a flexible tube[1.5] (as distinct from a pipe, which is rigid or semi-rigid), used for conveying liquids or gases (occasionally also particulate solids) typically to, from, and in equipment[1.5] (cf. e.g. PVC tubing in medical transfusion sets, PE tubing for conveying milk in milking equipment; PA tubing for compressed air in pneumatic brake systems, etc.).

TUBULAR FILM and TUBULAR FILM PROCESS

See *lay-flat film*.

ULTRA-HIGH-MOLECULAR-WEIGHT POLYETHYLENE (PE-UHMW or UHMWPE)

Substantially linear polymers and copolymers of polyethylene (the copolymers essentially with less than 50% by weight of 1-olefin monomer(s) - BS ISO 11542-1:1994) with nominal weight-average molecular weight in the range of 1 million to about 5 million, and density typically in the range 920 kg m^{-3} to 940 kg m^{-3}. The density range is lower than that of ordinary high-density polyethylene (PE-HD) because the very large molecular size of PE-UHMW hinders crystallisation. The same factor is responsible for the fact that typical values for the rigidity and yield strength of PE-UHMW are slightly below those of PE-HD. At the same time, the ultra-high molecular weight of the former polymer makes for greater toughness (impact resistance), and greater resistance to creep and environmental stress cracking in comparison with PE-HD. Because its molecular weight is so high, resulting also in very high solution and melt viscosities, the standard ways of determining and expressing the molecular weight of PE-UHMW moulding and extrusion materials are either in terms of the relative viscosity (q.v.) of a hot, very dilute solution (in decahydronaphthalene - cf. ASTM D 4020-92), or in terms of a "flow value" (ISO 11542-2). The latter is defined in the ISO standard as the tensile stress which - in prescribed test conditions - will increase by a specified factor the length of a standard test specimen. PE-UHMW materials for use in surgical implants are covered in ISO 5834:Parts 1 and 2:1985.

Commercial versions of PE-UHMW include *Hostalen GUR, Lupolen UHM, NCPE 22, Stamylan UH, Tivar* and *Ultrac.*

"ULTRA-MICROCELLULAR" FOAM

See under *"supermicrocellular" foam.*

UNIAXIALLY ORIENTED FILM

Film which in the course of its production has been deliberately stretched in one of the two principal directions (machine direction or transverse direction). Cf. also OPP-U in ASTM D 2673-94 (a standard specification for oriented polypropylene film).

UNSATURATED POLYESTER (UP); UNSATURATED POLYESTER RESIN

A polyester (q.v.) crosslinkable by reaction with an unsaturated monomer (e.g. styrene) or prepolymer[1.1,1.5], by virtue of the presence of carbon-to-carbon double bonds in its polymer chain (cf. also ISO 4618-3:1984).

UP resin systems incorporating various additives, including catalysts (also termed hardeners or curing agents), activators, accelerators, inhibitors, fillers, flame retardants, and others, find their widest use in the manufacture of a great variety of products from thermoset composites (reinforced laminates), in which the resin system is combined with a fibrous reinforcement. Some examples are: tanks (including petrol-storage tanks and the containers of bulk-transport tankers for various liquids), pipes for conveyance of water and liquid chemicals, chemical storage vessels, boat bodies, lamp posts, vehicle panels, swimming-pool linings, and many others. Some composite products are manufactured from fibre-reinforced UP resin moulding compounds, including dough moulding compounds (DMC), bulk moulding compounds (BMC), sheet moulding compounds (SMC); and moulding compounds in pellet (granular) form (PMC; GMC). UP resins and/or compounds are available in the following commercial ranges: Alpolit; *Arapol; Beetle; Chempol; Crystic; Dielectrite; Distitron; Durapol; Durez; Durodet; Esteform; Fiberite; Flomat; Freeflow; Haysite; Hetron; Hi-Flex; Illandur; Insultruc; Keripol; Legupren; Leguval; Menzolit; Oldapal; Palatal; Plenco; Polyrite; Premi-Glas; Premi-Jet; Resartherm; Resotherm; Resipol; Rigolac; Rosite; Rovilux; Selectron; Setapol; Shimoco; Synolit; Varimat; Vestopal; Uromix.*

Unreinforced UP moulding and casting systems are also used in some applications, in particular those in which the good electrical properties and heat resistance of the products are beneficial (as e.g. in the potting of electronic and electrical components). Non-plastics applications of UP resins include their use as the polymeric binders in some surface coatings.

The following standards are concerned with UP resins and compounds used in plastics.

ISO 3672, Parts 1 and 2 : basic standard for the designation and determination of properties of UP resins. Inter alia refers to ISO standards for the determination of reactivity at 80°C (ISO 584:1982); gelation time at 25°C (ISO 2535:1974); acid value (ISO 2114:1974); hydroxyl value (ISO 2554:1974); shrinkage of casting resins (ISO 3521:1976); chlorine content (ISO 4615:1975); and methods of other ISO standards not exclusive to UP resins, but relevant to them.

BS 3532:1990(1995): Basic standard for specifying UP resin systems. Like ISO 3672, this standard refers to a number of relevant (British standard) methods for the determination of properties of UP resins, inter alia acid value (identical with ISO 2114); hydroxyl value (identical with ISO 2554); gelation time - manual method (BS 2782:Method 835B:1980(1986)); and gel-timer method (BS 2782:Method 835C:1980(1986); and some relevant general methods of other BS standards.

BS 5734:1990: Basic standard (in five parts) for UP moulding compounds (DMC and SMC) for electrical and other purposes, including relevant test methods.

ASTM D 1201-81 (1992): Standard specification for UP moulding compounds (DMC; SMC and GMC).

ASTM PS 15-94: Provisional standard specification for thermosetting moulding compounds, including UP compounds inter alia.

UREA-FORMALDEHYDE (UF) RESIN

The kind of amino resin (q.v.) that results from polycondensation of urea with formaldehyde[1.1,1.5]. The ISO and ASTM plastics terminology standards also define *urea plastic(s)*. The ASTM standard[1.7] confines its definition to plastics based on resins which are condensation products of urea and aldehydes, whilst the ISO standard[1.5] defines urea plastic (in the singular) as one based on those amino resins (in the plural) in which urea is the component predominant (by mass) among the amine or amide reactants used in the production of the resins.

UF resin precondensates, capable of further curing (crosslinking) serve as bases of UF moulding compounds used mainly to produce white or light-coloured thermoset mouldings. Non-plastics uses of the resins include their application in certain types of water-based adhesives, in surface finishes for wood, and as non-crease finishes for cellulosic textiles.

The moulding compounds contain curing catalysts for the resin, and fillers which often serve also as reinforcements (fibrous cellulose is a common example of this dual function). Other additives may include flow promoters, plasticisers, white pigments, and colorants.

Among the useful characteristics of UF mouldings (enhanced in some cases by appropriate fillers) are generally good electrical and mechanical properties, hardness, heat resistance, high deflection temperature under load, and relatively low flammability. Limitations include susceptibility to boiling water, many oxidising and reducing agents, and strong acids and alkalis. Moulding shrinkage is comparatively high, as is water absorption, and weathering resistance is not outstanding. The main applications of UF moulding compounds are in the production of electrical goods, such as plugs, sockets, lights mountings, switch boxes, and the like. Trade names under which the compounds are available include *Bakelite; Beetle; Carbaicar; Legupren; Scarab; Urochem;* and *Uroplast* (see also Section 1 of Part 3 of this book).

There are few major standards dealing directly with UF resins or plastics. Most of those cited here relate to them as thermosetting or aminoplastic materials.

ASTM D 705-94	Standard specification for UF moulding compounds.
ASTM D 1490-93	Determination of non-volatile content of UF resin solutions.
BS 5617:1985 (1996)	UF foam - Cavity wall insulation.
BS 5618:1985 (1996)	UF foam - Cavity wall insulation - Code of practice.
BS 1322:1992	Aminoplasts - Moulding materials.
ISO 295:1991	Thermosetting plastics materials - Compression moulding of test specimens.
ISO 2112:1990	Aminoplastic moulding materials - Specification
ISO 2577:1984	Thermosetting moulding materials - Determination of shrinkage.
ISO 3671:1976	Aminoplastic moulding materials - Determination of volatile matter.
ISO 7808:1992	Thermosetting moulding materials - Determination of transfer flow.
ISO 10724:1993	Thermosetting moulding materials - Injection moulding of multipurpose test specimens.
ISO 11248:1993	Thermosetting moulding materials - Evaluation of short-term performance at elevated temperatures.
ISO 11402:1993	Condensation resins - Determination of free formaldehyde.

UREA PLASTIC(S)

See under *urea-formaldehyde resin*.

URETHANE PLASTICS

Plastics whose base polymers are polyurethanes - q.v. (i.e. polymers with molecular chains made up of repeat units of the urethane type[1.1.1.5.1.7]) or copolymers with molecular chains containing repeated structural units of both the urethane and some other type(s)[1.1.1.5.1.7].

Urethane plastics form a large, diversified group comprising materials and products ranging from elastomeric and non-elastomeric thermoplastics, through elastomeric or rigid solid thermosets, to flexible or rigid thermoset foams.

The following commercial ranges are representative of thermosetting reaction systems for urethane plastics, their isocyanate components (for polyol components see under polyol), and thermoplastic urethane materials.

Isocyanates (simple and/or prepolymer) : *Hypol; Lupranate; Luxate; Mondur; Multrathene; Tedimon.*

Thermosetting urethane systems : *Adiprene: Autofroth* (foam systems); *Autopak* (foam systems); *Baydur* (RIM systems); *Bayfit* (foam systems); *Bayflex* (elastomeric RIM systems); *Baygal; Baymidur; Baytec; Baytex; Chempol* (RIM systems); *Coro-Foam* (foam systems); *Cyanaprene* (elastomer and foam systems); *Diprane; Elastopan; Elastopan Aqua* (system for water-blown foam); *Hyperlast; Isostac; Maxcels; Maxflex, Maxpak* (three foaming systems); *Mistrabond* (foam systems); *Rimline* (RIM systems); *Spectrim* (RIM systems); *Stepanfoam* (foam systems); *Tediflex; Tedilast; Tedipur; Tedirim; Teditherm; Urestyl* (rigid foam system); *Vibrathane* (casting systems); *Voralast* (RIM systems); *Vulkollan*.

Thermoplastic urethane materials

Elastomeric: *Elastollan; Estaloc; Estane; Lastane; Morthane; Oldopren; Pellethane; Polypur; Rimplast; Tecoflex; Tecophilic; Tecoplast; Texin; Urepan.*

Miscellaneous : *Caprolan; Desmopan; Isoplast.*

The following standards are relevant to various aspects of urethane plastics.

Urethane foams

Flexible foams and foam products

Foam specifications : ISO 5999:1982 (foam for load-bearing applications - except carpet underlay); ISO 6915:1991 (foam for laminate use); BS 3379:1991 (cellular materials for load-bearing applications - this standard differs from ISO 5999); BS 4021:1971 (1984) (foam sheeting for use in laminates); ASTM D 3453-91 (materials for furniture, automotive cashioning, bedding, and the like); ASTM D 3490-80 (1990) (bonded foam); ASTM D 3770-91 (high-resilience foam); ASTM D 3851-84 (1990) (microcellular soling materials).

Test method standards (see also under cellular plastic) : ASTM D 3489-85 (1990) (test methods for microcellular urethane); ASTM D 3574-95 (test methods for slab, bonded, and moulded foams); ASTM D 3768-85 (1990) (test for flexural recovery); ASTM D 3769-85 (1990) (test for high-temperature sag); ASTM D 5308-92 (test methods for semi flexible cellular urethanes for automotive and similar energy-absorption applications).

Products : BS 5223:Part 2:1992 (mattress cores); BS 5223:Part 3:1976 (1991) (pillows); BS 5340:1976 (1991) (pillows for domestic use).

Rigid foams and foam products (see also ASTM Book 09.01 and under cellular plastic)

Sprayed or dispensed foam applications : ISO 8873:1987 (spray-applied foam for thermal insulation of buildings); BS 5241:Part 1:1994 (foam insulation sprayed on a construction site); BS 5241:Part 2:1991 (1996) (specification for dispensed foam for thermal insulation or buoyancy applications); BS 7456:1991(1996) (code of practice for wall cavity filling with urethane foam systems - BS 7457 provides a specification for suitable foam systems); BS 6586:Part 1:1993 (foam for insulation panels for transport containers, produced by press-injection method); BS 7021:1989 (1996) (code of practice for insulation of roofs with sprayed PUR and PIR foams).

Preformed foam applications : BS 5608:1993 (preformed PUR and PIR foams for thermal insulation of pipework and equipment); BS 4841 (PUR and PIR foam for building applications - Part 1:1993, laminated board for general purposes; Part 2:1975, laminated board for wall and ceiling insulation; Part 3:1994, laminated roof boards).

Non-foam applications

Electrical insulation : BS 5664 (polymerisable resinous compounds for electrical insulation - Part 3 : Section 3.3:1995, unfilled PUR compounds; Part 3:Section 3.4:1995, filled PUR coumpounds); BS EN 60317 : various parts and sections (PUR - enamelled copper wires).

PUR-coated fabrics : ISO 7617-3:1988 (PUR-coated woven fabrics for upholstery); ISO 8096-2:1989 (fabrics coated with PUR and silicone elastomers); BS 5223:Part 4:1976 (1996) (mattress covers of PUR-coated nylon); BS 5455:1977 (1996) (specification for PUR-coated nylon fabrics for mattress covers).

PUR footwear : ISO 5423:1992 (moulded PUR boots for industrial use); ISO 6910:1992 (moulded PUR boots resistant to animal fats and vegetable oils).

VICAT SOFTENING TEMPERATURE

The temperature at which, in standard test conditions, a flat-ended indentor of specified dimensions and bearing a specified lod penetrates a prescribed thermoplastic test specimen to a specified depth. The following standards give methods of determination of Vicat softening temperature. ISO 306:1987 (identical with BS 2782:Part 1:Methods 120A, B, D, and E:1990); BS 2782:Part 1:Method 120C (determination of the so-called "1/10 Vicat softening point", in which the prescribed depth of penetration is one-tenth of that given in ISO 306; the method is envisaged for use on PMMA sheet); ASTM D 1525-95. It should be noted that, as the standard test methods allow two possible loads on the indentor viz. either 10N or 50N, the load used should always be stated when a Vicat softening-temperature value is being quoted.

VINYL CHLORIDE PLASTIC

A plastic based on polymers of vinyl chloride or copolymers of vinyl chloride with other monomers, the vinyl chloride being in the greatest amount by mass.

Reproduced, with permission, from ISO 472:1988.

The same definition appears in the BS and ASTM terminology standards[1.1,1.7]. In informal, but widely used terminology, vinyl chloride plastics are generally styled "PVC plastics".

See also *paste (1)*, *poly(vinyl chloride) plastic*, *PVC resin*, *PVC plastic*, and *vinyl resin*.

Some commercial vinyl chloride (PVC) materials (compounds and/or resins) are available under the following trade names (cf. also Section 1 of Part 3 of this book).

Advex; Artevyl; Benvic; Corvic; Decelith; Duflex; Dural; Egeplast; Episol; Epivyl; Evicom; Evipol; Fiberloc; Garaprene; Gary PVC; Geloy; Geon; Helm; Hy-Vin; Ixan; Kohinor; Lacovyl; Lucolene; Lucorex; Lucovyl; Lutofan; Marvylex; Mistique; Nakan; Nipolit T; Norvinyl; Novablend; Novacycle (recycled materials); Oxyblend; Oxyclear; Pantalast; Petkim; Pevicon; Plastilit; Polytron; Scovinyl; Solvic; Sunprene; Trivin; Tuf-Stif; Vampstyr; Varlan; Vestolit; Vicom; Vinakon; Vinidur; Vinnol; Vinoflex; Vinychlon; Vipla; Vista; Vistel, Welvic.

Noteworthy among the commercial PVC plastics compounds are chemically crosslinkable grades (cf. e.g. *Norvinex*), electrically conductive compounds (cf. e.g. *PVC-Semicon*), and compounds incorporating glass-fibre reinforcement, in which adhesion between the reinforcement and the polymer matrix is promoted by special coupling agents (cf. e.g. *Fiberloc*).

In commercial PVC compounds in which the PVC resin is blended with other polymers, the polymeric modifier is typically incorporated to increase toughness (i.e. to act as impact modifier) or to modify melt properties. Many polymeric modifiers tend to reduce the hardness, modulus and strength of the PVC material, and to lower its softening point and deflection temperature under load (some modified styrene/maleic anhydride copolymers - e.g. the *Elix* polymeric modifiers - can have the opposite effect).

Polymeric modifiers incorporated in PVC plastics as toughening agents include ABS and MBS terpolymers, acrylonitrile/butadiene copolymers (nitrile rubber), EVA and graft-modified EVA (e.g. *Elvaloy*), all-acrylic and modified acrylic modifiers, some grades of chlorinated polyethylene, and certain polyurethane elastomers. Polymeric modifiers which primarily affect melt properties (often called processing aids) are mainly acrylic polymers.

In PVC plastics, the base polymers are always associated with additives, some of which may even be present in predominant proportions (e.g. the plasticisers in certain soft PVC materials). Indeed, among plastics, PVC is remarkable for the great variety and range of properties realised through planned formulation of individual compositions for particular purposes (cf. e.g. Wickson, 1993, and/or Titow, 1990). One small illustrative example is the contrast between the soft, flexible PVC coating on an upholstery fabric, and the rigid, hard material of a PVC pipe. This scope for wide variation of material properties through appropriate formulation, which is the basis of PVC's great applicational versatility, is unmatched by any other plastic. Because of this versatility - responsible inter alia for the fact that world-wide consumption of PVC materials is second only to that of the polyolefins - it is not practicable adequately to summarise in a short passage all the varied applications of these materials. By way of a general indication it may be mentioned that PVC pipes and conduit, profiles and cladding, and film and sheeting account for the highest rigid-product tonnages, whilst cable coverings, and flexible film and sheeting are the highest-tonnage outlets for plasticised PVC.

Many standards dealing with PVC building and related materials products (including pipes) are assembled in Annual Book 08.04 of ASTM standards. Standards concerned with PVC pipes and fittings are featured in ISO Standards Handbook 28:1986 (Pipes, fittings and valves - Plastics products). Other standards relevant to vinyl chloride plastics materials and products include the following (see also *PVC resin*).

Standard material specifications

1. *Rigid materials*

ISO 1163	Unplasticised PVC moulding and extrusion materials. Part 1:1995 Designation system and basis for specification. Part 2:1995 Preparation of test specimens and determination of properties.
ASTM D 1784-95	Specification for rigid PVC and CPVC compounds.
ASTM D 3010-71 (1981)	Preparation of compression-moulded test plaques from rigid PVC compounds.
ASTM D 3915-92	PVC and related compounds for plastics pipe and fittings (pressure applications).
ASTM D 4216-92	Rigid PVC and related compounds (including CPVC) for plastics building products.

ASTM D 4396-92	Rigid PVC and related compounds for non-pressure piping products.
ASTM D 5260-92	Standard classification for chemical resistance of PVC and CPVC compounds.

2. *Flexible materials*

ISO 2898	Plasticised compounds of vinyl chloride homopolymers and copolymers. Part 1:1986 Designation. Part 2:1989 Preparation of test specimens and determination of properties.
BS 2571:1990	Flexible PVC compounds.
ASTM D 2287-92	Specification for non-rigid vinyl chloride polymer and copolymer moulding and extrusion compounds.
ASTM D 1565-81 (1990)	Specification for flexible, open-cell foam based on vinyl chloride polymers and copolymers.

Temperature effects

1. *Physical*

BS 2782:Part 1	
Method 122 A:1976 (1993)	Determination of deformation under heat of flexible PVC compound.
Method 131 B:1983 (1989)	Determination of extensibility of flexible PVC sheet after heat ageing.
Method 150 B:1976 (1988)	Determination of cold-flex temperature of flexible PVC compound (Clash and Berg test).
Method 150 C:1983 (1989)	Determination of low-temperature extensibility of flexible PVC sheet.
Method 151 A:1984 (1991)	Determination of cold-bend temperature of flexible PVC extrusion compound.
Method 153 B:1991 (identical with ISO 458-2:1985)	Determination of stiffness in torsion (at different temperatures) of flexible vinyl chloride polymer compounds.
Method 641 A:1983 (1991)	Determination of dimensional stability of flexible poly(vinyl chloride) sheet at 100°C.

2. *Thermal stability*

ISO 182	Thermal stability of PVC compounds and products.
Part 1:1990 (identical with Method 130 A: 1991 of BS 2782:Part 1)	Congo Red method.

Part 2:1990 (identical with Method 130 B: 1991 of BS 2782:Part 1)	pH method.
Part 3:1993	Conductometric method.
Part 4:1993	Potentiometric method.
ISO 305:1990	Determination of thermal stability of PVC polymers and copolymers and their compounds, discoloration method.
ASTM D 793	Test for short-time thermal stability of plastics containing chlorine. Standard discontinued in 1983.
ASTM D 2115-92	Oven-heat stability of PVC compositions.

Miscellaneous tests and analysis

1. *Rheology and fusion*

ASTM D 2538-94	Fusion test for PVC compounds in a torque rheometer.
ASTM D 3364-94	Flow rates of PVC and rheologically unstable thermoplastics in an extrusion plastometer.

2. *Analysis*

ASTM D 2124-70 (1994)	Analysis of components of PVC compounds.
ASTM D 3421-76	Extraction and analysis of plasticiser mixtures from vinyl chloride plastics. Standard discontinued in 1987.

See also under *composition*.

3. *Other properties and tests*

BS 2782	
Part 2:Method 241A:1984 (1992)	Determination of the effect of PVC compound on electrical loss tangent of polyethylene.
Part 11:Method 1110:1989	Tensile properties of dumbbell specimens from PVC gutter profiles or pipes for non-pressure applications.
ASTM D 2151-95	Test for staining of PVC compositions by rubber compounding ingredients.
ASTM D 3351-93	Gel count method for PVC film.
ASTM D 5226-92	Standard practice for dissolving polymer materials (solvent data for PVC in the Annex, Section A1.2.63).

VINYLIDENE CHLORIDE PLASTICS

Plastics whose base polymers are homopolymers of vinylidene chloride, or copolymers in which vinylidene chloride is the major monomeric component, predominant by mass[1.1,1.5,1.7].

Virtually all commercial vinylidene chloride plastics, including those sometimes referred to as poly(vinylidene chloride) plastics, are based on vinylidene chloride copolymers containing significant proportions of vinyl chloride comonomer (see poly(vinylidene chloride) and poly(vinylidene chloride) plastics). Products include pipes, mouldings, films and barrier coatings. Filaments and fibres are also produced (e.g. *Saran*).

Two relevant standards are: BS 7291-4:1990 (1995) - specification for PVDC pipes and fittings for hot and cold water, and for applicable solvent cement; ASTM D 729-95 - specification for vinylidene chloride/vinyl chloride copolymer moulding compounds.

VINYL ESTER RESINS

A group of thermosetting resins, crosslinkable through the reactive vinyl groups of their molecular chains. They are sometimes referred to as "epoxy vinyl ester resins" because a typical vinyl ester resin is produced by esterification of an epoxy resin (via its terminal epoxy groups) with an unsaturated acid, usually methacrylic acid[1.4], or sometimes acrylic acid, whose double bonds confer polymerisability[1.4] upon the resulting vinyl ester resin. The resins are dissolved in a reactive monomer (normally styrene)[1.4]. In this feature, and in the kind of catalysts used for their curing, they are similar to unsaturated polyester resins. However, in the cured state, vinyl ester resins are superior to the unsaturated polyesters in strength, toughness, thermal stability, and resistance to water and chemicals at room and elevated temperatures. It is these properties that make glass-fibre-reinforced vinyl ester resin laminates the materials of choice for the construction of various kinds of chemical plant. Cf. also ASTM C 904-95.

The trade names under which vinyl ester resins and/or their compounds are available include *Atlac; Derakane; Dion; Fiberite VE;* and *Varimat.*

VINYL RESIN

According to the formal standard definition, a resin resulting from polymerisation of monomers containing the vinyl group[1.1,1.5] (CH_2=CH–). This definition thus applies to each of the polymers within a large group containing, for example, such diverse materials as polystyrene, poly(vinyl fluoride), poly(vinyl pyrrolidone), and poly(vinyl chloride), among others. Whilst the definition correctly reflects an important feature of the polymers within its ambit, it is relevant to note that in the vocabulary of the plastics industry, as well as in wide general non-technical usage, the term "vinyl" (as a noun or an adjective) is a virtual synonym for "PVC" in that abbreviation's widest popular sense, viz. of, or appertaining to, the family of polymers or copolymers of vinyl chloride or their derivative plastics. This is illustrated by the practical equivalence of such phrases as "vinyl upholstery" and "PVC upholstery", "vinyl flooring" and "PVC flooring", or "vinyl foam" and "PVC foam". Conversely, no polymer technologist would normally refer to expanded polystyrene as "cellular vinyl" or "vinyl foam", or to, say, a hair lacquer based on poly(vinyl pyrrolidone) as a "vinyl preparation".
See also *PVC resin* and *vinyl chloride plastic*.

VIRGIN MATERIAL[1.1]; VIRGIN PLASTIC[1.5,1.7]

A plastics material, in a form (e.g. pellets[1.5,1.7], prills, powder[1.5,1.7], flakes, floc[1.5,1.7], liquid[1.7], etc.) suitable for use as feedstock in the relevant processing into products, but that has not yet been used or processed[1.5,1.7] (other than in the course of its own production[1.5,1.7]).

NOTE - The definition in the BS terminology standard[1.1], additionally refers to the material's composition as "polymer and appropriate ingredients" with no reprocessed material, but allows for inclusion of a proportion (to be specified) of "arisings" from the manufacturing process.

"VISBREAKING"

A colloquial term occasionally used for the action of some additives which modify the melt flow of a polymer in processing. An example is the modification of the melt viscosity of polypropylene by certain peroxides.

VISCOSITY

The property of a substance manifested as resistance to steady flow if the substance is a fluid, and - in a solid - as resistance to gradual yielding to a deforming force. Thus in its effect viscosity may be likened to internal friction. For a fluid, a unit of viscosity can be defined as unit tangential force acting per unit area per unit velocity gradient between two adjacent layers of the fluid. Viscosity defined in this way is termed *dynamic viscosity* (symbol η).

NOTE - In the SI system dynamic viscosity has the units $Pa.s$ (i.e. $N.m^{-2}.s$: the force of one newton per area of one square metre, per velocity gradient of one metre per second per metre, i.e.

$$\frac{N}{m^2} \Big/ \left(\frac{m}{s} \cdot \frac{1}{m} \right) = N.m^{-2}.s)$$

In the cgs system the unit of dynamic viscosity is the *poise* (the force of one dyne per area of one centimetre, per velocity gradient of one centimetre per second per centimetre, which reduces to $g.cm^{-1}.s^{-1}$). The usual symbol for the poise is *P*.

The quotient of the dynamic viscosity of a fluid by its density is called *kinematic viscosity* (SI units : $m^2.s^{-1}$). The cgs system unit of kinematic viscosity is the *stoke* (usual symbol *St*; units $cm^2.s^{-1}$).

In viscosity tests on liquid polymeric systems (solutions, dispersions, melts) the viscosity is defined, and determined, as the ratio of the shearing stress (τ) to the rate of shear (γ)[1.1,1.5,1.7]. In that context, the shearing stress required to produce a unit velocity flow gradient in the liquid is termed the *viscosity coefficient* (with the dynamic viscosity units of $N.m^{-2}.s$ in the SI system and *poises* in the cgs system).

Liquids in which the shearing stress is directly proportional to the shear rate ($\tau = \gamma \times a$ constant) are called "Newtonian". For such liquids, the proportionality constant, i.e. the ratio of shearing stress to the shear rate, is often termed "Newtonian viscosity". That is, for "Newtonian" liquids

$$\tau / \gamma = \eta$$

where η is the "Newtonian viscosity", and the plot of τ versus γ (often termed the flow curve) is a straight line of slope η, passing through the origin. Liquids for which the shear stress is some other, more complex, function of the shearing rate (i.e. whose viscosity is not constant, but varies with the shearing stress) are called "non-Newtonian". For such liquids, viscosity as the ratio of τ to γ can only be validly quoted for a particular pair of values of these variables (a particular point on the plot of τ versus γ), and it is then termed "apparent viscosity"[1.1,1.5,1.7]. Similarly, the corresponding viscosity coefficient is an "apparent" value[1.1,1.5,1.7].

The viscosity of a dilute solution of a polymer in a suitable solvent can, under certain assumptions, be expressed as a function of the polymer's average molecular weight (relative molecular mass). A general expression of the relationship is the Mark-Houwink equation (also called Mark-Houwink-Sakurada equation[1.5], M–H–S[1.5] equation, or Staudinger-Kühn equation)

$$[\eta] = K(\overline{M}_r)^{\alpha}$$

where K and α are constants for a given polymer/solvent system at a given temperature, $\overline{M}_r$ is the relevant molecular weight average (sometimes referred to as *viscosity-average* molecular weight) and $[\eta]$ is the limiting viscosity number (intrinsic viscosity) of the polymer.

The expression provides a link between the values commonly determined in practice in viscosity measurements on dilute solutions of polymers, and the average molecular weights of the solutes. The values, and the relevant terms, are shown in the table below. See also *K value*. The following standards deal with such determinations.

ISO 1628	Methods of determination of viscosity number and limiting viscosity number of polymers in dilute solution
Part 1:1984	General conditions.
(Identical with Method 732 A:1991 of BS 2782)	
Part 2:1988	PVC resins. (This part replaces the former ISO 174:1974).
(Identical with Method 732 B:1991 of BS 2782)	
Part 3:1991	Polyethylenes and polypropylenes.
(Identical with Method 732 C:1991 of BS 2782)	
Part 4:1986	Polycarbonate moulding and extrusion materials.
(Identical with Method 732 D:1991 of BS 2782)	
Part 5:1986	Poly(alkylene terephthalates). (This part undergoing revision in 1997).
(Identical with Method 732 E:1991 of BS 2782)	
Part 6:1990	Methyl methacrylate polymers.
(Identical with Method 732 F:1991 of BS 2782)	
ISO 307:1984	Determination of viscosity number of polyamides.
(Identical with Method 733 A:1991 of BS 2782)	
ISO 1157:1990	Determination of viscosity number and viscosity ratio of cellulose acetates.
(Identical with Method 733 C:1991 of BS 2782)	
BS 6306:Parts 1 and 2:1982 (1992)	Determination of limiting viscosity number of cellulose in dilute solutions.
(Technically equivalent to ISO 5351: Parts 1 and 2:1981)	

ASTM D 2857-93	Standard practice for determination of dilute-solution viscosity of polymers.
ASTM D 1601-86 (1991)	Test method for dilute-solution viscosity of ethylene polymers.
ASTM D 789-94 (Section 9)	Test method for relative viscosity of polyamides.
ASTM D 1243-95	Test method for dilute-solution viscosity of vinyl chloride polymers.
ASTM D 3591-77 (1991)	Determination of logarithmic viscosity number of polyvinyl chloride in formulated compounds.

Other standards generally relevant to viscosity determinations include the following.

ISO 2555:1989	Determination of apparent viscosity of resins by the Brookfield test method.
(Identical with Method 730C:1992 of BS 2782)	
BS EN ISO 3219:1995	Determination, with a rotational viscometer, of viscosity of polymer emulsions or dispersions.
BS EN ISO 9371:1995	Determination of viscosity of phenolic resins in the liquid state or in solution.
(Identical with Methods 734 A to D:1995 of BS 2782)	

VOLATILE MATTER (VOLATILE MATERIAL; VOLATILE MATERIAL CONTENT; VOLATILE CONTENT; VOLATILES; VOLATILE LOSS)

The amount, commonly expressed as percentage by weight, of gaseous and vaporised matter given off, and lost, by a material in specified test conditions. This amount is thus dependent to a substantial extent upon the particular conditions, especially the temperature and time of heating, and the presence or absence of absorbents (such as activated carbon, which surrounds the solid specimen in some standard test methods). Some test standards do not include water (i.e. the moisture content of the material concerned) in the definition of volatile matter in their particular contexts. A large proportion of the volatile content of PVC plasticised with monomeric plasticisers typically consists of the fraction of plasticiser(s) volatilised in the conditions of the test.

The following standards deal with some methods of determination of volatile matter in various plastics materials:

ISO 176:1976 (identical with Methods 465 A and B:1979 of BS 2782) - Determination of loss of plasticisers (activated carbon method).
ISO 1269:1975 (identical with Method 454 D:1978 (1983) of BS 2782) - Determination of volatile matter (including water) in PVC resins.
ISO 3671:1976 (identical with Method 461 A:1978 (1992) of BS 2782) - Volatile matter in aminoplastics moulding materials.
ISO 8618:1987 (identical with Method 451 M:1991 of BS 2782) - Volatile content of liquid phenolic resins.

TERMS AND EXPRESSIONS OF INTEREST IN CONNECTION WITH DETERMINATIONS OF SOLUTION VISCOSITY OF POLYMERS

TERMS AND SYMBOLS				MATHEMATICAL EXPRESSION(a)(b)	REMARKS
ORIGINAL		RECOMMENDED BY TERMINOLOGY STANDARDS[1.1,1.5] (AFTER IUPAC)			
TERM	SYMBOL	TERM	SYMBOL		
Relative viscosity	η_r or η_{rel}	Viscosity ratio; solution/ solvent viscosity ratio	η_r	η/η_s	
Specific viscosity (Term discouraged by ISO[1.5])	η_{sp}	Relative viscosity increment; viscosity ratio increment	η_i	$(\eta-\eta_s)/\eta_s$	Represents viscosity increase attributable to the dissolved polymer.
Reduced viscosity(c)	RV	Viscosity number	η_i/c	$(\eta-\eta_s)/(\eta_s.c)$	Provides a measure of the capacity of the polymer to increase the viscosity ratio. The ISO terminology standard[1.5] directs that the units of viscosity number must be specified, and recommends them as cm^3/g.
Inherent viscosity(c)	IV; η_{inh}	Logarithmic viscosity number	$\eta_{\ln}$	$(\ln \eta_r)/c$	
Intrinsic viscosity(c) (Sometimes also referred to as *Staudinger index*[1.5])	$[\eta]$	Limiting viscosity number	$[\eta]$	$\lim_{c \to o} \eta_i/c = \lim_{c \to o} \eta_{\ln}$	The reduced and inherent viscosities both tend to the same limit as the concentration of dissolved polymer tends to zero (i.e. at infinite dilution). The value of the limit is different for the same polymer in different solvents because of differences in polymer/solvent interactions.

(a) In practical determinations, the relationships given in this column are valid only when the viscosities concerned are determined under the same standard conditions.

(b) Key to the symbols used in the expressions : η = viscosity of dilute polymer solution (or its time of flow in standard conditions); η_s = viscosity of the solvent alone (or its time of flow in standard conditions); c = mass concentration of the polymer in grams per cubic centimetre[1.5] (or per millilitre[1.1]) of solution.

(c) The ISO terminology standard[1.5] points out that the mathematical expressions of these terms are neither viscosities nor pure numbers, and recommends that the terms be regarded simply as traditional names. The standard also recognises that any attempt at their prescriptive replacement by consistent terminology would produce unnecessary confusion in polymer literature.

BS 2782:Method 107 B:1970	Loss of mass on heating at 100°C phenolic resin in solution.
BS 3424:Part12:Method 14A:1996	Percentage mass loss on heating by PVC-coated upholstery fabrics (the relevant numerical requirement is given in BS 5790:Part 1:1996).
ASTM D 1203-94	Volatile loss from plastics: activated carbon method.
ASTM D 3030-84(1990)	Volatile matter (including water) of vinyl chloride resins.
ASTM D 4526-85(1991)	Determination of volatiles in polymers by headspace gas chromatography.
ASTM D 706-82(1989), D 707-84(1990), D 729-95, D 787-81(1995) and D1562-82(1989)	Specifications for the following moulding and extrusion compounds: cellulose acetate, cellulose acetate butyrate, vinylidene chloride, ethyl cellulose and cellulose propionate, with - in each case - a test for volatile loss (inter alia).

The oxygen complexes found on the surfaces of carbon black particles are also sometimes referred to, collectively, as "volatile content".

VOLUME CONDUCTANCE

The reciprocal of volume resistance (q.v.).

VOLUME CONDUCTIVITY

The reciprocal of volume resistivity (q.v.).

VOLUME RESISTANCE

The resistance between two electrodes set into, or in surface contact with, an appropriate test specimen. This resistance is given by the quotient resulting when the direct voltage applied to the electrodes is divided by that part of the current which flows through the volume of the specimen[1.1,1.5] (i.e. exclusive of any current flowing along the surface[1.5]). Cf. also IEC Publication 93-1980 and ASTM D 1711-94. Volume resistance is expressed in ohms. For methods of determination see under *volume resistivity*.

VOLUME RESISTIVITY

In relation to its measurement in practice by standard methods, the volume resistivity of a plastics material may be regarded as the volume resistance (q.v.) between the opposite sides of a unit-side cube, expressed in ohms multiplied by the appropriate unit of length (e.g. Ω.cm in the case of a centimetre cube[1.1,1.5]). Ohm-centimetre is in fact the usual unit of volume resistivity; ohm-metre is also sometimes used. According to a more formal definition, the volume resistivity of a material is the quotient resulting when the potential gradient (parallel to the direction of the current flow in the material[1.1]) is divided by the current density[1.1,1.5]. The following standards are relevant to the concept and methods of determination of volume resistivity. IEC Publication 93-1980; BS 4618:Section 2.3:1975; BS 2782:Part 2:Method 230A:1982; ASTM D 257-93.

VULCANISED FIBRE

A dense cellulosic material available in the form of sheets, tubes and rods, used in a number of applications, including electrical insulation (where high resistance to water is not of primary importance), containers (originally widely used as the material of suitcases), backing for abrasive discs, and some industrial and consumer products. Vulcanised fibre is produced from absorbent paper of high alpha-cellulose (q.v.) content (and preferably made from rag stock). The paper is gelatinised by immersion in a swelling agent (originally aqueous zinc chloride), and the appropriate number of layers is combined under pressure to produce a welded, compact mass in which the individual layers are largely merged together. The original fibre structure may be retained in the mass to a degree depending on the severity of the swelling treatment and the quality of the fibers. The consolidated material is washed free of the swelling agent (with zinc chloride the maximum residual content should not exceed 0.2%), and heat-dried; sheets are sometimes finished by calendering. Vulcanised fibre is a relatively hard, rigid material, but a flexible version is produced by impregnation with glycerine: this post-treatment imparts resistance to petroleum, some liquid fuels, oils, greases, and some organic liquids (BS 3964:1980). Susceptibility of vulcanised fibre to water vapor and liquid - which stems from the material's cellulosic nature - may be reduced by suitable surface coatings.

Vulcanised fibre is available in three general grades:

1. High hardness and stiffness grade (the "bone" grade of ASTM D 710-89), and Type A of BS 6091:Part 1:1981 and IEC 667-1:1980.
2. General purpose grade, also known as mechanical grade ("commercial" grade of ASTM D 710, and Type B of BS 6091:Part 1 and IEC 667-1).
3. Electrical insulation grade, also used in applications involving difficult bending and forming operations. Thin material of this grade is sometimes referred to as "fishpaper" (ASTM D 710); it is the Type C of BS 6091:Part 1 and IEC 667-1, also known as "leatheroid" (BS 6091:Part 1).

Standard tests are available for determining the properties of vulcanised fibre - cf. the three standards quoted above.

Among the original trade names of vulcanised-fibre products are *Amite* (Spaulding Fibre Co., USA); *Diamond* and *Vulcoid* (Budd Co., Polychem Div., USA); and *Taylorite* (Synthane-Taylor Corp. - an Alco Standard company, USA).

WATER ABSORPTION

The amount of water absorbed by a material[1.5] (typically determined in terms of weitht gain by a prescribed test specimen[1.1]) under specified test conditions[1.5]

(including specified temperature and duration of exposure[1.1]). In a typical standard water-absorption test the exposure is by immersion[1.1,1.5]. A test in which the specimen is exposed to a humid atmosphere, so that it absorbes water from the vapour phase, is more accurately described as a *water vapour absorption* test[1.5]. In standard water absorption tests, the arbitrary test conditions (in particular the duration of immersion at the test temperature) are not specifically directed to achieving equilibrium absorption (one of the methods of ASTM D 570 is a partial exception here), and the test methods do not make provision for following the sorption kinetics. The diffusional and sorption-kinetics aspects of water absorption by some polymers and plastics, including some pratical implications, are outlined in papers by Braden (1963; 1968) and Stafford and Braden (1968).
The standard water-absorption tests in frequent use are described in the following specifications.

ISO 62:1980 (identical with BS 2782:Part 4:Methods 430 A to 430 D:1983 (1992). ASTM D 570-95. ASTM D 2842-94 (water absorption by rigid cellular plastics).

WATER VAPOUR ABSORPTION

see under *water aborption.*

WATER VAPOUR TRANSMISSION RATE (WVT)

See under *permeability.*

WEAR

The cumulative effect (or action[1.5]) of the mechanical factors responsible for deterioration which tends to impair the serviceability[1.5] or appearance of a material or product. See also *abrasive wear*.

WEATHERING (NATURAL WEATHERING)

Depending on the context, this term may have one of the following two interrelated meanings.

(1) The *effects* on a material of exposure out of doors[1.5].

These effects, virtually always detrimental in plastics and rubbers[1.2], are brought about by the action of environmental factors such as sunlight (especially its UV component), atmospheric oxygen, ozone (particularly aggressive towards some rubbers under stress), temperature and temperature variation, moisture, and some atmospheric pollutants. The effect are typically manifested in deterioration of some property or properties (e.g. impact resistance, tensile strength and elongation), and very often in the altered state and appearance of the surface, which may exhibit chalking, cracking, crazing[1.3], blistering, discoloration, or - e.g. with plasticised PVC - tackiness caused by exudation of plasticiser.

(2) The *exposure itself*, i.e. the act of submitting a material to the action of the environmental factors operating outdoors (as e.g. in the sentence "the tests included three year's weathering").

It may be noted that the only standard plastics-terminology definition of natural weathering[1.5], and the two corresponding definitions relating to rubber[1.2,1.3], are confined to the first of the above meanings. At the same time, the standard definitions of *artificial weathering* (q.v.)[1.1,1.5,1.7] are all concerned with the second meaning, i.e. the action and conditions of exposing materials in the laboratory to treatments designed to cause changes of the kind brought about by outdoor exposure (usually at a somewhat increased rate).

The following standards are relevant to natural weathering of plastics:

Outdoor weathering, general

BS 4618: Section 4.2: 1972 (discussion of resistance to natural weathering)
Section 4.3: 1974 (discussion of resistance of plastics to colour changes produced by exposure to light).
ISO 877: 1994 (methods of exposure to direct weathering inter alia)
ISO 4607: 1979 (methods of exposure).
ASTM D 1435-94 (standard practice for outdoor weathering).
ASTM D 5272-92 (outdoor exposure testing of photodegradable plastics).
ASTM G 7-89 (outdoor exposure testing of non-metallic materials).

Accelerated outdoor weathering

ISO 877: 1994 (methods of exposure to Fresnel-mirror intensified daylight (inter alia))
ASTM D 4141-93 Procedure C (Fresnel-reflector rack exposure).
ASTM D 4364-94 (accelerated outdoors weathering with Fresnel-reflector concentrated natural sunlight).
ASTM E 838-81: Discontinued in 1992 (accelerated outdoor weathering with natural sunlight).
ASTM G 90-91 (accelerated outdoor weathering with concentrated natural sunlight).

Outdoor weathering under glass

ISO 877: 1994 (methods of exposure to glass-filtered daylight (inter alia)).
ISO 4582-1980 (1986) Determination of changes in colour and other properties of plastics after exposure to daylight under glass, natural weathering, or artificial light.
BS 2782 : Part 5 : Method 540 B : 1982 (1987) (identical with the superseded (1981) edition of ISO 877).
ASTM G 24-87 (conducting exposures to glass-filtered natural light).

WHITE-POINT TEMPERATURE

In latex (q.v.) technology, the limiting temperature (as determined in a test - cf. e.g. ISO 2115) below which the latex particles left after evaporation of water from a layer of latex deposited on a substrate form an opaque, white laminar mass. The whiteness and opacity are the optical consequences of the fact that the mass is discontinuous on the microscopic scale because the temperature has not been sufficiently high to promote coalescence of the latex particles. The white-point temperature is always a few degrees Celsius lower than the minimum film-forming temperature (q.v.).

WINDOW FRAMES (plastics)

Frames produced from extruded, hollow-section rigid PVC profiles ("lineals"), often of considerable structural complexity. The profiles are made into a frame by heat-welding at the frame's corners. Rigid PVC formulations, devised for good resistance to weathering (including photochemical degradation) and to impact are available (cf. e.g. ASTM D 4216-92, and Titow 1990). Some compositions made to these formulations are suitable for service in hot, sunny climates: such compositions typically incorporate titanium dioxide which contributes to UV stabilization, and hence are white in colour. The whiteness also tends to reduce absorption of sunlight.

Several English-language standards have been published, providing information on, and requirements for, PVC-framed windows: cf. e.g. (for rigid PVC profiles) BS 7413:1991 and BS 7414: 1991; ASTM D 4726-90; Canadian Standard 41-GP-19 Ma; and (for rigid-PVC-framed windows) BS 7412: 1991; BPF/GGF *Trade Standard for uPVC Windows*: 1986; ASTM D 4099-93.

WLF EQUATION (Williams, Landel and Ferry 1955)

In its standard form, the equation is

$$\log_{10} a_T = -C_1 (T - T)_g / (C_2 + T - T_g)$$

where a_T is a shift factor relating the state of a polymer system at a temperature T to the state at T_g (serving as a reference temperature), and C_1 and C_2 are numerical factors. The shift factor a_T may be thought of as representing the variation with temperature of a notional "friction coefficient" of polymer-chain segments in any mechanical relaxation of the polymer. The WLF equation thus applies, inter alia, to the temperature dependence of stress relaxation time (Θ) and of viscosity (η), the relevant forms being

$$\log_{10}\left(\Theta_T / \Theta_{Tg}\right) = -C_1\left(T - T_g\right) / \left(C_2 + T - T_g\right)$$

$$\text{and } \log_{10}\left(\eta_T / \eta_{Tg}\right) = -C_1\left(T - T_g\right)\left(C_2 + T - T_g\right)$$

The WLF equation (which is essentially an empirical relationship) has been found to fit the behavior of many non-crystalline polymers in the temperature range between the T_g and about (T_g + 100°C). In this range the values of C_1 and C_2 are nearly constant ($C_1 = 17.44$; $C_2 = 51.6$).

YELLOWNESS INDEX

A numerical expression of the yellow tinge or tint whereby the colour of a material differs from complete whiteness or a colourless state (cf. also Ref. 1.14 and ASTM D 123-96). The index can be determined by measurements of reflectance or transmittance (as appropriate) at two or three of the relevant wavelengths such as blue and red, or blue, red and green. For the latter case, which provides a more precise value of yellowness index, BS 2782:Part 5:Method 530A:1976 (1983) defines the index (YI) as

$$YI = \left(T_{700} - T_{420}\right)/T_{560}$$

where T_{700}, T_{420} and T_{560} are, respectively, the transmittance values for red light (at 700 nm), blue light (at 420 nm) and green light (at 560 nm). The companion BS method, Method 530 B, deals with determination of colour of near-white or near-colourless materials. ASTM D 1925 (discontinued in 1995) also provided a method of determination of yellowness index of plastics : this standard is replaced in part by ASTM E 313-73 (1993) which is concerned with determination of whiteness and yellowness of near-white opaque materials.

Determination of yellowness index of white or colourless plastics materials before and after exposure to light or to weathering can be used as a way of quantifying visual colour effects of the exposure.

YIELD POINT

In a plastics material (or test specimen) under stress, that value of the stress at which strain first begins to increase without a further rise of stress[1.1,1.5]. In a stress/strain plot this stress value is marked by the point at which the initial ascending part of the curve first levels off, or reaches a maximum. For a material whose stress/strain curve does not exhibit a yield point, *yield strength* is sometimes defined[1.10] in terms of a specified limiting stress value for the departure from linearity of the initial ascending portion of the curve. See also *offset yield point* and *offset yield stress*.

ZERO STRENGTH TIME (ZST)

In prescribed test conditions, the time in seconds for a notched strip specimen of a fluoropolymer to break on heating at 250°C while weighted with a load of 7.5 g. ZST correlates with the molecular weight for a number of fluoropolymers (e.g. some PCTFE resins) and is thus a measure of the relative molecular weight of such materials (cf. ISO 12086-2:1995 and ASTM D 1430-95).

Littera docet - Littera nocet
(Anon.)

PART 2

ABBREVIATED TERMS (LETTER SYMBOLS) FOR THE BASE POLYMERS AND OTHER COMPONENTS OF PLASTICS MATERIALS, AND SOME MISCELLANEOUS ABBREVIATIONS AND ACRONYMS OF RELEVANCE TO PLASTICS

PART 2

SECTION 1

ABBREVIATED TERMS (LETTER SYMBOLS) FOR POLYMERS AND COPOLYMERS

The abbreviated terms (letter symbols) for the polymers and copolymers used in, or as, plastics materials are almost invariably made up of the initials of the names (or components of names) of the polymers concerned. This is so both in the standard abbreviations[2.1.,2.2,2.3,2.11] and in those cases where no standard versions are available, or where common usage differs from the standard. Some illustrative examples are: CA for cellulose acetate; PPSU for poly(phenylene sulfone); and PEUR for polyester urethane.

There is not at present complete uniformity in standard presentation of the names of copolymers (including terpolymers, etc.), either in full or by means of abbreviated terms (letter symbols).

The former edition of the relevant ISO standard[2.1], and the corresponding British Standard[2.3], make up the abbreviated term for a copolymer (or terpolymer, etc.) from the letter symbols for the individual monomeric components, appearing - from left to right - in the order of decreasing molar percentage or mass percentage, and separated by oblique strokes: e.g. *VC/E/MA* for *vinyl chloride/ethylene/methyl acrylate*. Provision is made for omitting the oblique strokes where the abbreviated symbol concerned is traditionally widely used without them (e.g. *ABS*, not A/B/S, for *acrylonitrile/butadiene/styrene*), but in the full names of all copolymers the oblique strokes are mandatory.

The new, revised ISO standard[2.11] now uses hyphens (not oblique strokes) in full names of copolymers, and does not call for the use of oblique strokes in the abbreviated terms. Some alterations in the letter symbols themselves have also been made: e.g. *vinyl chloride/ethylene/methyl acrylate* (see above) should, according to this standard, be abbreviated to *VCEMAK*.

The ASTM standard for abbreviated terms[2.2] subscribes to the principle that in the full name of a copolymer (or terpolymer, etc.) the names of the monomeric components should be separated by hyphens (e.g. *vinyl chloride-ethylene-methyl acrylate*) ; an oblique stroke (termed "virgule" or "solidus" in the standard) is acceptable only where the use of a hyphen could lead to confusion (cf. e.g. *styrene-ethylene/butylene-styrene* block copolymer). At the same time, no separation is required in the abbreviated terms: thus, according to the standard, such terms for the copolymers just mentioned are, respectively, VCEMA and SEBS.

It may be noted that in the case of the standard letter-symbol systems for rubber polymers - contained in the relevant major specifications (ASTM, ISO and BS)[2.4,2.5,2.6] - the letter combinations used are mainly somewhat arbitrary material-class designations, and only in some cases do they relate directly to actual material names: cf. CO for polychloromethyl oxirane rubbers; MQ for silicone rubbers having only methyl substituent groups on the polymer chain (e.g. dimethyl polysiloxane); NBR for acrylonitrile/butadiene rubbers; EU for polyether urethane rubbers.

There are also in existence standard code designation systems for such commercial plastics materials as moulding and extrusion compounds. In these systems the abbreviation for the base polymer is typically preceded by the number of the standard concerned, and followed by various symbols (letters and/or numbers). The symbols are arranged in groups representing "data blocks" prescribed by the code; they indicate levels of certain selected properties, type of application, the material's physical form - e.g. pellets or powder (only in some systems, e.g. that of DIN 7749), the presence of fillers or reinforcements, and other features. For example, in accordance with ISO 2580-1:1990, an ABS compound for pipe extrusion could be coded: ISO 2580 ABS 1 EN 085-04-25-30. The information conveyed by this particular coding (in addition to the obvious terms, i.e. the number of the specification concerned and the nature of the polymer) is that the percentage acrylonitrile content of the polymer is between 10 and 30 (code"1"), that the intended application is pipe extrusion (code "E"), that the material is not coloured (code "N"), and that it has a Vicat softening temperature of 90°C or less (code "085"), a melt flow rate of 5 g/10 min or less (code "04"), an Izod impact resistance in the range 20-30 kJ/m^2 (code "25") and flexual modulus above 2800 MPa (code "30").

It would be impracticable to include such detailed, extended code designations among the abbreviations explained in this part of the dictionary, as the possible combinations of the various symbols are too numerous; moreover, there are differences among the coding systems promoted by different standards (e.g. those of ISO and ASTM). The relevant standards should be consulted in specific cases: they can be located - normally via the name of the base polymer of the compound concerned - through the appropriate indexes of standards. The indexes are available in English not only for those standards which are published in English (ISO, BS and ASTM standards) but also for the German DIN standards, and certain others. A few examples, taken at random, of the standards dealing with designatory coding of plastics materials are given in the table.

POLYMER, OR BASE POLYMER OF PLASTICS MATERIAL	STANDARDS RELEVANT TO MATERIAL CODING	
	NUMBER AND DATE	SUBJECT (NOT NECESSARILY THE FULL OR EXACT TITLE)
ACRYLONITRILE/ BUTADIENE/STYRENE (ABS)	BS ISO 2580-1:1990	Designation of ABS moulding and extrusion materials
	ASTM D 4673-87 (1993)	Standard specification for ABS moulding and extrusion materials
	DIN 16772: Part 1:1987	ABS materials: classification and designation
POLYETHYLENE (PE)	ISO 1872-1:1993	PE and ethylene copolymer thermoplastics: designation
	ASTM D 1248-84 (1989)	Standard specification for PE plastics moulding and extrusion materials
	BS 3412:1976 (Amended 1982 and 1985)	PE materials for moulding and extrusion
	DIN 16776 Part 1:1987	PE and ethylene copolymer thermoplastics: classification and designation
POLYSTYRENE (PS)	BS ISO 1622-1: 1994	Designation of PS moulding and extrusion materials
	ASTM D 4549-85:(1994)	Standard specification for PS moulding and extrusion materials
	DIN 7741: Part1:1986	PS moulding materials: classification and designation
POLYVINYL CHLORIDE (PVC)	ISO 1060-1:1982	Homopolymer and copolymer resins of vinyl chloride: designation
	ISO 1163-1:1985	Unplasticised compounds of homopolymers and copolymers of vinyl chloride: designation
	ISO 2898-1:1986	Plasticised compounds of homopolymers and copolymers of vinyl chloride (PVC-P): designation
	ASTM D 1755-92	Standard specification for PVC resins
	ASTM D 2474-81 (1987)	Standard specification for vinyl chloride copolymer resins
	ASTM D 1784-95	Standard specification for rigid PVC and chlorinated PVC (CPVC) compounds
	ASTM D 2287-92	Standard specification for non-rigid vinyl chloride polymer and copolymer moulding and extrusion compounds
	DIN 7746-1:1986	Vinyl chloride (VC) polymers: classification and designation
	DIN 7748-1:1985	Unplasticised PVC (PVC-U) moulding materials: classification and designation
	DIN 7749-1:1985	Plasticised PVC (PVC-P) moulding materials: classification and designation

AB[2.2]	Acrylonitrile/butadiene plastics
ABA[2.2] **or A/B/A**[2.1,2.3]**; ABAK**[2.11]	Acrylonitrile/butadiene/acrylate terpolymer
ABR[2.4,2.5,2.6]	Acrylate-butadiene rubber
ABS[2.1,2.2,2.3,2.11]	Acrylonitrile/butadiene/styrene
ACM[2.4,2.5,2.6]	A rubber, copolymer of an acrylate (typically ethyl acrylate) and a minor proportion of a monomer which facilitates vulcanisation. (Often termed "acrylic rubber"[2.5,2.6])
ACPES[2.2] **or A/CPE/S**[2.1,2.3]**; ACS**[2.11]	Acrylonitrile/chlorinated polyethylene/styrene
ADC[2.2]	Allyl diglycol carbonate
AEM[2.4]	A rubber. Copolymer of an acrylate (typically ethyl acrylate) and ethylene. See also EEA and EEAK
AEPDMS[2.2]	Acrylonitrile/ethylene-propylene-diene/styrene plastic[2.2]. See also AEPDS
A/EPDM/S[2.1,2.3]	Acrylonitrile/ethylene-propylene-diene/styrene[2.1,2.3]. See also AEPDS
AEPDS[2.11]	Acrylonitrile/ethylene-propylene-diene/styrene[2.11]. See also A/EPDM/S
AES[2.2]	Acrylonitrile-ethylene-styrene plastic[2.2]
AMMA[2.2,2.11]	Acrylonitrile/methyl methacrylate[2.1,2.3,2.11]
A/MMA[2.1,2.3]	Acrylonitrile-methyl methacrylate plastics[2.2]
APET	Abbreviation sometimes used for amorphous poly(ethylene terephthalate).
ARP[2.2]	Aromatic polyester, or thermoplastic polyester (polyaryl terephtalate).
AS	Styrene/acrylonitrile copolymer. Abbreviation, sometimes used in the USA and Japan in commercial contexts. (Standard form: SAN)
ASA[2.1,2.2,2.3,2.11]	Acrylonitrile/styrene/acrylate terepolymer
AMS	See ASM
ASM; AMS	Abbreviated terms sometimes used for acrylonitrile/styrene/methacrylate terpolymers and their plastics
BOPP	Abbreviation sometimes encountered for "biaxially oriented polypropylene" (in the context of film, sheeting, or blow-moulded containers)
BR[2.4,2.5,2.6]	Letter symbol for butadiene rubber (polybutadiene) in the relevant terminology standards. See also PBD
B/S/MMA	Butadiene/styrene/methyl methacrylate. Abbreviation not listed in, but according with the priciples of, references 2.1 and 2.3
CA[1.1,1.5,2.1,2.2,2.3,2.11]	Cellulose acetate
CAB[1.1,1.5,2.1,2.2,2.3,2.11]	Cellulose acetate butyrate
CAP[1.1,1.5,2.1,2.2,2.3,2.11]	Cellulose acetate propionate
CE[2.2]	General abbreviation for cellulose plastics[2.2]
CEF[2.2]	Cellulose formaldehyde
CF[1.1,1.5,2.1,2.2,2.3,2.11]	Cresol-formaldehyde resin
CFM[2.4,2.6]	Polychlorotrifluoroethylene rubber (polymer abbreviation also PCTFE-q.v.)
CIIR[2.4,2.5,2.6]	Chloro-isobutene-isoprene rubber (commonly known as chlorobutyl rubber)

CM[2.4,2.5,2.6]	Chloropolyethylene rubber. See also CPE and PE-C (the standard letter symbols for chlorinated polyethylene in the plastics context)
CMC[1.1,1.5,2.1,2.2,2.3,2.11]	Carboxymethyl cellulose
CN[1.1,1.5,2.1,2.2,2.3,2.11]	Cellulose nitrate
CO[2.4,2.5,2.6]	Polychloromethyloxirane rubber (commonly known as epichlorohydrin rubber)
COC	Abbreviation sometimes used for metallocene-catalysed, amorphous, transparent copolymers of ethylene with cyclic olefins (cf. TOPAS)
CP[1.1,1.5,2.1,2.2,2.3,2.11]	Cellulose propionate
CPE[2.2]	Chlorinated polyethylene. See also CM (standard abbreviation for CPE rubber), and PE-C
CPVC[2.2]	Chlorinated poly(vinyl chloride). See also PVC-C
CR[2.4,2.5,2.6]	Chloroprene rubber
CS[2.2]	Casein
CSF[2.2,2.11]	Casein-formaldehyde resin
CSM[2.4,2.5,2.6]	Chlorosulfonylpolyethylene rubber (chlorosulfonated polyethylene)
CTA[2.1,2.2,2.3]	Cellulose triacetate
CTFE[2.2]	Chlorotrifluoroethylene
DAIP[2.2]	Diallyl isophthalate (used as monomer in thermosetting DAIP compositions - see PDAIP, and as plasticiser with crosslinking potential)
DAP[2.2]	Diallyl phthalate; diallyl orthophthalate (applications similar to those of DAIP - cf. PDAP)
EAA	Abbreviation sometimes used for ethylene/acrylic acid copolymer (see e.g. *Escor*)
E/BA; EBA	Abbreviation sometimes used for ethylene/butyl acrylate copolymer
EC[1.1,1.5,2.1,2.2,2.3,2.11]	Ethyl cellulose[2.1,2.2,2.3,2.11]; ethylcellulose[1.1,1.5]
ECTFE	Copolymer of ethylene and chlorotrifluoroethylene (cf. ISO/DIS 12086-1)
EEA[2.2] **OR E/EA**[2.1,2.3]**; EEAK**[2.11]	Ethylene/ethyl acrylate copolymer (in the plastics context; for EEA rubbers see AEM). Cf. the commercial version: *NCPE5868* in Table 3.1
EFEP **E/FE/P**	Ethylene/perfluoroethylene/propylene copolymer (cf. ISO/DIS 12086-1)
EMA[2.2,2.11]	Ethylene/methacrylic acid[2.1,2.3] copolymer;
E/MA[2.1,2.3]	Ethylene-methacrylic acid plastics[2.2]
EMA	Ethylene/maleic anhydride copolymer
EMA	Ethylene/methyl acrylate copolymer
EP[1.1,1.5,2.1,2.2,2.3,2.11]	Epoxy; epoxide (resin or plastic)
E/P[2.1,2.3,2.11]	Ethylene/propylene copolymer
EPD[2.2]	Ethylene/propylene/diene plastics
EPDM[2.1,2.3,2.4,2.5,2.6]	Ethylene/propylene/diene terpolymer rubber - see also entry in Part 1
EPM[2.2]	Ethylene-propylene polymer[2.2]
EPP	Abbreviation sometimes used for expandable propylene (beads), or the expanded product.

EPS	Abbreviation sometimes used for expandable polystyrene beads, or the expanded final product
ETFE[2.2,2.11] **E/TFE**[2.1,2.3]	Ethylene/tetrafluoroethylene copolymer (cf. also ISO/DIS 12086-1)
EVA[2.2] **or E/VAC**[2.1,2.3]; **EVAC**[2.11]	Ethylene/vinyl acetate copolymer
E/VAL[2.1,2.3]; **EVOH**[2.2,2.11]	Ethylene/vinyl alcohol copolymer
EVS	Abbreviation sometimes used for ethylene/vinyl silane graft copolymer
FCEA[2.2]	Fully crosslinked elastomeric alloy (see ASTM D 5046-92)
FEP[1.1,1.5,2.1,2.2,2.3]	Perfluoro (ethylene/propylene)[1.1,1.5,2.1,2.3]; perfluoro(ethylene-propylene) copolymer[2.2]; tetrafluoro/hexafluoropropylene[2.1,2.3] copolymer. (Cf. also ISO/DIS 12086-1) The letters "FEP" actually stand for "fluorinated ethylene/propylene".
FEVE	Fluorinated ethylene/vinyl ether polymers; thermosetting fluoropolymers.
FF[2.1,2.2,2.3,2.11]	Furane-formaldehyde[2.1,2.3]; furan-formaldehyde resin[2.2]; furan-formaldehyde[2.11]
GMT	Abbreviation which has been used for "glass-mat thermoplastic"
GPPS	Abbreviation sometimes used for general-purpose polystyrene
HDPE[1.7,2.2]	High density polyethylene plastics (standard density 0.941 g cm^{-3} or greater[1.7]). See also PE-HD
HIPS[2.2]	High-impact-resistant ("high impact") polystyrene. For other standard versions see IPS and PS-HI
HPDL	High pressure decorative laminate (cf. ISO 4586)
IPS[2.2]	Impact-resistant polystyrene. See also PS-HI and HIPS
LCP[2.2,2.11]	Liquid crystal polymer (liquid-crystalline polymer) - q.v.
LDPE[1.7,2.2]	Low density polyethylene plastics (standard density range given[1.7] as 0.910 g cm^{-3} to 0.925 g cm^{-3}). See also PE-LD.
LLDPE[1.7,2.2]	Linear low density polyethylene plastics (standard density range given[1.7] as 0.919 g cm^{-3} to 0.925 g cm^{-3}). See also PE-LLD
LMDPE[1.7,2.2]	Linear medium density polyethylene plastics (standard density range given[1.7] as 0.926 g cm^{-3} to 0.940 g cm^{-3}). See also PE-LMD
MABS	Methyl methacrylate/acrylonitrile/butadiene/ styrene copolymers (cf. ISO 10366: Parts 1 and 2)
MBS[2.1,2.2,2.3,2.11]	Methacrylate-butadiene-styrene[2.11] plastics[2.2]. Methacrylate/butadiene/styrene[2.1,2.3]
MC[2.1,2.2,2.3,2.11]	Methyl cellulose
MDPE[1.7,2.2]	Medium density polyethylene plastics (standard density range given[1.7] as 0.926 g cm^{-3} to 0.940 g cm^{-3}). See also PE-MD
MF[1.1,1.5,2.1,2.2,2.3,2.11]	Melamine-formaldehyde[1.1,1.5,2.1,2.3,2.11] resin[2.2]
MFA	Methylfluoroalkoxyethylene - see e.g. *Hyflon MFA*
M-LLDPE	Linear low-density polyethylene produced with the aid of metallocene catalysts
MMA[2.2]	Methyl methacrylate. See also PMMA
MMABS[2.2,2.11]	Methyl methacrylate-acrylonitrile-butadiene-styrene[2.11] resin[2.2]. See also MABS

MPF[2.1,2.2,2.3,2.11]	Melamine/phenol-formaldehyde[2.1,2.3,2.11] resin[2.2]
NBR[2.4,2.5,2.6]	Acrylonitrile-butadiene rubber. In the plastics context a standard letter symbol for this copolymer is PBAN[2.2]
OPP	Abbreviated term sometimes used for "oriented polypropylene" i.e. stretched polypropylene (PP) sheet, film or blow mouldings. In the context of ASTM D 2673-94, this abbreviation stands specifically for oriented PP film of a prescribed minimum tensile strength (68 MPa) in at least one of the two principal directions (machine or transverse)
OPP-B	Abbreviated term used in ASTM D 2673-94 for "balanced" oriented PP film, i.e. one characterised by a prescribed minimum strength (› 68 MPA) in both the machine and the transverse directions (the strengths not differing by more than 55 MPa), and with the corresponding elongations not differing by more than 60% (absolute)
OPP-U	Abbreviated term used in ASTM D 2673-94 for uniaxially oriented ("unbalanced") PP film, i.e. film stretched in one principal direction (machine or transverse), and having a tensile strength in that direction of at least 68 MPa.
OPP-X	Abbreviated term used in ASTM D 2673-94 for biaxially oriented PP film whose tensile strengths in both the machine direction and the transverse direction exceed 68 MPa.
PA[1.1,1.5,2.1,2.2,2.3,2.11]	Polyamide[1.1,1.5,2.1,2.2,2.3,2.11] (nylon)[2.2]
PA+[2.2]	Polyamide (amorphous) blend
PAA[2.2]	Poly(acrylic acid)
PADC[2.2]	Poly(allyl diglycol carbonate)
PAE[2.2]	Polyarylether
PAEK[2.2]	Poly(aryl ether ketone)[2.2]
PAESU (PAES)	Polyarylethersulfone
PAI[2.1,2.2,2.3,2.11]	Polyamide/imide[2.1,2.3]; polyamide-imide[2.2]; polyamideimide[2.11]
PAK[2.2]	Polyester alkyd
PAK[2.11]	Polyacrylate
PAN[1.5,2.1,2.2,2.3,2.11]	Polyacrylonitrile
PAni	Abbreviated term often used for polyaniline (q.v.)
PAR[2.2,2.11]	Polyarylate (i.e. an aromatic linear polyester).
PARA[2.2]	Polyaryl amide
PASU[2.2]	Polyarylsulfone
PAT[2.2]	Poly(aryl terephthalate); a thermoplastic polyester (polyarylate)
PAUR[2.1,2.2,2.3]	Poly(ester urethane)
PB[1.5,2.1,2.2,2.3,2.11]	Polybutylene[1.5]; polybutene[1.5,2.11]; polybutene-1[2.1,2.2,2.3]
PBA[2.1,2.2,2.3]	Poly(butyl acrylate). This abbreviation is also still occasionally encountered as a non-standard symbol for polyether block amide (see PEBA).
PBAK[2.11]	Poly(butyl acrylate)
PBAN[2.2]	Polybutadiene-acrylonitrile. (In abbreviated terminology for rubbers: NBR - q.v.)
PBD	Abbreviated term sometimes used for polybutadiene. See also BR.
PBI	Abbreviation sometimes used for polybenzimidazole
PBN	Poly(butylene naphthalate) (q.v.)
PBS[2.2]	Polybutadiene-styrene. See also SB.

PBT[1.5,2.1,2.2,2.3,2.11]	Poly(butylene terephthalate): name commonly used for poly(tetramethylene terephthalate)
PC[1.1,1.5,2.1,2.2,2.3,2.11]	Polycarbonate
PCCE[2.2]	Poly(cyclohexylenedimethylene cyclohexandicarboxylate), glycol and acid comonomer
PCT[2.2]	Poly(cyclohexylenedimethylene terephthalate)[2.2]
PCTA[2.2]	Poly(cyclohexylenedimethylene terephthalate), acid comonomer[2.2] - see *Kodar*, *Ektar*
PCTFE[1.1,1.5,2.1,2.2,2.3,2.11]	Polychlorotrifluroroethylene (cf. also CFM - abbreviation for PCTFE rubber)
PCTG[2.2]	Poly(cyclohexylenedimethylene terephthalate), glycol comonomer[2.2] - see *Ektar*, *Kodar*
PDAIP	Poly(diallyl isophthalate)
PDAP[1.1,1.5,2.1,2.2,2.3,2.11]	Poly(diallyl phthalate)
PDCPD[2.11]	Polydicyclopentadiene
PE[1.1,1.5,2.1,2.2,2.3,2.11]	Polyethylene (polyethene)[1.1,1.5]; Polyethylene[2.1,2.2,2.3,2.11]
PEBA[2.1,2.2,2.3,2.11]	Polyether block amide[2.1,2.3]. Poly(ether block amide)[2.2,2.11]
PEC	Polyestercarbonate; abbreviation also sometimes used for a blend of polycarbonate and a polyester
PE-C[1.5,2.1,2.3]	Chlorinated polyethylene. See also CM (the standard letter symbol for chloropolyethylene rubber), and CPE
PEDT	3,4-polyethylene dioxythiophene. Inherently antistatic polymer (Bayer)
PEEK[1.5,2.1,2.2,2.3,2.11]	Polypolyetheretherketone (q.v.)
PEEKK[2.2,2.11]	Polyetheretherketoneketone
PEEST[2.11]	Polyetherester
PE-HD	High-density polyethylene. Abbreviated term not listed directly in the relevant ISO and BS standards[2.1,2.3,2.11], but in full accord with their stated principles. See also HDPE.
PEI[2.1,2.2,2.3,2.11]	Polyetherimide
PEK[2.2,2.11]	Polyetherketone
PEKEKK[2.2,2.11]	Polyetherketoneetherketoneketone
PEKK[2.2,2.11]	Polyetherketoneketone
PE-LD	Low-density polyethylene. Abbreviated term not listed directly in the relevant ISO and BS standards[2.1,2.3,2.11], but in full accord with their stated principles. See also LDPE.
PE-LLD	Linear low-density polyethylene. Abbreviated term not listed directly in the relevant ISO and BS standards[2.1,2.3], but in full accord with their stated principles. The ISO draft standard[2.11] quotes PE-LLD as an example. See also LLDPE.
PE-LMD	Linear medium-density polyethylene. Abbreviated term not listed directly in the relevant ISO and BS standards[2.1,2.3], but in full accord with their stated principles. See also LMDPE.
PE-MD	Medium-density polyethylene. Abbreviated term not listed directly in the relevant ISO and BS standards[2.1,2.2,2.3], but in full accord with their stated principles. See also MDPE.
PEN	Polyethylenenaphthalate (q.v.)
PEOX[1.1,15,2.1,2.2,2.3,2.11]	Poly(ethylene oxide)
PEO	Version of the abbreviated term in an earlier (1994) edition of Ref 2.2
PES[1.5,2.1,2.2,2.3,2.11]	Polyethersulfone[1.5,2.11]. Poly(ether sulfone)[2.1,2.2,2.3] NOTE - It may be noted that in the standard abbreviations for polysulfone (PSU[2.1,2.2,2.3]) and poly(phenylene sulfone) (PPSU[2.1,2.2,2.3]) the symbol for "sulfone" is 'SU" (not "S" as in "PES")

PESTUR[2.11]	Polyesterurethane[2.11]
PET[1.5,2.1,2.2,2.3,2.11] **; PETP**	Poly(ethylene terephthalate). ("PETP" used by a former (1978) version of Ref. 2.3, and still by BS EN 60249-2-8:1995)
PETG[2.2]	Poly(ethylene terephthalate) glycol comonomer[2.2] - see *Kodar* and *Spectar*
PE-UHMW	Ultra-high-molecular-weight polyethylene. Abbreviated term not listed directly in the relevant ISO and BS standards[2.1,2.3,2.11], but in full accord with their stated principles. See also UHMWPE.
PEUR[2.1,2.2,2.3,2.11]	Poly(ether urethane); polyetherurethane[2.11]
PF[1.5,2.1,2.2,2.3]	Phenol-formaldehyde[2.1,2.3]. Phenol-formaldehyde resin[1.5,2.2]
PFA[2.1,2.2,2.3,2.11]	Perfluoro alkoxyl alkane[2.1,2.3] polymer[2.11] (cf. also ISO/DIS 12086-1). Perfluoro(alkoxy alkane)[2.2]
PFEP[2.11]	Perfluoro(ethylene/propylene)
PFF[2.2]	Phenol-furfural resin
PHB	Poly(3-hydroxybutyrate)
PHBV	Poly(hydroxybutyrate/valerate)
PI[2.1,2.2,2.3,2.11]	Polyimide
PIB[1.1,1.5,2.1,2.2,2.3,2.11]	Polyisobutylene[1.1,1.5,2.1,2.2,2.3,2.11]; polyisobutene[2.1,2.3,2.11]; poly-2-methylpropene[1.1,1.5]
PIR[2.1,2.2,2.3,2.11]	Polyisocyanurate
PISU[2.2]	Polyimidesulfone
PMCA[2.2]	Poly(methyl-α-chloroacrylate)
PMI[2.1,2.2,2.3,2.11]	Polymethacrylimide
PMMA[1.1,1.5,2.1,2.2,2.3,2.11]	Poly(methyl methacrylate)
PMMI[1.7,2.2,2.11]	Polymethylmethacrylimide[1.7]; Poly(methyl methacrylimide)[2.2]; poly(N-methylmethacrylimide)[2.11]. Cf. e.g. *Pleximid*
PMP[1.5,2.1,2.2,2.3,2.11]	Poly(4-methylpentene)[1.5]; poly-4-methylpentene-1[2.1,2.2,2.3]; poly(4-methylpent-1-ene)[2.11]
PMS[2.1,2.2,2.3,2.11]	Poly-α-methylstyrene[2.1,2.2,2.3]; poly(α-methyl styrene)[2.11]
PO	Abbreviation sometimes used for "polyolefin", or for "propylene oxide"
POB[2.2]	Poly-p-oxybenzoate
POE	Polyoxyethylene. An alternative letter symbol and term for poly(ethylene oxide). Both the latter term and the corresponding letter symbols PEOX or PEO (q.v.) are favoured by the relevant abbreviated-term standards.
POM[1.1,1.5,2.1,2.2,2.3,2.11]	Polyoxymethylene, also known as polyformaldehyde.
POP[2.2]	Abbreviation for poly(phenylene oxide) given, but deprecated, in the 1994 edition of Ref.2.2. The term itself is also deprecated in that reference, in favour of poly(phenylene ether) and its abbreviation PPE (q.v.), which appears also in Refs.2.1 and 2.3. Ref.1.5 features only poly(phenylene oxide), with the abbreviation PPO and a note to the effect that this abbreviation is not acceptable in the USA because it is a registered trademark.
PP[1.1,1.5,2.1,2.2,2.3,2.11]	Polypropylene (polypropene)[1.1,1.5]; polypropylene[2.1,2.2,2.3,2.11]
PPA[1.7,2.2]	Polyphthalamide
PPE[1.5,2.1,2.2,2.3]	The standard letter symbol for the polymer known as poly(phenylene ether) - q.v. (principally in the USA), or poly(phenylene oxide) - q.v.
PPG	Abbreviation sometimes used for polypropylene glycol

PPO[1.5]	Poly(phenylene oxide) - q.v. An abbreviation common outside the USA (where PPO is a registered trademark for a different, commercial product).
PPOX[1.5,2.1,2.2,2.3,2.11]	Poly(propylene oxide)
PPS[1.5,2.1,2.2,2.3,2.11]	Poly(phenylene sulfide) - q.v.
PPSS	Abbreviation sometimes used for poly(phenylene sulfide sulfone) - cf. *Amorvon*
PPSU[1.5,2.1,2.2,2.3]	Poly(phenylene sulfone) - q.v. NOTE - The abbreviation PPSU is also sometimes used for a line of commercial polysulfone-based moulding compounds (*Radel R* - q.v.) described by the producers as polyphenylsulfone
PPV	Poly(p-phenylenevinylene)
pPVC	See PVC-P
PS[1.1, 1.5,2.1,2.2,2.3, 2.11]	Polystyrene
PS-HI[2.1,2.3]	Impact-resistant ("high impact") polystyrene
PSU[2.1,2.2,2.3]	Polysulfone (q.v.)
PTES	Poly(thioether sulfone) - see PPSS. Cf. *Amorvon*
PTFE[1.1,1.5,2.1,2.2,2.3,2.11]	Polytetrafluoroethylene (cf. also ISO/DIS 12086-1)
PUR[1.5 ,2.1,2.2,2.3, 2.11]	Polyurethane (see also TPU and TPUR)
PVAC[1.1,1.5,2.1,2.2,2.3,2.11]	Poly(vinyl acetate)
PVAL[1.1,1.5,2.1,2.3,2.11]	Poly(vinyl alcohol). This abbreviated term was also featured in an earlier (1994) edition of Ref. 2.2 (replaced by PVOH in the1995 edition)
PVB[1.1,1.5,2.1,2.2,2.3,2.11]	Poly(vinyl butyral)
PVC[1.1,1.5,2.1,2.2,2.3,2.11]	Poly(vinyl chloride). The abbreviation is also widely used for a member of (or, collectively, the whole) of the large family of plastics materials based on homopolymers and some copolymers of vinyl chloride. Its adjectival use is illustrated by such expressions as "PVC pipe" or "PVC upholstery".
PVCA[2.2]	Poly(vinyl chloride-acetate). See also VCVAC
PVC-C[2.1,2.3]	Chlorinated poly(vinyl chloride). See also CPVC
PVC-P[2.1,2.3]	Plasticised PVC. The abbreviation pPVC, formerly favoured by ISO, is still sometimes encountered.
PVC-U[2.1,2.3]	Unplasticised PVC. The abbreviation uPVC, formerly favoured by ISO, is still sometimes encountered.
PVDC[1.1,1.5,2.1,2.2,2.3,2.11]	Poly(vinylidene chloride)
PVDF[1.1,1.5,2.1,2.2,2.3,2.11]	Poly(vinylidene fluoride)
PVF[1.1,1.5,2.1,2.2,2.3,2.11]	Poly(vinyl fluoride) (Cf. also ISO/DIS 12086-1)
PVFM[1.1,1.5,2.1,2.2,2.3,2.11]	Poly(vinyl formal)
PVK[1.1,1.5,2.1,2.2,2.3,2.11]	Polyvinylcarbazole[1.1,1.5,2.1,2.3]. Poly(vinyl carbazole)[2.2,2.11]
PVOH[2.2]	Poly(vinyl alcohol)
PVP[1.5,2.1,2.2,2.3,2.11]	Poly(vinylpyrrolidone)[1.5]; polyvinylpyrrolidone[2.1,2.3]; poly(vinyl pyrrolidone)[2.2,2.11]
RTPU[2.2]	Thermoplastic polyurethane, rigid
SAN[2.1,2.2,2.3,2.11]	Styrene/acrylonitrile[2.1,2.3,2.11]; styrene-acrylonitrile plastic[2.2]
SB[2.2,2.11]; **S/B**[1.1,1.5,2.1,2.3]	Styrene/butadiene[2.1,2.3,2.11] plastic(s)[1.1,1.5,2.2]
SBR[2.4, 2.5, 2.6]	Styrene-butadiene rubber
SBS[2.2]	Styrene-butadiene-styrene block copolymer
SEBS[2.2]	Styrene-ethylene/butylene-styrene block copolymer

SEPS[2.2]	Styrene-ethylene/propylene-styrene block copolymer
SI[1.1,1.5,2.1,2.2,2.3,2.11]	Silicone[2.1,2.3,2.11] plastic(s)[1.1,1.5,2.2]
SIS[2.2]	Styrene-isoprene-styrene block copolymer
SMA[21,2.3]; **S/MA**[2.2]; **SMAH**[2.11]	Styrene/maleic anhydride[2.1,2.3,2.11] plastics[2.2]
SMMA	Abbreviation sometimes used for styrene/methyl methacrylate copolymers
S/MS[2.1,2.3]	Styrene/ α-methylstyrene[2.1,2.3]
SMS[2.2]	Styrene- α-methylstyrene plastics[2.2]
SMS[2.11]	Styrene/ α-methylstyrene[2.11]
SP[2.1,2.2,2.3]	Saturated polyester[2.1,2.3] plastic[2.2]
SPS	Abbreviation sometimes used for "syndiotactic polystyrene" (as made e.g. by the metallocene "single-site" catalyst technique - Idemitsu Chemical Co., Japan, and Dow Plastics, USA).
SRP[2.2]	Styrene-rubber plastics
TE	Thermoplastic elastomer. In the ASTM letter-symbol standard[2.2] this abbreviated term appears only as part of a composite term (cf. e.g. TECEA). It has also been adopted by ISO, on completion of the revision of ISO 1043-1[2.11], among the "special characteristics" letter symbols, to be placed, after a hyphen, at the end of the main abbreviated term: e.g. PEBA-TE for "poly(ether block amide) thermoplastic elastomer".
TECEA[2.2]	Thermoplastic elastomer - chlorinated ethylene alloy (see ASTM D5021-89)
TEEE[2.2]	A poly(ether-ester) thermoplastic elastomer: Cf. also ASTM D 4550-92 (for TEEE materials for moulding, extrusion and thermal processing) and Ref. 2.13
TEO[2.2]	An olefinic thermoplastic elastomer (q.v.). See also ASTM D 5593-94
TES[1.7,2.2]	A styrenic thermoplastic elastomer (cf. also ASTM D 4474-91 for TES injection moulding and extrusion materials).
TESS[2.2]	A saturated styrenic thermoplastic elastomer
TESU[2.2]	An unsaturated styrenic thermoplastic elastomer
TFE[2.2]	Tetrafluoroethylene
TFE/PDD	Tetrafluoroethylene-perfluoro(dioxole) copolymer. See ISO/DIS 12086-1)
TPE[1.3,1.7,2.2]	Thermoplastic elastomer (q.v.). See also TE, and cf. ASTM D 5538-94
TPES[2.2]	Thermoplastic polyester. (See also ASTM D 4507-93a)
TPO	Non-standard abbreviation for "toughened polyolefin" fairly common in the technical and trade literature.
TPU[2.2,2.13]	Thermoplastic polyurethane. In Ref.2.2, this abbreviated term appears to be somewhat inconsistent with the one (PUR) which this reference gives for "polyurethane". In an earlier (1994) version of Ref. 2.2, the abbreviated term for thermoplastic polyurethane was "TPUR".
TPUR	Thermoplastic polyurethane. Although not actually quoted in Refs. 2.1 and 2.3, this abbreviation complies with the guidance these two terminology standards provide on the use of letter symbols for components of the abbreviated terms ("T" for "thermoplastic" and "PUR" for "polyurethane"). TPUR was also the abbreviated term for thermoplastic polyurethane in an earlier (1994) version of Ref. 2.2. Another ASTM standard (for the terminology and abbreviations of thermoplastic elastomers)[2.13] gives the abbreviation TPUR for "thermoplastic polyurethane, reinforced".

TPV	"Thermoplastic vulcanisate": non-standard but often used abbreviation for PP/EPDM thermoplastic elastomers (q.v.), in which the EPDM component is vulcanised.
TSPU[2.2]	Thermoset polyurethane
TSUR	Abbreviated term for "thermoset polyurethane" given in an earlier (1994) version of Ref. 2.2. See also TSPU
UF[1.1,1.5,2.1,2.2,2.3, 2.11]	Urea formaldehyde[2.1,2.3,2.11] resin[1.1,1.5,2.2]
UHMWPE[2.2]	Ultra-high molecular weight polyethylene. See also PE-UHMW
UP[1.1,1.5,2.1,2.2,2.3,2.11]	Unsaturated polyester
uPVC	See PVC-U
VC	Vinyl chloride
VCE[2.2,2.11]; **VC/E**[2.1,2.3]	Vinyl chloride-ethylene resin[2.2]. Vinyl chloride/ethylene[2.1,2.3,2.11]
VC/E/MA[2.1,2.3]	Vinyl chloride/ethylene/methyl acrylate[2.1,2.3]
VCEMA[2.2]	Vinyl chloride-ethylene-methyl acrylate resin[2.2]
VCEMAK[2.11]	Vinyl chloride/ethylene/methyl acrylate[2.11]
VC/E/VAC[2.1,2.3]	Vinyl chloride/ethylene/vinyl acetate[2.1,2.3,2.11]
VCEVAC[2.2,2.11]	Vinyl chloride-ethylene-vinyl acetate[2.11] resin[2.2]
VCM	Abbreviation often used for vinyl chloride ("vinyl chloride monomer")
VC/MA[2.1,2.3]	Vinyl chloride/methyl acrylate[2.1,2.3]
VCMA[2.2]	Vinyl chloride-methyl acrylate resin[2.2]
VCMAK[2.11]	Vinyl chloride-methyl acrylate[2.11]
VC/MMA[2.1,2.3]	Vinyl chloride-methyl methacrylate[2.1,2.3]
VCMMA[2.2]	Vinyl chloride-methyl methacrylate resin[2.2]
VCMMA[2.11]	Vinyl chloride-methyl methacrylate[2.11]
VC/OA[2.1,2.3]	Vinyl chloride/octyl acrylate[2.1,2.3]
VCOA[2.2]	Vinyl chloride-octyl acrylate resin[2.2]
VCOAK[2.11]	Vinyl chloride-octyl acrylate[2.11]
VC/VAC[1.1,1.5,2.1,2.3]	Vinyl chloride/vinyl acetate[2.1,2.3]; poly(vinyl chloride covinyl acetate)[1.1,1.5]
VCVAC[2.2,2.11]	Vinyl chloride-vinyl acetate resin[2.2]; vinyl chloride/vinyl acetate[2.11]. See also PVCA
VC/VDC[2.1,2.3]	Vinyl chloride/vinylidene chloride[2.1,2.3]
VCVDC[2.2]	Vinyl chloride-vinylidene chloride resin[2.2]
VCVDC[2.11]	Vinyl chloride-vinylidene chloride[2.11]
VDF/CTFE	Vinylidene fluoride-chlorotrifluoroethylene copolymer (cf. ISO/DIS 12086-1)
VDF/HFP	Vinylidene fluoride-hexafluoropropylene copolymer (cf. ISO/DIS 12086-1)
VDF/TFE	Vinylidene fluoride-tetrafluoroethylene copolymer (cf. ISO/DIS 12086-1)
VDF/TFE/HFP	Vinylidene fluoride-tetrafluoroethylene-hexafluoropropylene copolymer (cf. ISO/DIS 12086-1)
VE	Abbreviated term sometimes used for vinyl ester resin

PART 2

SECTION 2

ABBREVIATED TERMS (LETTER SYMBOLS) FOR POLYMER MIXTURES (BLENDS AND ALLOYS)

The standard principle for representing mixtures, i.e. blends or alloys, of polymers by abbreviated terms (letter symbols) is that the abbreviated term for a given mixture is made up of the letter symbols for the individual polymer components joined by a plus sign (in accordance with the ASTM standard[2.2]) or joined by a plus sign and enclosed in parentheses (as required by the ISO and BS standards[2.1,2.3,2.11]). The ISO and BS standards simply state the principle and illustrate it with a single example - the same in both cases - viz. (PMMA + ABS) as the abbreviated term for a blend of poly(methyl methacrylate) and acrylonitrile/butadiene/styrene terpolymer. The ASTM standard[2.2] lists abbreviated terms for a number of polymer mixtures, several of which are represented in commercial polymer ranges. The contents of the present section are partly based on that list.

It may be noted that in trade literature abbreviated terms for polymer blends and alloys are often given in the form which, according to two ISO and BS standards[2.1,2.3], should be employed for copolymers (see Section 1 above). Thus for instance a blend of polyetherimide with polycarbonate - such as e.g. *Ultem LTX 100A* - may be designated PEI/PC, whereas the standard abbreviated term is PEI + PC[2.2] or (PEI + PC)[2.1,2.3,2.11]. The non-standard abbreviations (with the oblique strokes) are also sometimes seen in technical publications. In the present section the abbreviated terms for the polymer mixtures listed are parenthesised in accordance with the ISO and BS practice.

ABBREVIATED TERM	POLYMER MIXTURE	REMARKS AND/OR COMMERCIAL EXAMPLES
(ABA + PC)[2.2]	Acrylonitrile/butadiene/acrylate + polycarbonate	
(ABA + PMMA)[2.2]	Acrylonitrile/butadiene/acrylate + poly(methyl methacrylate)	
(ABS + PA)[2.2]	Acrylonitrile/butadiene/styrene + polyamide	NOVALLOY A; TRIAX 1120
(ABS + PBT)	Acrylonitrile/butadiene/styrene + poly(butylene terephthalate)	CEVIAN; LATIBLEND; NOVALLOY B
(ABS + PC)[2.2]	Acrylonitrile/butadiene/styrene + polycarbonate	BAYBLEND; CYCOLOY; TERBLEND
(ABS + PP)	Acrylonitrile/butadiene/styrene + polypropylene	Compatibilised alloys
(ABS + PPSU)[2.2]	Acrylonitrile/butadiene/styrene + poly(phenylene sulfone)	
(ABS + PTFE)[2.2]	Acrylonitrile/butadiene/styrene + polytetrafluoroethylene	
(ABS + PVC)[2.2]	Acrylonitrile/butadiene/styrene + poly(vinyl chloride)	This term, with ABS in the first position, is properly applicable to blends containing ABS in substantial proportions (cf. e.g. RONFALOY V; NOVALOY 9000). Moderately ABS-modified PVC considered as a blend is more accurately designated (PVC + ABS) -q.v.

ABBREVIATED TERM	POLYMER MIXTURE	REMARKS AND/OR COMMERCIAL EXAMPLES
(ABS + SMA)[2.2]	Acrylonitrile/butadiene/styrene + styrene/maleic anhydride	
(ABS + TPU)[2.2] **(ABS + PUR)**[2.1,2.3,2.11]	Acrylonitrile/butadiene/ styrene + thermoplastic polyurethane	PREVAIL
(ASA + AES)	Acrylonitrile/styrene/acrylate + acrylonitrile/ethylene /styrene	CENTREX
(ASA + PC)[2.2]	Acrylonitrile/styrene/acrylate + polycarbonate	TERBLEND S KR
(ASA + PMMA)[2.2]	Acrylonitrile/styrene/acrylate + poly(methyl methacrylate)	
(PA + EMA)[2.2]	Polyamide + ethylene/methacrylic acid (ionomer)	
(PA + PE)[2.2]	Polyamide + polyethylene	
(PA + PP)	Polyamide + polypropylene	ENPNITE; ORGALLOY
(PA + PPE)[2.2]	Polyamide + poly(phenylene ether)	EMAREX; NORYL GTX; LURANYL
(PA + PPO)	Polyamide + poly(phenylene oxide)	PPO and poly(phenylene oxide) are the alternative abbreviation and name to PPE and poly(phenylene ether) - see the relevant entries in Part 1 and Section 1 of Part 2
(PA + RBR)[2.2]	Polyamide + a rubber	
(PA + SAN)[2.2]	Polyamide + styrene/acrylonitrile	
(PBT + PET)[2.2]	Poly(butylene terephthalate) + poly(ethylene terephthalate)	
(PBT + PPE)[2.2]	Poly(butylene terephthalate) + poly(phenylene ether)	
(PBT + PPO)	Poly(butylene terephthalate) + poly(phenylene oxide)	See comment to (PA + PPO)
(PBT + RBR)[2.2]	Poly(butylene terephthalate) + a rubber	
(PC + PBT)[2.2]	Polycarbonate + poly(butylene terephthalate)	LATIBLEND; XENOY
(PC + PE)[2.2]	Polycarbonate + polyethylene	
(PC + PET)[2.2]	Polycarbonate + poly(ethylene terephthalate)	STANULOY; STAPRONE
(PC + SMA)[2.2]	Polycarbonate + styrene/maleic anhydride	
(PC + TPU)[2.2] **(PC + PUR)**[2.1,2.3,2.11]	Polycarbonate + thermoplastic polyurethane	
(PE + PS)	Polyethylene + polystyrene	Compatibilised alloys
(PEI + PC)	Polyetherimide + polycarbonate	ULTEM LTX
(PEI + SI)	Polyetherimide + silicone	SILTEM STM 1500
(PET + PMMA)[2.2]	Poly(ethylene terephthalate) + poly(methyl methacrylate)	
(PET + PPSU)[2.2]	Poly(ethylene terephthalate) + poly(phenylene sulfone)	
(PET + RBR)[2.2]	Poly(ethylene terephthalate) + a rubber	

ABBREVIATED TERM	POLYMER MIXTURE	REMARKS AND/OR COMMERCIAL EXAMPLES
(PMMA + ABS)	Poly(methyl methacrylate) + acrylonitrile/ butadiene/styrene	
(POM + PTFE)[2.2]	Polyoxymethylene + polytetrafluoroethylene	POM is the standard letter symbol for acetal polymers
(POM + RBR)[2.2]	Polyoxymethylene + a rubber	
(PP + PE)	Polypropylene + polyethylene	FERROPAK TPP; POLYTROPE TPP
(PP + PS)	Polypropylene + polystyrene	Compatibilised alloys
(PPE + IPS)[2.2]	Poly(phenylene ether) + impact-resistant polystyrene	
(PPO + PS-HI)[2.1,2.3,2.11]	Poly(phenylene oxide) + high-impact polystyrene	See comment to (PA + PPO); PS-HI is the standard ISO symbol
(PPS + PTFE)[2.2]	Poly(phenylene sulfide) + polytetrafluoroethylene	
(PS + PPO)	Polystyrene + poly(phenylene oxide)	CARIL (expandable beads)
(PSU + ABS)	Polysulfone + acrylonitrile/butadiene/styrene	MONDEL A-670
(PVC + ABA)[2.2]	Poly(vinyl chloride) + acrylonitrile/butadiene/acrylate	
(PVC + ABS)	Poly(vinyl chloride) + acrylonitrile/ butadiene/styrene	GEON HTX
(PVC + ASA)[2.2]	Poly(vinyl chloride) + acrylonitrile/styrene/acrylate	GELOY
(PVC + CPE)[2.2] **(PVC + PE-C)**[2.1,2.3,2.11]	Poly(vinyl chloride) + chlorinated polyethylene	
(PVC + EPDM + PE-C)	Poly(vinyl chloride) + ethylene/ propylene/diene + chlorinated polyethylene	A ternary blend, in which the PE-C has a compatibilising action on PVC and EPDM (Lee and Chen 1987)
(PVC + EVA)	Poly(vinyl chloride) + ethylene/ vinyl acetate	PANTALAST 1162
(PVC + MBS)	Poly(vinyl chloride) + methacrylate/ butadiene/ styrene	
(PVC + NBR)[2.2]	Poly(vinyl chloride) + nitrile rubber	VYNITE
(PVC + PMMA)[2.2]	Poly(vinyl chloride) + poly(methyl methacrylate)	
(PVC + PU)* **(PVC + PUR)**[2.1,2.2,2.3,2.11]	Poly(vinyl chloride) + polyurethane	DURALEX 75; VYTHENE
(PVC + SMA)	Poly(vinyl chloride) + styrene/ maleic anhydride	
(SMA + IPS)[2.2] **(SMA + PS-HI)**[2.1,2.3,2.11]	Styrene/maleic anhydride + impact-resistant ("high-impact") polystyrene	

* Abbreviated term featured in an earlier (1994) edition of Ref. 2.2

PART 2

SECTION 3

ABBREVIATED TERMS (LETTER SYMBOLS) FOR PLASTICISERS

This section lists and identifies a large number of letter symbols (abbreviated terms) for plasticisers, including those recognised by the three main relevant terminology standards[2.2,2.7,2.8], as well as numerous other letter symbols encountered in general usage or in technical or trade literature.

Two of the three standards[2.7,2.8] additionally list letter symbols for some individual components of plasticiser names. These subsidiary lists are intended to assist in making up abbreviated terms for those plasticisers for which such terms are not actually quoted in the standards; they also illustrate the way in which the quoted terms have been put together. Thus, for example, the individual components of the abbreviated term DEGDB, given in the ISO standard[2.7] for diethylene glycol dibenzoate, are D for "di", E for "ethylene", G for "glycol", and B for "benzoate". It should be noted that some of the standard individual-component letter symbols stand for more than one component - e.g. B can denote not only benzoate, but also benzyl, bromo, butoxy, and butyl - and some symbols would appear not to be entirely mutually consistent; cf. e.g. HX for "hexyl", but CH for "cyclohexyl"[2.2,2.7,2.8]. All three standards sanction the use of either F or P as the letter symbol for "phosphate" (in the UK , P tends to be more widely used, except in the case of TCF for tricresyl phosphate, as "TCP" is a registered trademark). The ISO standard[2.7] and (with some exceptions) the BS standard[2.8] recognise the following principles as applicable to their letter symbols for plasticisers.

a) Although the letter N is accepted as the symbol for normal (*n*-) linear alkyl radicals, it is in fact used only in the case of the *n*-octyl group (designated NO, as in the symbol DNOP for di-*n*-octyl phthalate), because of the long established universal practice of referring to the 2-ethylhexyl group in the plasticisers as "octyl" (cf. "dioctyl phthalate" for "di-2-ethylhexyl phthalate") and the corresponding normal use of the letter "O" in the relevant abbreviated term (e.g. DOP for di-2-ethylhexyl phthalate).
b) In the absence of a special indication to the contrary, all alkyl groups are understood to be *n*-alkyl groups, and phthalates to be esters of *o*-phthalic acid. Thus e.g. DNP stands for di-*n*-nonyl orthophthalate.
c) The letter I is part of the symbol for an iso-branched group (as e.g. in DIBP for diisobutyl phthalate) except in certain special cases (cf. DTDP in the table).
d) In the case of an adipate or phthalate plasticiser produced by esterification of the acid with a mixture of alkyl alcohols, the abbreviated term may be given in the form of XXXA or XXXP, were the first digit stands for the number of carbon atoms in the shortest alkyl group and the second digit, or pair of digits, similarly represents the longest alkyl group: cf. e.g. 711P for HNUP (heptyl nonyl undecyl phthalate) and 610A for HXODA (hexyl octyl decyl adipate) in the table.

The table additionally contains items of amplifying information, such as names of commercial versions of many of the plasticisers listed (including some which are no longer current, but which are occasionally encountered in the literature).

PART 2

SECTION 3

ABBREVIATED TERMS (LETTER SYMBOLS) FOR PLASTICISERS

ABBREVIATION (STANDARD OR USUAL)	NAME (STANDARD OR USUAL)	OTHER NAMES AND/OR ABBREVIATIONS	CHEMICAL TYPE REF†	REMARKS, OR EXAMPLES OF SPECIFIC COMMERCIAL PRODUCTS
AAF	Alkyl aryl phosphate	AAP	2	See e.g. DPOF
AAP	Alkyl aryl phthalate		1	See e.g. BBP
ABP	Alkyl benzyl phthalate		1	SANTICIZER 261
ACA	Alkyl capryl adipate		4	E.g. alkyl(C_7-C_9) capryl adipate
ACO	Acetylated castor oil		7.1	
AGS esters			4	Mixed esters of the "nylon acids" (i.e. adipic, glutaric and succinic acids), hence also sometimes referred to as "nylonate esters" or "nylonates"
AIDP	Amyl isodecyl phthalate		1	STAFLEX CP
AIOP	Alkyl isooctyl phthalate		1	
ASE[2.2,2.7,2.8]	Alkylsulfonic acid esters[2.2,2.7,2.8]	Alkane sulfonates* or alkyl alkane sulfonates*	10	MESAMOLL and MESAMOLL II
ATSA	Acetyltoluene sulfonamide		10	
BAG	Benzoyl acetyl glycerol		9.1	MOLLIT A (originally produced by Bayer)
BAR[2.7]	Butyl O-acetylricinoleate[2.7]	Butyl(R)-12-acetoxyoleate*	7.1	STAFLEX BR
BAST	Butyl acetoxystearate			PARICIN 6
BBA	Benzyl butyl adipate	Butyl benzyl adipate	4	ADIMOLL BB
BBP[2.2,2.7,2.8]	Benzyl butyl phthalate[2.2,2.7,2.8]*	Butyl benzyl phthalate	1	SANTICIZER 160

† As identified in relation to trade names in Section 2 (Table 1) of Part 3.

* Name conforming to the nomenclature of the International Union of Pure and Applied Chemistry (IUPAC)

ABBREVIATION (STANDARD OR USUAL)	NAME (STANDARD OR USUAL)	OTHER NAMES AND/OR ABBREVIATIONS	CHEMICAL TYPE REF†	REMARKS, OR EXAMPLES OF SPECIFIC COMMERCIAL PRODUCTS
BBSA	Butylbenzene sulfonamide		10	
BCHP[2.7]	Butyl cyclohexyl phthalate[2.7]*		1	
BDDA	Butanediol diacrylate	1,4-butanediol diacrylate	13	
BDP	Butyl decyl phthalate		1	
BEL	Butoxyethyl laurate	Ethylene glycol monobutyl ether laurate; butyl cellosolve** laurate	7.1	
BEO	Butoxyethyl oleate	Ethylene glycol monobutyl ether oleate; butyl cellosolve oleate	7.1	
BEP	Butoxyethyl pelargonate	Ethylene glycol monobutyl ether pelargonate ; butyl cellosolve pelargonate	7.1	
BEST	Butyl epoxy stearate		7.1	
BEST	Butoxyethyl stearate	Ethylene glycol monobutyl ether stearate; butyl cellosolve stearate	7.1	
BGDG	Butylene glycol diglycolate		7.2.2	
BGDMA	Butylene glycol dimethacrylate	1,3-butylene glycol dimethacrylate	13	
BGS	Butyl glycerol sebacate		4	
BHXP	Benzyl hexyl phthalate		1	See also HXBP

** Cellosolve was originally the Union Carbide trade name for ethylene glycol monoethyl ether (ethoxyethanol). The name is now often used generically for monoalkyl ethers and ether esters of ethylene glycol; cf. e.g. ethyl cellosolve (the original "Cellosolve"); butyl cellosolve (butoxyethanol); ethyl cellosolve stearate (ethoxyethyl stearate), etc.

ABBREVIATION (STANDARD OR USUAL)	NAME (STANDARD OR USUAL)	OTHER NAMES AND/OR ABBREVIATIONS	CHEMICAL TYPE REF†	REMARKS, OR EXAMPLES OF SPECIFIC COMMERCIAL PRODUCTS
BIDP	Butyl isodecyl phthalate		1	
BIHXP	Butyl isohexyl phthalate		1	BIHP in some commercial plasticiser trade names
BNP[2.7]	Butyl nonyl phthalate[2.7,*]		1	
BO	Butyl oleate		7.1	
BOA[2.2,2.7,2.8]	Benzyl octyl adipate[2.2,2.7,2.8]	Benzyl 2-ethylhexyl adipate[2.2,*]	4	BEHXA These abbreviations (in the commerical version H usually replaces HX are justified
BOP[2.7,2.8]	Butyl octyl phthalate[2.7,2.8.]	Butyl 2-ethylhexyl phthalate*	1	BEHXP (though less commonly used), as the esterifying alcohol is indeed 2-ethylhexanol,
BOP[2.2]	Benzyl octyl phthalate[2.2]	Benzyl 2-ethylhexyl phthalate[2.2,*]	1	BEHXP not n-octanol.
BPAE	Butyl polyacetoxy ether		12	
BPBG	Butyl phthalyl butyl glycolate		7.2.2	SANTICIZER B-16
BPDPF	t-butylphenyl diphenyl phosphate	BPDPP	2	
BR	Butyl ricinoleate	n-butyl ricinoleate	7.1	The same abbreviation could stand for benzyl ricinoleate
BS	Benzyl succinate		4	
BST[2.7]	Butyl stearate[2.7,*]		7.1	
BTAF	Butyl triaryl phosphate	BTAP	2	
BTG	Butyl thioglycolate		7.2.2	
CDPF	Cresyl diphenyl phosphate	CDPP	2	See DPCF
CHGS	Cyclohexyl glutarate succinate		4	HEXAPLAS CGS
CIOP	Cycloisooctyl phthalate		1	
CO	Chlorinated olefins		11	E.g. the FLEXCHLOR range
CP	Chlorinated paraffins		11	E.g. the CERECLOR range

ABBREVIATION (STANDARD OR USUAL)	NAME (STANDARD OR USUAL)	OTHER NAMES AND/OR ABBREVIATIONS	CHEMICAL TYPE REF†	REMARKS, OR EXAMPLES OF SPECIFIC COMMERCIAL PRODUCTS
CPB	Cumylphenyl benzoate		9.1	KENPLAST ESB
DAA	Dialkyl adipate		4	E.g. dialkyl (C_7-C_9) adipate; $DA_{79}A$
DAC[2.2]	Diallyl chlorendate[2.2]	Diallyl ester of 1,4,5,6, 7,7-hexachlorobi-cyclo-(2,2,1)-5-heptene-2,3-dicarboxylic acid[2.2]	4	In some trade names of commercial plasticisers, "allyl" has also been abbreviated to "AL" - cf. e.g. BISOMER DALP for diallyl phthalate originally marketed by the Distillers Co. in the UK.
DAF[2.2]	Diallyl fumarate[2.2]		4	
DAGS	Dialkyl glutarate sebacate		4	E.g. dialkyl (C_7-C_9) glutarate sebacate; $DA_{79}GS$: HEXAPLAS AGS
DAIP[2.2]	Diallyl isophthalate[2.2]		1	
DAIP	Dialkyl isophthalate		1	
DAM[2.2]	Diallyl maleate[2.2]		4	
DAP[2.2]	Diallyl phthalate[2.2]	DALP	1; 13	BISOMER DALP
DAP	Dialkyl phthalate		1	E.g. dialkyl (C_7-C_9) phthalate; $DA_{79}P$. See also DNAP
DAP	Diamyl phthalate		1	
DAS	Dialkyl sebacate		4	E.g. dialkyl (C_7-C_9) sebacate; $DA_{79}S$
DAZ	Dialkyl azelate		4	E.g. dialkyl (C_7-C_9) azelate; $DA_{79}Z$
DBA[2.7]	Dibutyl adipate[2.7,*]		4	
DBEA	Di-(2-butoxyethyl) adipate*	Di-(ethylene glycol monobutyl ether) adipate; di-(butyl cellosolve) adipate	4	
DBEP[2.7]	Di-(2-butoxyethyl) phthalate[2.7,*]	Di-(ethylene glycol monobutyl ether) phthalate; di-(butyl cellosolve) phthalate	1	
DBES	Di-(2-butoxyethyl) sebacate	Di-(ethylene glycol monobutyl ether) sebacate; di-(butyl cellosolve) sebacate	4	
DBEZ	Di-(2-butoxyethyl) azelate	Di-(ethylene glycol monobutyl ether) azelate; di-(butyl cellosolve) azelate	4	

ABBREVIATION (STANDARD OR USUAL)	NAME (STANDARD OR USUAL)	OTHER NAMES AND/OR ABBREVIATIONS	CHEMICAL TYPE REF†	REMARKS, OR EXAMPLES OF SPECIFIC COMMERCIAL PRODUCTS
DBF[2.7]	Dibutyl fumarate[2.7,*]		4	
DBGA	Di-(butyl glycol) adipate		4	
DBM[2.7]	Dibutyl maleate		4	
DBP[2.2,2.7,2.8]	Dibutyl phthalate[2.2,2.7,2.8,*]		1	
DBS[2.2,2.7,2.8]	Dibutyl sebacate[2.2,2.7,2.8,*]		4	
DBS	Dibenzyl sebacate		4	The abbreviation "DBZS" also found in some trade names of commercial plasticisers
DBT	Dibutyl tartrate		4	
DBZ[2.7]	Dibutyl azelate[2.7,*]		4	
DCEEP	Di-(chloroethoxyethyl) phthalate		1	
DCGA	Dicresyl glyceryl acetate		7.2.1	
DCHP[2.2,2.7,2.8]	Dicyclohexyl phthalate[2.2,2.7,2.8,*]		1	UNIMOLL 66M
DCP[2.2,2.7,2.8]	Dicapryl phthalate[2.2,2.7,2.8]	Bis(1-methylheptyl) phthalate*	1	
DDA	Didecyl adipate		4	
DDP[2.2,2.7,2.8]	Didecyl phthalate[2.2,2.7,2.8,*]		1	
DEEP	Di-(ethoxyethyl) phthalate	Di-(ethylene glycol monoethyl ester) phthalate; di-(ethyl cellosolve) phthalate	1	
DEGDB[2.7]	Diethylene glycol dibenzoate[2.7]	Oxydiethylene dibenzoate*	9.1	BENZOFLEX 2-45
DEGDM	Diethylene glycol dimethacryalate		13	
DEGDP	Diethylene glycol dipelargonate		7.1	
DEGDST	Diethylene glycol distearate		7.1	The abbreviation "DGS" ("diglycol stearate") also found in some trade names of commercial plasticisers

ABBREVIATION (STANDARD OR USUAL)	NAME (STANDARD OR USUAL)	OTHER NAMES AND/OR ABBREVIATIONS	CHEMICAL TYPE REF†	REMARKS, OR EXAMPLES OF SPECIFIC COMMERCIAL PRODUCTS
DEHA	Di-2-ethylhexyl adipate	See DOA	4	In common usage these abbreviations, though more accurate, are not as frequent as DOA and DOP (DEHA and DEHP)
DEHP	Di-2-ethylhexyl phthalate	See DOP	1	
DES	Diethyl sebacate		4	
DHP[2.2,2.7,2.8]	Diheptyl phthalate[2.2,2.7,2.8,*]		1	
DHXA	Dihexyl adipate		4	
DHXP[2.2,2.7,2.8]	Dihexyl phthalate[2.2,2.7,2.8,*]		1	
DHXS	Dihexyl sebacate		4	
DHXZ	Dihexyl azelate		4	
DIBA[2.7,2.8]	Diisobutyl adipate[2.7,2.8,*]		4	
DIBM[2.7]	Diisobutyl maleate[2.7,*]		4	
DIBP[2.2,2.7,2.8]	Diisobutyl phthalate[2.2,2.7,2.8,*]		1	
DIBZ	Diisobutyl azelate		4	
DIDA[2.2,2.7,2.8]	Diisodecyl adipate[2.2,2.7,2.8]		4	
DIDETHP	Di-(isodecyl)-4,5-epoxy tetrahydro-phthalate		1	FLEXOL PEP (originally produced by Union Carbide)
DIDG	Diisodecyl glutarate		4	PLASTHALL DIDG
DIDP[2.2,2.7,2.8]	Diisodecyl phthalate[2.2,2.7,2.8]		1	PALATINOL Z
DIDS	Diisodecyl sebacate, or diisodecyl succinate		4	
DIDZ	Diisodecyl azelate		4	
DIHG	Diisoheptyl glycolate		7.2.2	
DIHP[2.2,2.7,2.8]	Diisoheptyl phthalate[2.2,2.7,2.8]		1	

ABBREVIATION (STANDARD OR USUAL)	NAME (STANDARD OR USUAL)	OTHER NAMES AND/OR ABBREVIATIONS	CHEMICAL TYPE REF†	REMARKS, OR EXAMPLES OF SPECIFIC COMMERCIAL PRODUCTS
DIHXP[2.2,2.7,2.8]	Diisohexyl phthalate[2.2,2.7,2.8]		1	
DINA[2.2,2.7,2.8]	Diisononyl adipate[2.2,2.7,2.8]		4	
DINP[2.2,2.7,2.8]	Diisononyl phthalate			DINP is available in several commercial versions differing in the number of positions of the branches of the esterifying isononyl alcohol chain. Some examples are: VESTINOL 9, JAYFLEX DINP, PALATINOL N (one and two branches); VESTINOL N, PALATINOL DN (two to three branches); WITAMOL 150 (three branches)
DIOA[2.2,2.7,2.8]	Diisooctyl adipate[2.2,2.7,2.8]		4	
DIOIP	Diisooctyl isophthalate		1	
DIOM[2.7]	Diisooctyl maleate[2.7]		4	
DIOP[2.2,2.7,2.8]	Diisooctyl phthalate[2.2,2.7,2.8]		1	
DIOS[2.7]	Diisooctyl sebacate[2.7]		4	
DIOZ[2.7]	Diisooctyl azelate[2.7]		4	
DIPP[2.2,2.7,2.8]	Diisopentyl phthalate[2.2,2.7,2.8]		1	
DITDP[2.2,2.8]	Diisotridecyl phthalate[2.2,2.8]	Ditridecyl phthalate (DTDP)	1	WITAMOL 190
DIUP	Diisoundecyl phthalate		1	
DLP	Dilauryl phthalate		1	PALATINOL DP
DMA	Dimethyl adipate		4	
DMCHP	Di-(methylcyclohexyl) phthalate		1	
DMEP[2.7]	Di-(2-methyloxyethyl) phthalate	Bis(2-methoxyethyl) phthalate *; di-(ethylene glycol monomethyl ether) phthalate; di-(methyl cellosolve) phthalate	1	The abbreviation "DMGP" ("dimethoxyglycol phthalate") also appears in some trade names of commercial plasticisers.
DMIP	Dimethyl isophthalate		1	

ABBREVIATION (STANDARD OR USUAL)	NAME (STANDARD OR USUAL)	OTHER NAMES AND/OR ABBREVIATIONS	CHEMICAL TYPE REF†	REMARKS, OR EXAMPLES OF SPECIFIC COMMERCIAL PRODUCTS
DMP[2.2,2.7,2.8]	Dimethyl phthalate[2.2,2.7,2.8,*]		1	
DMS[2.7,2.8]	Dimethyl sebacate[2.7,2.8,*]		4	
DNA	Dinonyl adipate		4	
DNAP	Di-n-alkyl phthalates	E.g. Di-(Linevol 79) phthalate; Di-(Alfol 810) phthalate	1	These phthalates are esters of o-phthalic acid with mixtures of predominantly straight-chain alcohols manufactured from synthetic hydrocarbon feedstocks, such as e.g. C_7-C_9 and C_9-C_{11} Linevol alcohols (Shell Chemicals), or C_6-C_{10}, C_7-C_{11}, C_8-C_{10}, C_9-C_{11} Alfol alcohols (Continental Oil Co.).
DNF[2.7]	Dinonyl fumarate[2.7,*]		4	
DNM[2.7]	Dinonyl maleate[2.7,*]		4	
DNNP	Di-n-nonyl phthalate		1	
DNOA	Di-n-octyl adipate	Dioctyl adipate*	4	
DNOP[2.2,2.7,2.8]	Di-n-octyl phthalate[2.2,2.7,2.8]	Dioctyl phthalate*	1	
DNP[2.2,2.7,2.8]	Dinonyl phthalate[2.2,2.7,2.8,*]		1	A commercial version is di-(3,5,5-trimethylhexyl) phthalate
DNS[2.7]	Dinonyl sebacate[2.7,*]		4	
DOA[2.2,2.7,2.8]	Dioctyl adipate[2.2,2.7,2.8]	Bis(2-ethylhexyl) adipate*; Di-2-ethylhexyl adipate	4	The abbreviation "DEHA" (diethylhexyl adipate) also fairly commonly encountered
DOF	Dioctyl fumarate	Bis(2-ethylhexyl) fumarate*	4	
DOHXHP	Dioctyl hexahydrophthalate		1	
DOIP[2.2,2.7,2.8]	Dioctyl isophthalate[2.2,2.7,2.8]		1	
DOM	Dioctyl maleate	Bis(2-ethylhexyl) maleate*	4	
DOP[2.2,2.7,2.8]	Dioctyl phthalate[2.2,2.7,2.8]	Bis(2-ethylhexyl) phthalate*; Di-2-ethylhexyl phthalate	1	The abbreviation "DEHP" (diethylhexyl phthalate) also fairly commonly encountered
DOPF	Dioctyl phenyl phosphate	DOPP; Bis(2-ethylhexyl) phenyl orthophosphate*	2	

ABBREVIATION (STANDARD OR USUAL)	NAME (STANDARD OR USUAL)	OTHER NAMES AND/OR ABBREVIATIONS	CHEMICAL TYPE REF†	REMARKS, OR EXAMPLES OF SPECIFIC COMMERCIAL PRODUCTS
DOS[2.2,2.7,2.8]	Dioctyl sebacate[2.2,2.7,2.8]	Bis(2-ethylhexyl) sebacate*	4	The abbreviation "DEHS" (diethylhexyl sebacate) also encountered
DOTP[2.2,2.7,2.8]	Dioctyl terephthalate[2.2,2.7,2.8]	Bis(2-ethylhexyl) terephthalate*	1	KODAFLEX DOTP
DOZ[2.2,2.7,2.8]	Dioctyl azelate[2.2,2.7,2.8]	Bis(2-ethylhexyl) azelate*	4	
DPA	Dipropyl adipate	Di-n-propyl adipate (DNPA)	4	
DPCF[2.2,2.7,2.8]	Diphenyl cresyl phosphate[2.2,2.7,2.8]	DPCP; tolyl diphenyl phosphate (TDP); cresyl diphenyl phosphate (CDP); diphenyl x-tolyl orthophosphate*	2	DISFLAMOLL DPK. In the IUPAC name "x" indicates a mixture of *o*, *p*, and *m*, or any single one of these
DPGDB[2.7]	Di-x-propylene glycol dibenzoate[2.7]	Dipropylene glycol dibenzoate (DPDB)	9.1	BENZOFLEX 9-88
DPOF[2.2]	Diphenyl octyl phosphate[2.7]	DPOP; octyl diphenyl phosphate (ODP); 2-ethylhexyl diphenyl orthophosphate* or octyl diphenyl orthophosphate*	2	DISFLAMOLL DPO
DPP[2.7,2.8]	Diphenyl phthalate[2.7,2.8,*]		1	
DPP	Dipropyl phthalate or Dipentyl phthalate		1	
DTDP[2.7]	Diisotridecyl phthalate[2.7]	DITDP	1	Di-n-tridecyl phthalate (DTDP or DNTDP) is not normally used as a plasticiser, so that the abbreviation DTDP is often written for the isodecyl compound (DITDP), which is so used
DTHFP	Di-(tetrahydro-furfuryl) phthalate		1	FLEXOL G-316 (originally produced by Union Carbide)
DTMCHP	Di-(trimethylcyclo-hexyl) phthalate		1	

ABBREVIATION (STANDARD OR USUAL)	NAME (STANDARD OR USUAL)	OTHER NAMES AND/OR ABBREVIATIONS	CHEMICAL TYPE REF†	REMARKS, OR EXAMPLES OF SPECIFIC COMMERCIAL PRODUCTS
DTMHXA	Di-(3,5,5-trimethyl-hexyl) adipate		4	HYLENE B } Formerly produced by Du Pont
DTMHXDG	Di-(3,5,5-trimethyl-hexyl) diglycolate		7.2.2	HYLENE C } Formerly produced by Du Pont
DUP[2.2,2.7,2.8]	Diundecyl phthalate[2.2,2.7,2.8]		1	PALATINOL 11
EAE	Epoxy-alkyl ester	E.g. octyl epoxy stearate (OEST)	6.3	EDENOL B35, B74 and D72; LANKROFLEX ED3 and ED6; MONOPLEX S-71; REOPLAST 38 and 42
EAR	Ethyl acetyl ricinoleate		7.1	
EBSA	Ethylbenzene sulfonamide		10	SANTICIZER 128
EEL	Ethoxyethyl laurate	Ethylene glycol monoethyl ether laurate; ethyl cellosolve laurate	7.1	
EEO	Ethoxyethyl oleate	Ethylene glycol monoethyl ether oleate; ethyl cellosolve oleate	7.1	
EEST	Ethoxyethyl stearate	Ethylene glycol monoethyl ether stearate; ethyl cellosolve stearate	7.1	
EGDA	Ethylene glycol diabietate		9.2	FLEXALYN C (originally produced by Hercules)
EGDA	Ethylene glycol diadipate		4	
EGDB	Ethylene glycol dibenzoate		9.1	BENZOFLEX E-60
EGDB	Ethylene glycol dibutyrate		7.2.1	
EGDP	Ethylene glycol dipropionate		7.2	
EGR	Ethylene glycol ricinoleate		7.1	

ABBREVIATION (STANDARD OR USUAL)	NAME (STANDARD OR USUAL)	OTHER NAMES AND/OR ABBREVIATIONS	CHEMICAL TYPE REF†	REMARKS, OR EXAMPLES OF SPECIFIC COMMERCIAL PRODUCTS
ELO[2.2,2.7,2.8]	Epoxidised linseed oil [2.2,2.7,2.8]		6.2	EPOXOL 9-5
EPEG	Ethyl phthalyl ethyl glycolate		7.2.2	SANTICIZER E-15
EPUR	Ethyl phenyl urethane	Phenyl ethyl carbamate	15	MOLLIT 41 (originally produced by Bayer)
ER	Ethyl ricinoleate		7.1	
ESO[2.2,2.7,2.8]	Epoxidised soya bean oil[2.2,2.7,2.8]	ESBO		EDENOL D 31; LANKROFLEX GE; PARAPLEX G 62; REOPLAST 39
ETS	Ethyltoluene sulfonate		10	
ETSA	Ethyltoluene sulfonamide		10	SANTICIZER 3
GA	Glycerol monoacetate	Acetin; GMA	7.2.1	
GB	Glycerol monobenzoate	GMB	9.1	MOLLIT B (originally produced by Bayer)
GDA	Glycerol diacetate	Diacetin	7.2.1	DIACETIN
GDHA	Glycerol di-(hydroabietate)		9.2	STAYBELITE ESTER No 10 (originally produced by Hercules)
GL	Glycerol monolaurate	GML	7.1	
GO	Glycerol monoleate	GMO	7.1	
GP	Glycerol phthalate	GMP	1	
GR	Glycerol monoricinoleate	GMR	7.1	
GST	Glycerol monostearate	GMST	7.1	
GTA[2.7]	Glycerol triacetate[2.7]*	Triacetin	7.2.1	TRIACETIN
GTAR	Glycerol tri-(acetal ricinoleate)		7.1	
GTB	Glycerol tributyrate		7.2.1	TRIBUTYRIN } Originally produced by Eastman Chemicals
GTP	Glycerol tripropionate		7.2.1	TRIPROPIONIN } Originally produced by Eastman Chemicals

ABBREVIATION (STANDARD OR USUAL)	NAME (STANDARD OR USUAL)	OTHER NAMES AND/OR ABBREVIATIONS	CHEMICAL TYPE REF†	REMARKS, OR EXAMPLES OF SPECIFIC COMMERCIAL PRODUCTS
HBP	Heptyl benzyl phthalate		1	
HE	Hydrocarbon extender	Hydrocarbon plasticiser extender	14	
HNP	Heptyl nonyl phthalate		1	
HNUA[2.2,2.7,2.8]	Heptyl nonyl undecyl adipate[2.2,2.7,2.8]	711 A[2.2,2.7,2.8]; alkyl (C_7-C_{11}) adipate ($A_{711}A$)	4	
HNUP[2.2,2.7,2.8]	Heptyl nonyl undecyl phthalate[2.2,2.7,2.8]	711 P[2.2,2.7,2.8]; alkyl (C_7-C_{11}) phthalate ($A_{711}P$)	1	
HXBP	Hexyl benzyl phthalate	See BHXP	1	
HXEST	Hexyl epoxy stearate		7.1	
HXODA[2.2,2.7,2.8]	Hexyl octyl decyl adipate[2.2,2.7,2.8]	610 A[2.2,2.7,2.8]; alkyl (C_6-C_{10}) adipate ($A_{610}A$)	4	
HXTDA	Hexanetriol diacetate		7.2.1	DIACETIN H
HXTTA	Hexanetriol triacetate		7.2.1	TRIACETIN H
IBAR	Isobutyl acetyl ricinoleate		7.1	
IBIDP	Isobutyl isodecyl phthalate		1	
IBNP	Isobutyl nonyl phthalate		1	
IBOB	Isobutyl oxybenzoate		9.1	
IDB	Isodecyl benzoate		9.1	BENZOFLEX 131
IDDDF	Isodecyl didecyl phosphate	IDDDP	2	
IDDPF	Isodecyl diphenyl phosphate	IDDPP	2	
IODA	Isooctyl decyl adipate		4	

ABBREVIATION (STANDARD OR USUAL)	NAME (STANDARD OR USUAL)	OTHER NAMES AND/OR ABBREVIATIONS	CHEMICAL TYPE REF†	REMARKS, OR EXAMPLES OF SPECIFIC COMMERCIAL PRODUCTS
IOEST	Isooctyl epoxy stearate		7.1	
IOET	Isooctyl epoxy tallate		6.3	The abbreviation "ETO" used in some commercial plasticiser trade names
IOIDA	Isooctyl isodecyl adipate		4	
IOIDP	Isooctyl isodecyl phthalate		1	
IOP	Isooctyl palmitate		7.1	MORFLEX 1125
IOT	Isooctyl tallate		7.1	MONOPLEX X-73
IPBSA	Isopropylbenzene sulfonamide		10	SANTICIZER 130
IPO	Isopropyl oleate		7.1	MORFLEX IPO
IPPDPF	Isopropylphenyl diphenyl phosphate	IPPDPP	2	
IPPF	Isopropylated phenyl phosphate	IPPP	2	KRONITEX 50 and 100; PLIABRAC 519, 521 and 524; REOFOS 50, 65 and 95
MADHXCH	Methylamine dihexyl cyclohexanone		12	
MAR	Methyl acetyl ricinolate		7.1	
MBA	Methoxybutyl acetate		7.2.1	
MBBSA	Methyl-butyl benzene sulfonamide		10	DELLATOL BBS
MBSA	Methylbezene sulfonamide		10	DELLATOL MMA
MCHO	Methyl cyclohexyl oleate		7.1	MOLLIT MCO (originally produced by Bayer)
MDHA	Methyl dihydroabietate		9.2	
MEAR	Methoxyethyl acetyl ricinoleate	Ethylene glycol monomethyl ether acetyl ricinoleate; methyl cellosolve acetyl ricinoleate	7.1	

ABBREVIATION (STANDARD OR USUAL)	NAME (STANDARD OR USUAL)	OTHER NAMES AND/OR ABBREVIATIONS	CHEMICAL TYPE REF†	REMARKS, OR EXAMPLES OF SPECIFIC COMMERCIAL PRODUCTS
MEL	Methoxyethyl laurate	Ethylene glycol monomethyl ether laurate; methyl cellosolve laurate	7.1	
MEO	Methoxyethyl oleate	Ethylene glycol monomethyl ether oleate; methyl cellosolve oleate	7.1	
MER	Methoxyethyl ricinoleate	Ethylene glycol monomethyl ether ricinoleate; methyl cellosolve ricinoleate	7.1	
MEST	Methoxyethyl stearate	Ethylene glycol monomethyl ether stearate; methyl cellosolve stearate	7.1	
MHST	Methyl hydroxy stearate		7.1	
MO	Methyl oleate		7.1	
MPCST	Methyl pentachloro stearate		7.1	
MPEG	Methyl phthalyl ethyl glycolate		7.2.2	SANTICIZER M - 17
MPST	Methyl phenylol stearate		7.1	SANTICIZER MPS
MPTMA	Methylol propane trimethacrylate		13	
MPUR	Methyl phenyl urethane	Phenyl methyl carbamate	15	MOLLIT 40 (originally produced by Bayer)
MR	Methyl ricinolate		7.1	
MTSA	Methyltoluene sulfonamide		10	
NBOP	n-butyl octyl phthalate		1	
NBST	n-butyl stearate		7.1	

ABBREVIATION (STANDARD OR USUAL)	NAME (STANDARD OR USUAL)	OTHER NAMES AND/OR ABBREVIATIONS	CHEMICAL TYPE REF†	REMARKS, OR EXAMPLES OF SPECIFIC COMMERCIAL PRODUCTS
NODA	n-octyl decyl adipate	See also ODA	4	MORFLEX 375
NODTM	n-octyl decyl trimellitate	See also ODTM	3	
NONDA	n-octyl n-decyl adipate	See also ODA	4	MONOPLEX NODA
NONDP	n-octyl n-decyl phthalate	See also ODP	1	MORFLEX 125
NONDTM	n-octyl n-decyl trimellitate	See also ODTM	3	MORFLEX 525
NONDZ	n-octyl n-decyl azelate	See also ODZ	4	MORFLEX 4
NPBSA	n-propylbenzene sulfonamide		10	SANTICIZER 130
NPGDR	Neopentyl glycol diricinoleate		7.1	Originally available as FERRICIN 30
NUA[2.2,2.7,2.8]	Nonyl undecyl adipate[2.2,2.7,2.8]	911A; alkyl (C_9-C_{11}) adipate; $A_{911}A$	4	
NUP[2.2,2.7,2.8]	Nonyl undecyl phthalate	911P; alkyl (C_9-C_{11}) phthalate; $A_{911}P$		
ODA[2.2,2.7,2.8]	Octyl decyl adipate[2.2,2.7,2.8]	Decyl octyl adipate*; see also NODA and NONDA	4	
ODCF	Octyl dicresyl phosphate	ODCP	2	SANTICIZER 142
ODP[2.2,2.7,2.8]	Octyl decyl phthalate[2.2,2.7,2.8]	Decyl octyl phthalate*; see also NONDP	1	MORFLEX 175
ODPF	Octyl diphenyl phosphate	ODPP; 2-ethylhexyl diphenyl phosphate (EHDPF)	2	DISFLAMOLL DPO
ODPS	Oxydiphenyl sulfone	p-oxydiphenyl sulfone	15	IGAMID-WEICHMACHER 8 (formerly produced by IG Farben)
ODTM[2.2,2.7,2.8]	n-octyl decyl trimellitate	Decyl octyl hydrogen benzene-1,2,4-tricarboxylate*; see also NODTM	3	

ABBREVIATION (STANDARD OR USUAL)	NAME (STANDARD OR USUAL)	OTHER NAMES AND/OR ABBREVIATIONS	CHEMICAL TYPE REF†	REMARKS, OR EXAMPLES OF SPECIFIC COMMERCIAL PRODUCTS
ODZ	Octyl decyl azelate	See also NONDZ	4	
OEST	Octyl epoxy stearate		6.3	LANKROFLEX ED
OET	Octyl epoxy tallate	2- ethylhexyl epoxy tallate	6.3	DRAPEX 4.4
OM	Octyl maleate		4	
OOB	Octyl *p*-oxybenzoate		9.1	
OST	Octyl stearate		7.1	
PCB	Pentaerithritol caprylate butyrate		7.1	
PEG	Polyethylene glycol		12	
PEGDO	Polyethylene glycol dioctoate	Polyethylene glycol di-2-ethylhexanoate	7.1	
PEL	Phenoxy ethyl laurate		7.1	
PEO	Phenoxy ethyl oleate		7.1	
PER	Pentaerithritol monoricinoleate		7.1	
PETH	Pentaerithritol tetraheptanoate		7.1	GARBEFLEX PE 47
PETR	Pentaerithritol tetraricinoleate		7.1	
PGDB	Propylene glycol dibenzoate		9.1	BENZOFLEX 400
PGO	Propylene glycol 1,2-oleate		7.1	
PGP	Propylene glycol 1,2-phthalate		1	
PGR	Propylene glycol ricinoleate		7.1	
PO[2.7]	Paraffin oil[2.7]		14	
PO	Phenyl oleate		7.1	
POEDB	Poly(oxyethylene) dibenzoate	Polyethylene glycol dibenzoate (PEGDB)	9.1	
POEDL	Poly(oxyethylene) dilaurate	Polyethylene glycol dilaurate (PEGDL)	7.1	

ABBREVIATION (STANDARD OR USUAL)	NAME (STANDARD OR USUAL)	OTHER NAMES AND/OR ABBREVIATIONS	CHEMICAL TYPE REF†	REMARKS, OR EXAMPLES OF SPECIFIC COMMERCIAL PRODUCTS
PPA[2.7]	Poly(propylene adipate)[2.7]		5	HEXAPLAS PPA (originally produced by ICI)
PPAL	Poly(propylene adipate/laurate)		5	HEXAPLAS PPL (originally produced by ICI)
PPGP	Phenyl polyglycol palmitate	Phenylpolyethylene glycol palmitate	7.1	
PPS[2.7]	Poly(propylene sebecate)[2.7]		5	PARAPLEX G 25
PPZ	Poly(propylene azelate)		5	
SOA[2.7]	Sucrose octa-acetate[2.7]	Sucrose octaacetate*	7.2.1	
TAC[2.8,2.10]	Triallyl cyanurate	2,4,6-triallyloxy-1,3,5-triazine*	13	
TBAC[2.7]	Tributyl *O*-acetylcitrate[2.7]*	Acetyl n-butyl citrate	8	CITROFLEX A-4
TBC	Tributyl citrate		8	CITROFLEX C-4
TBEP[2.7]	TBEF; tri-(2-butoxyethyl) phosphate[2.7]	Tris(2-butoxyethyl) orthophosphate*	2	
TBP[2.7]	TBF; tributyl phosphate[2.7]	Tributyl orthophosphate*	2	
TBPP	TBPF; tri-(p-*ter*-butylphenyl) phosphate		2	
TBTDS	Tetra-n-butyl thiodisuccinate	TNBTDS	4	FLEXOL TWS (originally produced by Union Carbide)
TCEF[2.2,2.7,2.8]	TCEP; trichloroethyl phosphate	Tris(2-chloroethyl) orthophosphate*	2	
TCF[2.2,2.7,2.8]	TCP; tricresyl phosphate[2.2,2.7,2.8]	Tritolyl phosphate (TTP); tri-x-tolyl orthophosphate* (where x stands for *o*, *m*, *p*, or a mixture)	2	In the UK, "TCP" is a registered trade mark for another product, so that the abbreviation "TTP" is used for tricresyl phosphate.

ABBREVIATION (STANDARD OR USUAL)	NAME (STANDARD OR USUAL)	OTHER NAMES AND/OR ABBREVIATIONS	CHEMICAL TYPE REF†	REMARKS, OR EXAMPLES OF SPECIFIC COMMERCIAL PRODUCTS
TCPF	TCPP; tri-(mono-chloropropyl) phosphate		2	
TDBPP[2.7]	TDBPF; tri-(2,3-dibromopropyl) phosphate[2.7]	Tris(2,3-dibromopropyl) orthophosphate*	2	
TDCPP[2.7]	TDCPF; tri-(2,3-dichloropropyl) phosphate[2.7]	Tris(2,3-dichloropropyl) orthophosphate*	2	
TEAC[2.7]	Triethyl *O*-acetyl citrate[2.7]*	Acetyltriethyl citrate	8	CITROFLEX A-2
TEC	Triethyl citrate		8	CITROFLEX C-2
TEGDA	Triethylene glycol diacetate	Triglycol diacetate	7.2.1	
TEGDB	Triethylene glycol dibenzoate	Triglycol dibenzoate	9.1	
TEGDC	Triethylene glycol dicaprylate	Triglycol dicaprylate	7.1	
TEGDEB	Triethylene glycol di-(2-ethylbutyrate)	Triglycol dihexoate (see TEGDHX)	7.2.1	FLEXOL 3 GH
TEGDEHX	Triethylene glycol di-(2-ethylhexoate)	Triglycol dioctoate (see TEGDO)	7.1	FLEXOL 360 (originally produced by Union Carbide)
TEGDHA	Triethylene glycol di-(hydroabietate)	Triglycol di-(hydroeabietate)	7.2.1	STAYBELITE ESTER NO. 3 (originally produced by Hercules)
TEGDHX	Triethylene glycol dihexoate	Triglycol dihexoate (see also TEGDEB)	7.1	
TEGDMA	Tetraethylene glycol dimethacrylate		13	
TEGDO	Triethylene glycol dioctoate	Triglycol dioctoate (see also TEGDEHX)	7.1	
TEGDP	Triethylene glycol dipelargonate	Triglycol dipelargonate	7.1	
TEGMEE	Triethylene glycol monoethyl ether	Ethoxytriglycol; trioxitol	7.1	
THFO[2.7]	Tetrahydrofurfuryl oleate[2.7]*		7.1	PLASTOLEIN THFO (originally produced by Emery Industries Inc.)

ABBREVIATION (STANDARD OR USUAL)	NAME (STANDARD OR USUAL)	OTHER NAMES AND/OR ABBREVIATIONS	CHEMICAL TYPE REF†	REMARKS, OR EXAMPLES OF SPECIFIC COMMERCIAL PRODUCTS
THODAC	Tri-(n-hexyl n-acetyl n-decyl *O*-octyl) citrate	Tri-(n-alkyl (Alfol 610) citrate	8	
THTM[2.2,2.7,2.8]	Triheptyl trimellitate[2.2,2.7,2.8]	Triheptyl benzene-1,2,4-tricarboxylate*	3	
THXAC	Tri-(n-hexyl) *O*-acetyl citrate	Acetyl tri-(n-hexyl) citrate	8	CITROFLEX A-6
THXBC	Tri-(n-hexyl) *O*-butyryl citrate	n-butyryl tri-(n-hexyl) citrate	8	
TIDTM	Triisodecyl trimellitate		3	
TINTM	Triisononyl trimellitate		3	
TIOTM[2.2,2.7,2.8]	Triisooctyl trimellitate[2.2,2.7,2.8]	Tris(6-methylheptyl) benzene-1,2,4-tricarboxylate*	3	
TIPPF	TIPPP; triisopropylphenyl phosphate		2	
TMETA	Trimethylol ethane triacetate		7.2.1	KESSCOFLEX MEA
TMPDB	2,2,4-trimethyl-1,3-pentanediol benzoate	Texanol monobenzoate	9.1	Texanol is a registered trade name of Eastman Chemicals Inc. for 2,2,4-trimethyl-1,3-pentanediol
TMPDDB	2,2,4-trimethyl-1,3-pentanediol dibenzoate	Texanol benzoate	9.1	BENZOFLEX 354
TMPDDIB	2,2,4-trimethyl-1,3-pentanediol diisobutyrate	Texanol isobutyrate	7.2.1	KODAFLEX TXIB
TMPDIBA	2,2,4-trimethyl-1,3-pentanediol isobutyrate adipate	Texanol monoisobutyrate adipate	4	
TMPDIBB	2,2,4-trimethyl-1,3-pentanediol isobutyrate benzoate	Texanol monobutyrate benzoate	9.1	

ABBREVIATION (STANDARD OR USUAL)	NAME (STANDARD OR USUAL)	OTHER NAMES AND/OR ABBREVIATIONS	CHEMICAL TYPE REF†	REMARKS, OR EXAMPLES OF SPECIFIC COMMERCIAL PRODUCTS
TMPDMA	Trimethylolpropane dimethacrylate		13	
TMPTA	Trimethylolpropane triacetate		7.2.1	KESSCOFLEX MEA
TMPTCA	Trimethylolpropane trichloroacetate		7.2.1	MOLLIT (originally produced by Bayer)
TMPTMA	Trimethylolpropane trimethacrylate	TMPTM	13	
TNOTM	Tri-n-octyl trimellitate		3	
TOAC	Trioctyl *O*-acetyl citrate	Tri-(2-ethylhexyl) acetyl citrate	8	CITROFLEX A-8
TODTM	Tri-(2-n-octyl n-decyl) trimellitate		3	
TOF[2.2,2.7,2.8]	TOP; trioctyl phosphate[2.2,2.7,2.8]	Tri-2-ethylhexyl phosphate[2.8]; tris(2-ethylhexyl) orthophosphate*	2	DISFLAMOLL TOF
TOPM[2.2,2.7,2.8]	Tetraoctyl pyromellitate[2.2,2.7,2.8]	Tetra-2-ethylhexyl pyromellitate[2.2,2.8]; tetrakis (2-ethylhexyl) benzene-1,2,4,5-tetracarboxylate*	3	
TOTM[2.2,2.7,2.8]	Trioctyl trimellitate[2.2,2.7,2.8]	Tri-2-ethylhexyl trimellitate[2.2,2.8]; tris(2-ethylhexyl) benzene-1,2,4-tricarboxylate*	3	
TPP[2.2,2.7,2.8]	TPF; triphenyl phosphate[2.2,2.7,2.8]	Triphenyl orthophosphate	2	
TSA	*p*-toluene sulfonamide		10	SANTICIZER 139
TTF[2.2]	Tritolyl phosphate[2.2] - see TCF[2.2]	Tricresyl phosphate	2	

ABBREVIATION (STANDARD OR USUAL)	NAME (STANDARD OR USUAL)	OTHER NAMES AND/OR ABBREVIATIONS	CHEMICAL TYPE REF†	REMARKS, OR EXAMPLES OF SPECIFIC COMMERCIAL PRODUCTS
TTHFF	Tris(tetrahydrofurfuryl) phosphate	THFP		KRONIFLEX THFP (formerly produced by FMC Corp.)
TTP[2.8]	Tritotyl phosphate[2.8] - see TCF[2.8]	Tricresyl phosphate	2	
TXF[2.7]	TXP; trixylyl phophate[2.7]	Tri-x-y-xylyl orthophosphate* (where x, y, denote *o*, *m*, *p* or a mixture)	2	
UDP	Undecyl dodecyl phthalate		1	
XDEGB	Xylyl diethylene glycol benzoate	Xylyl diglycol benzoate	9.1	
XTS	Xylyl *p*-toluene sulfonate		10	
XDPF	XDPP; xylyl diphenyl phosphate		2	

PART 2

SECTION 4

ABBREVIATED TERMS (LETTER SYMBOLS) FOR MISCELLANEOUS COMPONENTS OF PLASTICS

ABBREVIATED TERM (STANDARD OR USUAL)	NAME: STANDARD OR USUAL	NAME: OTHER NAMES, ABBREVIATED TERMS, OR DESCRIPTIONS	FUNCTION	REMARKS, AND/OR EXAMPLES OF COMMERCIAL PRODUCTS
AANA[2.8,2.12]	Aldol-α-naphthylamine		Antioxidant	
ACM[2.4,2.5,2.6]	Acrylic rubber		Impact modifier	HYTEMP
ADC[2.8,2.10,2.12]	Azodicarbonamide	C,C′-azodi- (formamide)*; azobisformamide (ABFA); ADA	Chemical blowing agent used in various plastics (e.g. PVC; polyolefins; PS; ABS; PA; TPE).	CELOGEN AZ and AZNP; UNICELL D series; POROFOR ADC.
ADPA[2.8,2.10]	Acetone-diphenylamine condensate		Antioxidant	
APPD[2.8,2.10]	N-alkyl-N′-phenyl-*p*-phenylenediamine		Antioxidant; retardant for UP resins	
ATH	Alumina trihydrate; alumina hydrate; $Al(OH)_3$	Hydrated alumina; $Al_2O_3.3H_2O$	Flame retardant for various plastics	HYDRAL; MICRAL
ATP	Alpha-tocopherol	Synthetic vitamin E	Antioxidant used in polyolefins	RONTEC ATP; UVINUL 2003 AO
AZDN	Azobis(isobutyronitrile)		Chemical blowing agent (mainly for PVC); free-radical polymerisation initiator.	MICROFINE AZDN; FICEL AZDN
BA[2.12]	Butyraldehyde-aniline condensate		Accelerator; crosslinking agent	
p-BBp 14[2.12]	4,4′-butylidene-bis(6-*t*-butyl-*m*-cresol)		Antioxidant	
BBPIB[2.12]	1,4-bis(t-butylperoxy-isopropyl) benzene		Crosslinking agent	

* Name conforming to the nomenclature of the International Union of Pure and Applied Chemistry (IUPAC).

ABBREVIATED TERM (STANDARD OR USUAL)	NAME: STANDARD OR USUAL	NAME: OTHER NAMES, ABBREVIATED TERMS, OR DESCRIPTIONS	FUNCTION	REMARKS, AND/OR EXAMPLES OF COMMERCIAL PRODUCTS
BCP	*tert*-butylcumyl peroxide		Crosslinking agent	TRIGONOX T
BDSH[2.8,2.10,2.12]	Benzene-1,3-disulfonyl-hydrazide	Benzene-1,3-di(sulfonohydrazide)*	Chemical blowing agent	
BEST	Butoxyethyl stearate	Butyl cellosolve stearate; BES	External lubricant (used in audio, video and computer tapes)	KESSCO BES
BHA[2.12]	Butyl hydroxyanisole		Antixoxidant	
BHEB	2,6-di-*tert*-butyl-4-ethyl-hydroxybenzene; butylated hydroxyethylbenzene		Antioxidant (for PE and other plastics)	PRODOX-500. For determination methods see ASTM D 1996-92; D5524-94
BHP	*tert*-butyl hydroperoxide		Curing agent for UP resins	
BHT[2.8,2.10,2.12]	2,6-di-*tert*-butyl-4-methylphenol; butylated hydroxytoluene	2,6-di-*tert*-butyl-*p*-cresol*	Antioxidant (for polyolefins; PS-HI; ABS; EVA)	For determination methods see ASTM D 1996-92; D5524-94
BHPF	bis(hydroxyphenyl fluorine)		A comonomer in the production of PC copolymers	Cf. INSPIRE
BMPPD[2.8,2.10]	N,N′-bis(1,4-dimethylpentyl)-*p*-phenylenediamine*	77PD[2.8,2.10,2.12]	Antioxidant; retardant for UP resins	Abbreviation 77PD preferred in ref. 2.8
BMTS[2.12]	bis-morpholino-thiocarbamyl sulfenamide		Accelarator (rubbers)	
BPA	*tert*-butyl peracetate	TBPA	Curing agent for UP resins	
BPB	*tert*-butylperoxy benzoate	*tert*-butyl perbenzoate (TBPB)[2.2]	Curing agent for UP resins; crosslinking agent	LUCIDOL S 50
BPO[2.12]	Benzoyl peroxide	BP, dibenzoyl peroxide(DBP)	Curing agent for UP resins; crosslinking agent	
BPV	n-butyl-bis(4,4′-*tert*-butyl peroxy) valerate		Curing agent for UP resins; crosslinking agent	
BSH[2.8,2.10,2.12]	Benzene sulfonylhydrazide	Benzene sulfohydrazide; benzene sulfono-hydrazide	Chemical blowing agent	
BTF	Borontrifluoride		Catalyctic curing agent for epoxy resins	LEECURE B series

ABBREVIATED TERM (STANDARD OR USUAL)	NAME: STANDARD OR USUAL	NAME: OTHER NAMES, ABBREVIATED TERMS, OR DESCRIPTIONS	FUNCTION	REMARKS, AND/OR EXAMPLES OF COMMERCIAL PRODUCTS
CBP	bis(*p*-chlorobenzoyl) peroxide		Curing agent for UP resins; crosslinking agent	PERKADOX PC
CBS[2.12]	N-cyclohexyl-2-benzothiazole sulfenamide		Accelarator (rubbers)	
CEA[2.12]	Cyclohexylethylamine		Accelerator (rubbers)	
CHDA	1,4-cyclohexane dicarboxylic acid	1,4-CHDA; CHDM	Speciality monomer (inter alia for polyamides and liquid-crystal polymers)	Produced by Eastman Chemical Co.
CHDI[2.12]	1,4- cyclohexane diisocyanate		Isocyanate reactant for PUR production	
CHP	Cumene hydroperoxide		Curing agent for UP resins	PEROXIMON CH; TRIGONOX K
CPB	Cumyl phenol benzoate		Processing aid, used in PVC, PA, and RIM PUR	KENPLAST ESB
CPPD[2.8,2.10,2.12]	N-cyclohexyl-N′-phenyl-*p*-phenylenediamine*		Antioxidant, retardant for UP resins	Highly effective antioxidant for polyamides (but can cause discoloration)
CTP[2.12]	N-cyclohexyl-thiophthalamide		Retarder (rubbers)	
CUTH	Cyanuric trihydrazide	THTA (q.v.)	Chemical blowing agent	
DAHQ[2.12]	2,5-di-*tert*-amylhyroquinone		Antioxidant (for polyether polyols, inter alia)	
DBA[2.12]	Dibutylamine		Accelarator (rubbers)	
DBDPO	Decabromodiphenyl oxide	Decabromodiphenyl ether (DBDPE)	Flame retardant	Used in engineering plastics (in conjuntion with antimony oxides). E.g. SATEX 102 E)
DBHQ	2,5-di-*tert*-butylhydro-quinone		Antioxidant (for polyether polyols, inter alia)	
DBNPG	Dibromoneopentyl glycol		Reactive flame retardant for UP resins and PUR systems	
DBODH	N,N′-dibenzyloxalyl-dihydrazide		Antioxidant (metal deactivator) for polyolefins	EASTMAN INHIBITOR OABH

ABBREVIATED TERM (STANDARD OR USUAL)	NAME STANDARD OR USUAL	OTHER NAMES, ABBREVIATED TERMS, OR DESCRIPTIONS	FUNCTION	REMARKS, AND/OR EXAMPLES OF COMMERCIAL PRODUCTS
DBP	Dibenzoyl peroxide	Benzoyl peroxide; BPO	Curing agent for UP resins; crosslinking agent	
DBPC[2.12]	1,1′-bis(*tert*-butyl peroxy)-3,5,5- trimethylcyclohexane		Curing and crosslinking agent	
DBPC	Di-*tert*-butyl-*p*-cresol		Antioxidant	
DBS	Dibromostyrene		Reactive flame retardant	Can replace styrene monomer
DBTDA	Dibutyltin diacetate	Dibutyltin acetate (DBTA)	Catalyst for PUR systems	CATA-CHEK 867
DBTDEH	Dibutyltin diethyl hexoate		Catalyst for PUR systems	
DBTDL	Dibutyltin dilaurate	Dibutyltin laurate (DBTL)	Heat stabiliser for PVC; catalyst for PUR systems; catalyst in chemical crosslinking for PE.	CATA-CHEK 820
DBTDM	Dibutyltin dimaleate	Dibutyltin maleate (DBTM)	Heat stabiliser for PVC; catalyst for PUR systems.	
DBTO	Dibutyltin oxide		Co-catalyst for PUR systems	
DCA	Erucamide	cis-13-docosenamide	Slip agent for polyolefins	KEMAMIDE-E. For determination method in PE see ASTM D 1996-92
DCBT	2,4-dichlorobenzoyl peroxide	DCBP[2.12]	Curing agent for UP resins; crosslinking agent	PERKADOX PD
DCDA	Dicyanodiamide		Curing agent for epoxy resins; costabiliser for POM	DYHARD 100 S
DCND	Dicinnamylidene hexamethylenediamine	Bis(cinnamylidene) hexamethylenediamine	Curing agent (blocked-diamine type) for fluororubbers	DIAK-3
DCP[2.8,2.10,2.12]	Dicumyl peroxide	Bis(1-methyl-1-phenylethyl) peroxide*	Curing agent and crosslinking agent	PERKADOX DI-CUP R
DDSA	Dodecenyl succinic anhydride		Curing agent for epoxy resins	
DEA[2.12]	Diethanolamine		Activator (rubber); processing aid	
DEA	Diethylaniline		Cure promoter for UP resins	

ABBREVIATED TERM (STANDARD OR USUAL)	NAME: STANDARD OR USUAL	NAME: OTHER NAMES, ABBREVIATED TERMS, OR DESCRIPTIONS	FUNCTION	REMARKS, AND/OR EXAMPLES OF COMMERCIAL PRODUCTS
DEG	Diethylene glycol		Activator (rubbers); processing aid	
DET	Diethylenetriamine	DETA	Curing agent for epoxy resins	
DLTDP[2.8,2.10,2.12]	Dilauryl thiodipropionate	Didodecyl-3,3′-thiodipropionate*; LTDP	Antioxidant (thioester type) used in polyolefins; styrenics, and TPE (styrenic type). FDA-approved for food packaging.	CYANOX LTDP; CARSTAB TLTDP
DMA	Dimethylaniline		Cure promoter for UP resins	
DMAE	Dimethylaminoethanol		Catalyst for PUR foam systems	Can be combined with tin-based catalysts.
DMAEE	Dimethylaminoethoxy-ethanol		Catalyst for PUR foam systems	TEXACAT ZR-70
DMAPA	Dimethylaminopro-pylamine	DMAP	Curing agent for epoxy resins	
DMBDPA	4,4′-di-(α,α-dimethylbenzyl) diphenylamine		Antioxidant	Widely used in polyether/ester TPE
DMBPHa[2.12]	2,5-dimethyl-2,5-di-(*tert*-butyl peroxy) hexane		Curing and crosslinking agent	
DMBPHy[2.12]	2,5-dimethyl-2,5-di-(*tert*-butyl peroxy) hexyne-3		Curing and crosslinking agent	
DMBPPD	See 6 PPD			
DMBS	Dimethylbenzylidene-sorbitol		Nucleating agent for polyolefins	GENISET MD
DMCHA	Dimethylcyclohexylamine	N,N′-dimethylcyclo-hexylamine	Catalyst for PUR systems	DIME
DMDNT	Dimethyldinitroso terephthalamide	N,N′-dimethyl-N,N′-dinitrosoterphthalamide	Chemical blowing agent used mainly in PVC	NITROSAN
DMEA	Dimethylethanolamine	N,N′-dimethylethanol-amine; DME	Catalyst in gas blowing of PUR foam systems	TEXACAT DME

ABBREVIATED TERM (STANDARD OR USUAL)	NAME: STANDARD OR USUAL	NAME: OTHER NAMES, ABBREVIATED TERMS, OR DESCRIPTIONS	FUNCTION	REMARKS, AND/OR EXAMPLES OF COMMERCIAL PRODUCTS
DMHPD[2.8,2.10,2.12]	N,N′-bis(1-methylheptyl)-*p*-phenylenediamine*		Antioxidant; retardant in UP resin systems	
DMI	Dimethylimidazole	1,2-dimethylimidazole	Catalyst for PUR systems	
DMMP	Dimethyl methylphosphonate	DMMP	Flame retardant	FYROL DMMP; REOFLAM DMMP
DMP	Dimethylpiperazine	N,N′-dimethylpiperazine	Catalyst for PUR foam systems	TEXACAT DMP
DMTDP	Dimyristyl thiodipropionate		Antioxidant	
DNPD[2.8,2.10,2.12]	N,N′-di-2-naphthyl-*p*-phenylenediamine*		Antioxidant	Highly effective in polyamides (but can cause discoloration)
DNPT[2.8,2.10,2.12]	Dinitrosopentamethylene-tetramine	3,7-dinitroso-1,3,5,7-tetraazobicyclo(3.3.1)-nonane*	Chemical blowing agent	Used mainly in rubber compositions
DOPD[2.8,2.10,2.12]	N,N′-dioctyl-*p*-phenylenediamine	N,N′-bis(1-ethyl-3-methylpentyl)-*p*-phenylenediamine*; 88PD[2.8,2.10,2.12]	Antioxidant; retardant in UP resin systems	DOPD is the symbol prefered (to 88PD) in ref. 2.8
DPA[2.8,2.10,2.12]	Diphenylamine*		Antioxidant (used as stabiliser in PUR systems)	
DPPD[2.8,2.10,2.12]	N,N′-diphenyl-*p*-phenylenediamine		Antioxidant; retardant in UP resin systems	
DPSDSH	Diphenylsulfone-3,3′-disulfonyl-hydrazide		Chemical blowing agent	Suitable for use in PVC, EVA, and some polyolefins
DSTDP	Distearyl thiodipropionate	Dioctadecyl-3,3′-thiodipropionate	Antioxidant (thioester type) used in polyolefins	HOSTANOX SE 10
DTBP[2.8,2.10]	Di-*tert*-butyl peroxide		Curing agent for UP resin systems; crosslinking agent	INTEROX DTPBP; TRIGONOX B
DTBPB	2,2-di-*tert*-butylperoxybutane	2,2-bis(*tert*-butyl peroxy) butane; TBPB	Curing agent for UP resin systems; crosslinking agent	INTEROX BU; TRIGONOX DB
DTDTDP	Ditridecyl thiodipropionate		Antioxidant (thioester type) used in polyolefins	
DTPD[2.8,2.10,2.12]	N,N′-ditolyl-*p*-phenylenediamine		Antioxidant; retardant in UP resin systems	

ABBREVIATED TERM (STANDARD OR USUAL)	NAME: STANDARD OR USUAL	NAME: OTHER NAMES, ABBREVIATED TERMS, OR DESCRIPTIONS	FUNCTION	REMARKS, AND/OR EXAMPLES OF COMMERCIAL PRODUCTS
EBPB	Ethyl-3,3-bis(*tert*-butyl peroxy) butane		Curing agent for PUR resin systems; crosslinking agent	
ECO[2.4,2.5,2.6]	Epichlorohydrin rubber	Copolymer of ethylene oxide (oxirane) and chloromethyloxirane	Impact modifier	HYDRIN
EDTBP	Ethylene-bis (tetrabromo-phthalimide)		Flame retardant for PS, polyolefins and polyesters	SAYTEX BT-93
EDTMQ[2.12]	6-ethoxy-1,1-dihydro-2,2,4-trimethylquinoline		Antioxidant	
EM	Ethylmorpholine	N-Ethylmorpholine (NEM)	Catalyst assisstant in PUR foam systems	TEXACAT NEM
EPDM[2.4,2.5,2.6]	Ethylene-propylene-diene terpolymer		An olefinic elastomer in its own right; used as impact modifier in some plastics	KELTAN EPDM See also entry for EPDM in Part 1
EPM	Ethylene-propylene copolymer	Ethylene-propylene rubber (EPR)	Used as impact modifier in some plastics	KELTAN EPR; VISTALON
GMS	Glycerol monostearate	GMST	Lubricant for PVC; antistatic agent for polyolefins	MYVAPLEX GMS
GTMA	Glyceryl trimethacrylate		Activator (rubbers)	
HBCD	Hexabromocyclododecane		Flame retardant	Used in expanded PS
HDI[2.12]	1,6-hexamethylene diisocyanate		Isocyanate reactant in PUR production	
HHP	Hydroxyheptyl peroxide		Curing agent for UP resin systems	
HHPA	Hexahydrophthalic anhydride		Curing agent for epoxy resins	
HMD[2.8,2.10,2.12]	Hexamethylenediamine	HMDA	Crosslinking agent	
HMDC[2.12]	Hexamethylenediamide carbamate		Curing agent for fluororubbers	DIAK-1
HMDI[2.12]	4,4′-dicyclohexylmethane diisocyanate		Isocyanate rectant in PUR production	

ABBREVIATED TERM (STANDARD OR USUAL)	NAME: STANDARD OR USUAL	NAME: OTHER NAMES, ABBREVIATED TERMS, OR DESCRIPTIONS	FUNCTION	REMARKS, AND/OR EXAMPLES OF COMMERCIAL PRODUCTS
HMMA[2.12]	N,N′-hexamethylene-bis-methacrylamide		Accelerator; curing agent (mainly in rubbers)	
HMT[2.8,2.10]	Hexamethylenetetramine	Hexamine; "Hexa"; HMTA	Crosslinking agent	Used inter alia in the production of phenolic resins
HNBR[2.4,2.5]	Hydrogenated nitrile rubber	Hydrogenated acrylonitrile-butadiene rubber	Synthetic rubber used as impact modifier in plastics	
HSAA	4-hydroxystearic acid anilide	HSTAA	Antioxidant	Widely used in polyamides
I**	Isophthalic acid		Reactant in the production of some polyamides	See under *polyamide* in Part 1 of this book
IA	Isatoic anhydride		Chemical blowing agent used in a number of plastics	KEMTEC
IND**	1,6-diamino-2,4,4-trimethylhexane		Reactant in the production of some polyamides	See under *polyamide* in Part 1 of this book
IPDA	Isophorone diamine	IPD	Hardener for epoxy resins	DEGAMIN
IPDI[2.12]	Isophrone diisocyanate		Isocyanate rectant in PUR coatings	
IPPD[2.8,2.10,2.12]	4-isopropylaminodiphenyl-amine	*N*-isopropyl-*N*′-phenyl-*p*-phenylenediamine*	Antioxidant	
MACM**	3,3′-dimethyl-4,4′-diamino-dicyclohexyl methane		Reactant in the production of some polyamides	See under *polyamide* in Part 1 of this book
MBI[2.12]	2-mercaptobenzimidazole		Antioxidant	
MBPP[2.12]	4-methyl-2,2-bis(*tert*-butyl peroxy) pentane		Curing agent aid in UP resin systems	
MC**	1,3-bis(aminomethyl) cyclohexane		Reactant in the production of some polyamides	See under *polyamide* in Part 1 of this book
MDA	Methylenedianiline		Curing agent for epoxy resins	TONOX MDA

** Abbreviation applying specifically in connection with use of this chemical as a component in the production of certain polyamides - cf. ISO 1874-1 : 1992 (Annex A).

ABBREVIATED TERM (STANDARD OR USUAL)	NAME STANDARD OR USUAL	OTHER NAMES, ABBREVIATED TERMS, OR DESCRIPTIONS	FUNCTION	REMARKS, AND/OR EXAMPLES OF COMMERCIAL PRODUCTS
MDI[2.12]	Diphenylmethane diisocyanate; diisocyanato-4,4′-diphenyl methane	Methylene-bis(4,4′-phenylisocyanate); methylene diphenyl isocyanate	Common isocyanate reactant in PUR production	
MEKP	Methylethylketone peroxide		Initiator for UP resin systems	THERMACURE; PEROXIMON K1
MHHPA	Methylhexahydrophthalic anhydride		Curing agent for epoxy resins	
MI	Methylimidazole	1-methylimidazole	Catalyst for PUR systems	
MIBKP	Methylisobutylketone peroxide		Curing agent for UP resin systems	PEROXIMON K2
MM	Methylmorpholine	N-methylmorpholine (NMM)	Catalyst for PUR foam systems	TEXACAT NMM
MMBI[2.12]	Methyl-2-mercapto-benzimidazole		Antioxidant	
MPDA	*m*-phenylenediamine	MPD	Curing agent for epoxy resins. Aromatic component of some polyamides.	
MTD**	*m*-toluylenediamine		Reactant in the production of some polyamides	See under *polyamide* in Part 1 of this book
MXD	*m*-xylylenediamine		Amine component of semi-aromatic polyamide PA MXD 6 (poly(*m*-xylylene adipamide))	
N**	2,6-naphthalene-dicarboxylic acid		Reactant in the production of some polyamides	See under *polyamide* in Part 1 of this book
ND**	1,6-diamino-2,2,4-tri-methylhexane		Reactant in the production of some polyamides	See under *polyamide* in Part 1 of this book
NDBC[2.8,2.10]	Nickel dibutyldithiocarbamate; NBC[2.12]	Nickel bis(dibutyldithio-carbamate)*	UV stabiliser; antioxidant	RYLEX NBC

** Abbreviation applying specifically in connection with use of this chemical as a component in the production of certain polyamides - cf. ISO 1874-1 : 1992 (Annex A).

ABBREVIATED TERM (STANDARD OR USUAL)	NAME STANDARD OR USUAL	OTHER NAMES, ABBREVIATED TERMS, OR DESCRIPTIONS	FUNCTION	REMARKS, AND/OR EXAMPLES OF COMMERCIAL PRODUCTS
NDI[2.12]	Naphthalene-1.5-diisocyanate		Isocyanate reactant in PUR production	
NEM	See EM			
NMA	Nadic methyl anhydride		Curing agent for epoxy resins	
NMM	See MM			
OABH	N,N′-dibenzyloxalyl-dihydrazide		Antioxidant (metal deactivator) for polyolefins	EASTMAN INHIBITOR OABH
OBBSH	4,4′-oxybis(benzene sulfonylhydrazide)	4,4′-oxybis(benzene-sulfone hydrazide); OBBS; OBSH	Chemical blowing agent (main use in PVC-P; PE-LD and some TPE)	CELOGEN OT; UNICELL OH; SULFOCEL
OBDPO	Octabromodiphenyl oxide	Octabromodiphenyl ether (OBDPE)	Flame retardant, similar in action to DBDPO, used mainly in PS, ABS and PA	SAYTEX 111
ODDBHPP	Octadecyl-3-(3,5-di-*tert*-butyl-4-hydroxyphenyl) propionate	ODDTBHPF	Antioxidant	Widely used in styrenics
ODPA[2.8,2.10,2.12]	Octylated diphenylamine	4,4′-di-*tert*-octyldiphenylamine	Antioxidant (used in some PUR systems inter alia)	
OPPD[2.8,2.10]	N-octyl-N-phenyl-*p*-phenylene diamine	8PPD[2.8,2.10,2.12]; N-(1-methylphenyl)-N-phenyl-*p*-phenylene-diamine*	Antioxidant; retardant for UP resin systems	
PABM**	Diphenylmethane-4,4′-diamine		Reactant in the production of some polyamides	See under *polyamide* in Part 1 of this book
PACM**	Bis(*p*-aminocyclohexyl) methane		Reactant in the production of some polyamides	See under *polyamide* in Part 1 of this book
PACP**	2,2′-bis(*p*-aminocyclo-hexyl) propane		Reactant in the production of some polyamides	See under *polyamide* in Part 1 of this book

** Abbreviation applying specifically in connection with use of this chemical as a component in the production of certain polyamides - cf. ISO 1874-1 : 1992 (Annex A).

ABBREVIATED TERM (STANDARD OR USUAL)	NAME: STANDARD OR USUAL	NAME: OTHER NAMES, ABBREVIATED TERMS, OR DESCRIPTIONS	FUNCTION	REMARKS, AND/OR EXAMPLES OF COMMERCIAL PRODUCTS
PAN[2.8,2.10] PANA[2.8,2.10,2.12]	N-phenyl-α-naphthylamine	*N*-(1-naphthyl) aniline*	Antioxidant	Symbol PAN preferred in ref. 2.8
PBBA	Pentabromobenzyl acrylate	Pentabromobenzyl monoacrylate	Reactive monomeric flame retardant	FR-1025 M
PBDPO	Pentabromodiphenyl oxide	Pentabromodiphenyl ether	Flame retardant	
PBEB	Pentabromoethyl benzene		Flame retardant	SAYTEX 105
PBGD**	Polybutylene glycol-diamine		Reactant in the production of some polyamides	See under *polyamide* in Part 1 of this book
PBN[2.8,2.10] PBNA[2.8,2.10,2.12]	N-phenyl-β-naphthylamine	*N*-(2-naphthyl) aniline*	Antioxidant	Symbol PBN preferred in ref. 2.8
77PD[2.8,2.10,2.12]	N,N′-bis(1,4-dimethylpentyl)-*p*-phenylenediamine*	BMPPD (q.v.)	Antioxidant; retardant for UP resin systems	
88PD[2.8,2.10,2.12]	See DOPD		Antioxidant; retardant for UP resin systems	Symbol DOPD preferred in ref. 2.8
PDBS	Polydibromostyrene		Flame retardant for styrenics and some other plastics	
PDEA PEEA	Phenyldiethanolamine Phenylethylethanolamine		Curing agents for PUR systems	
PEG[2.12]	Polyethylene glycol	Polyoxyethylene	Activator; process aid (mainly for rubbers)	
PGA PGDA	Polyglycolamine Polyglycoldiamine		Curing agents for epoxy resins	Supplied by Union Carbide Chemicals and Plastics Co., Specialty Chemicals Div., USA
PHP	Pinene hydroperoxide		Curing agent in UP resin systems	
PMDETA	Pentamethyldiethylene-triamine	PMDET	Catalyst for water-blown PUR foam systems	TEXACAT PMDETA
PMP	Phenylmercuric propionate		Catalyst for PUR foam systems	METASOL 57

** Abbreviation applying specifically in connection with use of this chemical as a component in the production of certain polyamides - cf. ISO 1874-1 : 1992 (Annex A).

ABBREVIATED TERM (STANDARD OR USUAL)	NAME STANDARD OR USUAL	NAME OTHER NAMES, ABBREVIATED TERMS, OR DESCRIPTIONS	FUNCTION	REMARKS, AND/OR EXAMPLES OF COMMERCIAL PRODUCTS
PMPPI[2.12]	Polymethylene polyphenyl isocyanate		Isocyanate reactant in PUR production	
PMS	*p*-methylstyrene		Used as comonomer in some impact modifiers	Cf. BROMO XP
POPD[2.12]	N-phenyl-N′-2-octyl-*p*-phenylenediamine		Antioxidant	
POPDA	Polyoxypropylenediamine	POPD	Curing agent for epoxy resins	
POPTA	Polyoxypropylene-tetramine	POPT	Curing agent for epoxy resins	JEFFAMINE T-403
6 PPD[2.8,2.10,2.12]	N-1,3-dimethylbutyl-N′-phenyl-*p*-phenylene-diamine*	DM3PPD	Antioxidant; retardant for UP resin systems	
7 PPD[2.8,2.10,2.12]	N-1,4-dimethylpentyl-N′-phenyl-*p*-phenylene-diamine		Antioxidant; retardant for UP resin systems	
8 PPD[2.12]	N-octyl-N′-phenyl-*p*-phenylene-diamine		Antioxidant; retardant for UP resin systems	
PPDI[2.12]	*p*-phenylene diisocyanate		Isocyanate reactant in PUR production	
PPDPA[2.8,2.10,2.12]	*p*-isopropoxydiphenyl-amine	4-isopropoxy-N-phenylaniline*	Antioxidant; retardant for UP resin systems	
PPG[2.12]	Polypropylene glycol	Polyoxypropylene	Activator; process aid (mainly for rubbers)	
PPGD**	Polypropylene glycoldiamine		Reactant in the production of some polyamides	See under *polyamide* in Part 1 of this book
PTAZ	5-phenyltetrazole	5-FT	Chemical blowing agent used in several plastics	EXPANDEX 5-PT
PTD**	*p*-toluylenediamine		Reactant in the production of some polyamides	See under *polyamide* in Part 1 of this book
PVME[2.12]	Polyvinyl methyl ether		Activator; process aid (mainly for rubbers)	

** Abbreviation applying specifically in connection with use of this chemical as a component in the production of certain polyamides - cf. ISO 1874-1 : 1992 (Annex A).

ABBREVIATED TERM (STANDARD OR USUAL)	NAME: STANDARD OR USUAL	NAME: OTHER NAMES, ABBREVIATED TERMS, OR DESCRIPTIONS	FUNCTION	REMARKS, AND/OR EXAMPLES OF COMMERCIAL PRODUCTS
PVP	Poly(*p*-vinyl phenol)		Curing agent for epoxy resins	LINCUR M
SA	Succinic anhydride	SAA	Curing agent for epoxy resins	
SB/CA		Combination of sodium bicarbonate with citric acid	Chemical blowing agent for PVC and other thermoplastics	HYDROCEROL BIH
SBR[2.4,2.5,2.6]	Styrene-butadiene rubber		Impact modifier for thermoplastics (especially styrenics)	STEREON; COPO SBR
SDPA[2.8,2.10]	Styrenated diphenylamine	SPDA[2.12]	Antioxidant; stabiliser in PUR systems	
SPH[2.12]	Styrenated phenol		Antioxidant	
T**	Terephthalic acid		Reactant in the production of some polyamides and polyesters	See under *polyamide* in Part 1 of this book
TAC[2.8,2.10,2.12]	Triallylcyanurate	2,4,6-triallyloxy-1,3,5-triazine*; triallylisocyanurate	Crosslinking agent (inter alia for radiation crosslinking of PE)	
TBBA	Tetrabromobisphenol-A		Reactive flame retardant for epoxy and UP resins	SAYTEX RB-100
TBCO	Tetrabromocyclooctane		Flame retardant	
TBCP[2.12]	See BCP			
TBLS	Tribasic lead sulfate	$3PbO.PbSO_4.H_2O$	Heat stabiliser for PVC	
TBPA	Tribromophthalic anhydride		Flame retardant additive for UP resin systems	SAYTEX RB-49
TBPAE	Tribromophenylallyl ether		Reactive flame retardant for PUR foam systems	
TBPB[2.12]	See BPB			
TBS	Tribromostyrene		Reactive flame retardant for styrenic materials	FR-803

** Abbreviation applying specifically in connection with use of this chemical as a component in the production of certain polyamides - cf. ISO 1874-1 : 1992 (Annex A).

ABBREVIATED TERM (STANDARD OR USUAL)	NAME: STANDARD OR USUAL	NAME: OTHER NAMES, ABBREVIATED TERMS, OR DESCRIPTIONS	FUNCTION	REMARKS, AND/OR EXAMPLES OF COMMERCIAL PRODUCTS
TCEP	Tris(2-chloroethyl) phosphate		Chlorophosphate flame retardant	FYROL CEF
TCPA	Tetrachlorophthalic anhydride		Flame retardant additive for UP resin systems	TETRAHAL
TCT[2.12]	Tricrotonylidenetetramine		Accelerator; crosslinking agent (rubbers)	
TDI[2.12]	Toluene diisocyanate; tolylene diisocyanate	1-methyl-2,4-diisicyanatobenzene	Widely used isocyanate reactant in PUR production	BAYTEC TS
TEA[2.12]	Triethanolamine		Activator; process aid (rubbers)	
TEA	Triethylamine		Catalyst for PUR systems	
TEDA	Triethylenediamine	TED	Catalyst for PUR elastomer and foam systems	TEXACAT TD-33A
TEPA	Tetraethylenepentamine	TEP	Curing agent for epoxy resins	D.H.E. 26
TETA	Triethylenetetramine	TET	Curing agent for epoxy resins	D.H.E. 24
THTA	Trihydrazinotriazine	Cyanuric trihydrazide (CUTH)	Chemical blowing agent for polyolefins, PA and ABS	
TIPP	Tri(isopropyl)phosphite		Antioxidant; costabiliser for PVC	
TIPT[2.12]	tris-(*p*-isocyanato phenyl)-thiophosphate		Isocyanate reactant in PUR production	
TMBDA	Tetramethyl-1,3-butanediamine		Catalyst used in PUR foam systems	
TMDI[2.12]	2,2,4-trimethyl-hexa-methylene diisocyanate or 2,4,4- trimethyl-hexa-methylene diisocyanate		Isocyanate reactant in PUR production	
TMEDA	Tetramethylethylene-diamine	TMD; TMED	Curing agent for epoxy resin and PUR systems	BULAB 600
TMG	Tetramethyl guanidine		Catalyst used in PUR foam systems	
TMP	Trimethylphosphite		Antioxidant	
TMPTMA	Trimethylolpropane trimethacrylate	TMPTM	Chemical crosslinking agent for PVC	ANCOMER 11

ABBREVIATED TERM (STANDARD OR USUAL)	NAME: STANDARD OR USUAL	NAME: OTHER NAMES, ABBREVIATED TERMS, OR DESCRIPTIONS	FUNCTION	REMARKS, AND/OR EXAMPLES OF COMMERCIAL PRODUCTS
TMQ[2.12]	2,2,4-trimethyl-1,2-dihydroquinone		Antioxidant	
TMXDI[2.12]	*m*-tetramethylxylylene diisocyanate		Isocyanate reactant in PUR production	WESTON TNPP
TNPP[2.8,2.10,2.12]	Tri(nonylphenyl) phosphite	Tris(*x*-nonylphenyl) phosphite* (where *x* denotes *o*, *m*, or *p*, or mixture)	Antioxidant (widely used in styrenics including styrenic TPE)	
TSA[2.12]	Salicylic acid		Retarder (rubbers)	UNICELL TSH
TSH	Toluene sulfonylhydrazide	Toluenesulfono-hydrazide	Chemical blowing agent for PVC, polyesters and some TPE	UNICELL TS; MICROFINE TSSC
TSSC	Toluene sulfonylsemicarbazide	Toluenesulfonosemi-carbazide	Chemical blowing agent for high-melting thermoplastics	
TTI[2.12]	4,4′,4″-triphenylmethane triisocyanate		Isocyanate reactant for PUR production	
VBR	Vinyl bromide	VB	Reactive flame-retardant monomer	SAYTEX VBR
ZDBC[2.8,2.10]	Zinc dibutyldithiocarbamate	ZnDBC[2.12]; Zinc bis(di-butyldithiocarbamate)*	Antioxidant (for polyolefins and some other plastics); vulcanising agent for rubbers	AGERITE
ZDBeC[2.8,2.10]	Zinc dibezyldithiocarbamate	ZBeDC[2.10]; ZnDBzC[2.12]; Zinc bis(dibezyldithio-carbamate)*	Antioxidant (for polyolefins and some other plastics); vulcanising agent for rubbers	
ZDBP[2.8,2.10]	Zinc dibutyldithiophosphate	ZnDBP[2.12]; Zinc bis(*O*,*O*-dibutyl phosphoro-dithioate)*	Antioxidant (for polyolefins and some other plastics); vulcanising agent for rubbers	

ABBREVIATED TERM (STANDARD OR USUAL)	NAME STANDARD OR USUAL	OTHER NAMES, ABBREVIATED TERMS, OR DESCRIPTIONS	FUNCTION	REMARKS, AND/OR EXAMPLES OF COMMERCIAL PRODUCTS
ZDEC[2.8,2.10]	Zinc diethyldithiocarbamate	ZEDC[2.10]; ZnDEC[2.12] Zinc bis(diethyldithiocarbamate)*	Antioxidant (for polyolefins and some other plastics); vulcanising agent for rubbers	
ZDMC[2.8,2.10]	Zinc dimethyldithiocarbamate	ZMDC[2.10]; ZnDMC[2.12] Zinc bis(dimethyldithiocarbamate)*	Antioxidant (for polyolefins and some other plastics); vulcanising agent for rubbers	
ZEPC[2.8,2.10]	Zinc ethylphenyldithiocarbamate	ZEPDC[2.10]; ZnEPC[2.12] Zinc bis[ethyl(phenyl) dithiocarbamate]*	Antioxidant (for polyolefins and some other plastics); vulcanising agent for rubbers	
ZnBX[2.12]	Zinc butyl xanthate		Activator, process aid (mainly rubbers)	
ZnEH[2.12]	Zinc 2-ethylhexanoate	Zinc octoate	Activator, process aid (mainly rubbers)	
ZnEHBP[2.12]	Zinc ethylhexyl-n-butyldithiophosphate		Activator, process aid (mainly rubbers)	
ZnEX[2.12]	Zinc ethyl xanthate		Activator, process aid (mainly rubbers)	
ZnIX[2.12]	Zinc isopropyl xanthate		Activator, process aid (mainly rubbers)	
ZnMBI[2.12]	Zinc-2-mercaptobenzimidazole	ZMBI	Antioxidant	
ZnMBT[2.12]	Zinc-2-mercaptobenzothiazole		Antioxidant	
Zn5MC[2.12]	Zinc pentamethylenedithiocarbamate	ZPMC	Antioxidant; vulcanising agent (rubbers)	
ZnMMBI[2.12]	Zinc-methyl-2-mercaptobenzimidazole		Antioxidant	
ZPMC[2.8,2.10]	Zinc pentamethylenedithiocarbamate	ZPMDC[2.10]; Zn5MC[2.12] Zinc bis(piperidine-1-carbodithioate)*	Antioxidant; vulcanising agent (rubbers)	

PART 2

SECTION 5

OTHER ABBREVIATIONS, OF RELEVANCE TO PLASTICS

ACORD	Automotive Consortium on Recycling and Disposal (UK).
AI	Acid index (q.v.).
ANSI	American National Standards Institute (originally known as ASA and USASI[1.11]).
BIBRA	The British Industrial Biological Research Association. The organisation which, inter alia, carries out health-hazard evaluations on plastics materials and their constituents.
BMC	Bulk moulding compound (q.v.).
BPF	The British Plastics Federation: the trade association for the UK plastics industry, grouping companies active in the fields of plastics materials and their constituents, plastics processing, and processing equipment.
"CAMPUS"	Computer-aided Material Preselection by Uniform Standards A system and database established by an international consortium of polymer suppliers to secure uniformity of methods of material testing and data reporting, so that the data from different sources are fully and directly comparable. Members of the consortium include BASF, Bayer, Huels, Hoechst, Akzo, Dow, Du Pont, ICI, and GE Plastics. (See also *FUNDUS*).
CEI	Commission Electrotechnique Internationale : the French version of the abbreviation IEC (q.v.).
CEN	European Committee for Standardization.
CENELEC	European Committee for Electrotechnical Standardization.
CIE	The initials of the French version of the name of the International Commission on Illumination (Commission Inter-national d'Eclairage).
DCL	Decorative continous laminate(s). Cf. BS 7332:1990.
DMA	Dynamic mechanical analysis (q.v.).
DMC	Dough moulding compound (q.v.).
DSC	Differential scanning calorimetry (q.v.).
DSS	Double-skin sheet (q.v.).
DTA	Differential thermal analysis (q.v.).
DTL	Deflection temperature under load (q.v.).
DTMM	Dynamic thermomechanical measurement (q.v.).
EFSG	European Fire and Security Group : an Agreement Group recognised by EOTC (q.v.).
EMCIT	Recognition Arrangement for European Testing of Electro-magnetic Compatibility of Information Technology Products : an Agreement Group recognised by EOTC (q.v.).
EMI	Electromagnetic interference (q.v.).
EMS	Electromagnetic shielding (q.v.).
EOTC	European Organisation for Testing and Certification (q.v.).

ETCOM Recognition Arrangement for European Testing for Certi-fication of Office and Manufacturing Protocols : an Agreement Group recognised by EOTC (q.v.).

EUROMAP The European Association of Machinery Manufacturers for the Plastics and Rubber Industries.

FRP Fibre-reinforced plastic(s) - q.v.

"FUNDUS" Material system and database for long-fibre-reinforced plastics, analogous in concept and operation to CAMPUS (q.v.); set up by a consortium of producers of bulk moulding compounds (BMC), sheet moulding compounds (SMC), and glass-mat thermoplastic compounds (GMT). The system was initiated by the German Reinforced Plastics Association (AVK) in Frankfurt am Main.

GGF Glass and Glazing Federation (UK).

GMC Granular moulding compound (q.v).

GMT Glass-mat-reinforced thermoplastic sheet. See under *reinforced thermoplastics.*

GR Abbreviation often used for the phrase "glass reinforced".

GRP The letters stand for "glass-reinforced polyester(s)" or "glass-reinforced plastic(s)". However, even in the latter, ostensibly very general meaning, the abbreviation is in fact customarily restricted to *glass-fibre-reinforced* plastics based on *thermosetting resins* (this usage is acknowledged e.g. in BS 3502:Part 1:1978). The general abbreviation for reinforced thermoplastics is "RTP" (q.v.).

GTR Gas transmission rate (see *transmission rate* under *permeability*).

HALS Hindered-amine light stabiliser(s) (q.v.).

HDT Heat distortion temperature (q.v.).

HPDL High-pressure decorative laminate (q.v.).

HPL High-pressure laminate(s), i.e. laminate(s) - q.v., produced by laminating at high pressure (regarded by one terminology standard[17] as pressure exceeding 1 400 kPa (200 psi)). See ISO 4586, Part 1 (1987) and Part 2 (1988).

ICTA International Confederation for Thermal Analysis.

IEC International Electrotechnical Commission (see also *CEI*).

IIEG International Instrumentation Evaluation Group : an Agree-ment Group recognised by EOTC (q.v).

IUPAC International Union of Pure and Applied Chemistry.

IUPAP International Union of Pure and Applied Physics.

kp kilopond (q.v.).

LOI Limiting oxygen index (see under *oxygen index*).

LOVAG Low Voltage Agreement Group : an Agreement Group recognised by EOTC (q.v.).

MFI Melt flow index (q.v.).

MFR Melt flow rate (q.v.).

MVR Melt volume-flow rate (see under *melt flow rate).*

NSF National Sanitation Foundation (q.v.).

NTP See *normal temperature and pressure.*

OI	Oxygen index (q.v.).
OIT	Oxidative induction time (q.v.).
OSTC	Open Systems Testing Consortium : an Agreement group recognised by EOTC (q.v.).
RIM	Reaction injection moulding (q.v.).
RRIM	Reinforced reaction injection moulding (q.v.).
RTI	Relative temperature index (q.v.).
RTM	Resin-transfer moulding.
RTP	Reinforced thermoplastic(s) - q.v.
SMC	Sheet moulding compound (q.v.).
SRIM	Structural reaction injection moulding.
STLA	Short-Circuit Testing Liaison Agreement Group : recognised by EOTC (q.v).
STP	See *standard temperature and pressure.*
TA	Thermal analysis (q.v.).
TG	Thermogravimetry (q.v.).
TGA	Thermogravimetric analysis : alternative term for "thermogravimetry" (deprecated by ICTA).
TI	Temperature index (q.v.).
$\mathbf{T_f}$	Letter symbol sometimes used for deflection temperature under load (q.v.).
$\mathbf{T_g}$	Glass-transition temperature (q.v.).
$\mathbf{T_m}$	Melting temperature (q.v.).
TMA	Thermomechanical analysis (q.v.).
TMC	Thick moulding compound (q.v.).
TSS	Triple-skin sheet (q.v.).
VAMAS	Versailles Project on Advanced Materials and Standards. A co-operative international operation promoting collaborative research projects in the field of materials, aimed at providing technical basis for standard specifications and codes of practice. Originated in 1987 by the Economic Summit countries (Canada, France, Italy, Japan, the UK, USA and West Germany) jointly with the Commission of the European Communities. VAMAS is led by a steering committee with current (1991) secretariat at the UK National Physical laboratory.
WECC	Western European Calibration Cooperation : an Agreement Group recognised by EOTC (q.v.).
WVT	Water-vapour transmission rate (see *transmission rate* under *permeability).*
XMC	Extra-thick moulding compound (q.v.).

I would to God thou and I knew where a commodity of good names were to be bought.
William Shakespeare
(King Henry IV, Part 1)

PART 3

TRADE NAMES

Informaton provided on trade names of commercial materials in Section 1, Table 2 of Section 2, and Section 3, includes the names of suppliers of materials under the trade names listed, and indications of their principal locations. Where the supplier concerned is a well known major company whose name (or its distinctive element) is used by the company as well as by its subsidiaries and representatives in all the main Western trading areas (Europe, UK, USA), only that name or element is given in the entry concerned, with no reference to any location. For example, only "DSM" for DSM Engineering Plastics Europe, The Netherlands; DSM Resins BV, The Netherlands; DSM Engineering Plastics, USA; DSM Thermoplastic Elastomers Inc., USA; DSM United Kingdom Ltd. Similarly only "Dow" for Dow Europe SA, Switzerland; Dow Plastics, USA; Dow Chemical Co. Ltd., UK; etc.

Where the name of a major manufacturing and/or trading subsidiary is different from that of its multinational principal, both names are given, with an indication of the subsidary's location - e.g. "Bayer; Miles Inc., USA", or "Hoechst; Hoechst Celanese Corp., USA" In all other cases, a fuller name of the company concerned is normally given, together with the principal location. It is possible that, in consequence to the current (1997) world-wide reorgansiation of Hoechst companies "Celanese" may not be retained as a company name, although it could be featured in the names of some products.

PART 3

SECTION 1

TRADE NAMES OF POLYMERS AND POLYMER COMPOUNDS

TRADE NAME	NATURE	PRODUCER OR SUPPLIER
ABSTAT	ABS	Mitech Corp., USA
A-C	Low-molecular-weight PE	AlliedSignal Inc., USA
ACCPRO	PP materials (various grades, including nucleated compounds)	Amoco Chemical
ACCTUF	Impact-resistant propylene copolymers	Amoco Chemical
ACCUREL	Porous-particle polymeric carriers for liquid additives (held in the pores)	Akzo Nobel
ACHIEVE	Isotactic PP (produced with metallocene catalysts)	Exxon Chemical
ACLAR	Fluoropolymer film	AlliedSignal Inc., USA
ACLON	PCTFE	AlliedSignal Inc., USA
ACLYN	Low-molecular-weight ionomers	AlliedSignal Inc., USA
ACRALDON	EVA	Bayer
ACRIFIX	PMMA	Rohm GmbH, Germany
ACRYLITE	Acrylic reins (PMMA)	CYRO Industries, USA
ACRYLOID	See PARALOID	
ACRYPOLY	Extruded acrylic sheeting	Chi Mei Corp.,Taiwan
ACRYREX	Acrylic resins (PMMA)	Chi Mei Corp.,Taiwan
ACRYSTEEL	Acrylic sheeting	Aristech Chemical Corp., USA
ACRYTHENE	Ethylene/methyl acrylate copolymers (for film production and additives)	Quantum Chemical Corp., USA
ACUMIST	Finely comminuted polyolefins	AlliedSignal Inc., USA
ADELL	Engineering thermoplastics	Adell Plastics Inc., USA
ADFLEX	PP (for calendering and extrusion)	Montell Italia, Italy; Himont, USA
ADIPRENE	PUR systems	Uniroyal Chemical Co. Inc., USA
ADPRO	PP	Novacor Chemicals Inc., USA
ADSTIF	PP (high-modulus grades)	Montell Italia, Italy; Himont USA
ADVEX	Rigid PVC	Geon Co., USA
AESNO	PA 12	Elf Atochem
AEROLAM	Epoxy resins and compounds	Ciba-Geigy
AEROLITE	Epoxy resins and compounds	Ciba-Geigy
AFFINITY	PE-ULD; PE-LD; PE-LLD (produced with metallocene catalysts)	Dow
AFLAS	TFE/propylene fluoroelastomer	3M Co.
AGEFLOC	Poly (hydroxyalkylene dimethyl ammonium) chloride: cationic polymer used as flocculant	CPS Chemical Co. Inc., USA
AIM	Proprietary styrenic resins (claimed intermediate in performance between ABS and PS-HI)	Dow
AKULON	PA 6 and 66	DSM
AKULOY	(PA + polyolefin) blends	DSM
ALATHON	PE-HD	Occidental Chemical Corp., USA
ALBIS	PA (various types); PP	Albis
ALCRYN	TPE (chlorinated olefin type)	Du Pont
ALCUDIA	PE (various grades); EVA	Repsol Quimica SA, Spain
ALCUPOL	Polyols	Repsol Quimica SA, Spain
ALEMID	PA 6 and 66 materials	Sebi SpA, Italy
ALGOFLON	PTFE polymers and compounds	Ausimont

TRADE NAME	NATURE	PRODUCER OR SUPPLIER
ALPHA 300	PVC-P	Alpha Chemical and Plastics Corp., USA
ALPHA GARY	PVC	Alpha Chemical and Plastics Corp., USA
ALPHASEAL	Styrenic blends	Alpha Chemical and Plastics Corp., USA
ALPHATEC	Styrenic blends	Alpha Chemical and Plastics Corp., USA
ALPHATHENE	Recycled PE	Alida Recycling, UK
ALPOLIT	UP resins	Hoechst
ALTAIR	Acrylic sheeting	Aristech Chemical Group, USA
ALTON	(PPS + fluoropolymer) blends	International Polymer Corp., USA
ALTUGLAS	PMMA	AtoHaas
AMILAN	PA 6; 66 and 610	Toray Industries, Japan, USA, Germany
AMILON	Glass-filled PP compounds	Amoco Chemical
AMOCO	PP (8000 and other series) PS	Amoco Chemical
AMODEL	Polyphthalamide (PPA) including fibre-production grades	Amoco Chemical
AMORVON	Block copolymer of poly (phenylene sulfide) and sulfide/sulfone (described as poly (thioether sulfone) - PTES)	Dainippon Ink and Chemicals Inc., Japan
AMPAL	UP and MF moulding compounds	Raschig AG, Germany
ANTRON	PA fibres (for upholstery and carpets)	Du Pont
APEC	Modifier PC (high-temperature-resistant)	Bayer; Miles Inc., USA; Albis
APEL	A cyclo-olefin polymer	Mitsui Sekka, Japan
API	ABS; PS-HI	American Polymers Inc., USA
APO	A cyclo-olefin polymer	Mitsui Sekka, Japan
APPRYL	Propylene homopolymers and copolymers	Elf Atochem (joint venture with BP Chemicals)
APYEIL	A meta-armid fibre	Unitica Ltd., Japan
AQUADRO	PVAL biodegradable resins	Planet Polymer Technologies Inc., USA
AQUA KEEP	High-absorbency polymers	Elf Atochem
AQUALIC CA	Crosslinked polyacrylate of high absorbency	BASF
AQUALOY	PA 66 and 612	Chem Polymer, Div. of T & N Industries, USA
AQUANYL	PA 612 (cast polymer)	Nylacast Ltd., Nylon Technics Div., UK
AQUATHENE	PE-LD; copolymer of ethylene and vinyltrimethoxysilane (a crosslinking agent for PE)	Quantum Chemical Co., USI Div., USA
ARALDITE	Epoxy resins	Ciba-Geigy
ARALL	Laminate of fibre (glass or aramid) plastic composite, and aluminium layers (used for aircraft parts and in armoured vehicles)	Akzo Chemie
ARCEL	Expanded PE beads	JSP International Co., USA
ARCOL	Polyols	Arco Chemical
ARDEL	Polyacrylate	Amoco Chemical
ARIMAX	Acrylamate RIM systems	Ashland Chemical Inc., USA
ARLEN	Modified PA 6 (containig terephthalate units in the molecular chains)	Mitsui & Co., UK and USA
ARMITE	Vulcanised fibre	Spaulding Fiber Co., USA
ARNITE	PET; PBT	DSM
ARNITEL	TPE (copolyester type)	DSM
AROPOL	UP resins	Ashland Chemical Inc., USA
ARPRO	Expandable PP beads	JSP International Co., USA
ARRHADUR	ABS	Elf Atochem
ARTEVYL	PVC resins	DSM

TRADE NAME	NATURE	PRODUCER OR SUPPLIER
ARTHOLENE	PP	Quality Additives Ltd., Ireland
ARTON	A cyclo-olefin	Japan Synthetic Rubber Co., Japan
ARYLON	Polyarylates	Du Pont
ASHLENE	Engineering thermoplastics (including various PA and polyester types, POM, ABS and PPO (PPE)	Ashley Polymers Inc., USA and Belgium
ASOTA	PP	Chemie Linz, Austria
ASPUN	Fibre-production-grade resins	Dow North America
ASTERITE	Acrylic dispersions	ICI Acrylics
ASTREL	PES	The Carborundum Co., USA
ASTRYN	PP; aesthetic polymers	Montell Italia, Italy; Himont USA
ATLAC	Vinyl ester resin compounds	Reichhold Chemicals Inc., Reactive Polymers Div., USA
ATTANE	Ultra-low-density PE (PE-ULD)	Dow
AURUM	Thermoplastic polyimide	Mitsui Toatsu Chemicals Inc., Japan and USA
AUTOFROTH	PUR foaming systems	BASF
AUTOPAK	PUR foaming systems	BASF
AUTOTHANE	Microcellular PUR elastomer system (for casting or RIM)	Kemira Polymers, UK and The Netherlands
AVP RESIN	Recycled thermoplastics	Polymerland Inc., USA and Europe
AVRON	Acrylic dispersion	ICI Acrylics
AZDEL	PP/glass mat formable sheet	GE Plastics
AZLOY	(PC + PBT) blend/glass mat sheet	GE Plastics
AZMET	(PC + PET) blend/glass mat sheet	GE Plastics
BAKELITE	Thermosetting resins and compounds (including phenolic and alkyd materials)	Bakelite AG, Germany; Bakelite Polymers, USA
BAPOLAN	PS materials, including PS-HI	Bamberger Polymers Corp., USA
BAPOLENE	Polyolefin materials	Bamberger Polymers Corp., USA
BAPOLON	PA 6 and 66	Bamberger Polymers Corp., USA
BAREX	Acrylonitrile copolymers (particularly for barrier applications)	BP Chemicals
BAREX RC	Barex containing recycled material	BP Chemicals
BASOFIL	Flame-resistant MF fibres (offered, inter alia, as asbestos substitute)	BASF
BASOSET	Epoxy resins and compounds	BASF
BAYBLEND	(PC + ABS) blends and blend compounds	Bayer; Miles inc., USA
BAYDUR	PUR systems for structural-foam and solid products (by RIM)	Bayer; Miles inc., USA
BAYER CM	Chlorinated polyolefins (including PE-C)	Bayer; Miles inc., USA
BAYFIT	PUR foaming systems	Bayer; Miles inc., USA
BAYFLEX	Elastomeric PUR systems (for RIM)	Bayer; Miles inc., USA
BAYFOL	PC films	Bayer; Miles inc., USA
BAYGAL	PUR casting resins	Bayer
BAYLON	PA 6 and 66	Bay Resins Inc., USA
BAYMIDUR	PUR casting resins	Bayer
BAYMOD A	Impact modifiers for PVC (ABS type)	Bayer; Miles inc., USA
BAYMOD L	Impact modifiers for PVC (EVA type)	Bayer; Miles inc., USA
BAYSILONE	Silicone resins	Bayer; Miles inc., USA
BAYTEC	PUR prepolymers (MDI-based)	Bayer; Miles inc., USA
BAYTEC RE	PUR liquid systems for resin-transfer moulding	Bayer; Miles inc., USA
BAYTEC RTM	PUR liquid spray systems	Bayer; Miles inc., USA

TRADE NAME	NATURE	PRODUCER OR SUPPLIER
BAYTEX	PUR elastomers (cast, heat-cured) similar to VULCOLLAN	Bayer
BECKOPOX	Epoxy resins and compounds	Hoechst
BECNO	PA 11	Elf Atochem
BECVO	PA 11	Elf Atochem
BEETLE	General name for a whole range of thermoplastic and thermosetting moulding materials including PA; PBT; PET; POM; UF and UP resins)	BIP Ltd., UK BIP Chemicals Inc., USA
BENVIC	PVC compounds	Solvay
BESNO	PA 11	Elf Atochem
BETA ABS	ABS	Beta Polymers Inc., USA
BETA 66	PA 66	Beta Polymers Inc., USA
BETA PP	PP	Beta Polymers Inc., USA
BEXLOY	Ionomers; TPE (copolyester type); PPO (PPE)	Du Pont
BIAPEN	PPO (PPE)	Chemolimpex, Hungary
BICOR	PP	Mobil Chemical Co., USA
BIFAN	PP	Showa High Polymer Co., Ltd., Japan
BIOCETA	CA (plasticised for biodegradability)	Tubize Plastics, Belgium (subsidiary of Nyltech)
BIONOLLE	Biodegradable aliphatic polyester	Showa High Polymer Co., Ltd., Japan
BIOPOL	Biodegradable copolymer of hydroxybutyrate and hydroxyvalerate	Zeneca (process being sold to Monsanto)
BLENDAW	(PC + ABS) blends	Jackdaw Polymers Ltd., UK
BLENDEX	ABS-type impact modifiers for PVC	GE Specialty Chemicals Inc., USA
BOALEN	PE-LD; PE-HD	Petrimex, Czech Republic
BOLATRON	Acrylic-modified PVC sheeting for thermoforming	Gen-Corp Polymer Products, USA
BROMO XP	Elastomeric impact modifiers (see Section 3 of Part 3)	Exxon Chemicals
BULEX	Styrene-butadiene rubber (SBR) and latex	Neftochim, Bulgaria
BUNA	EPDM	Huels
BUPLEN	PP	Neftochim, Bulgaria
BUSTREN	PS	Neftochim, Bulgaria
BUTACITE	Polyvinyl butyral sheet for laminated motor-car windscreens	Du Pont
BUTAFORM	Polyvinyl butyral sheet for laminated motor-car windscreens	Du Pont
CAB	CAB	Albis
CABELEC	Conductive (carbon-filled) polymer compounds	Cabot Plastics
CADON	SMA copolymers	Monsanto
CALIBRE	PC (including high-clarity and UV-resistant grades)	Dow
CALPRENE	Synthetic rubbers	Repsol Quimica SA, Spain
CAP	CAP	Albis
CAPILENE	Propylene polymers and copolymers	Carmel Olefins, Israel and UK
CAPRAN	PA films	AlliedSignal Inc., Engineering Plastics Div., USA
CAPRON	PA 6	AlliedSignal Inc., Engineering Plastics Div., USA
CAPRON AB	(PA 6 + amorphous PA) blends	AlliedSignal Inc., Engineering Plastics Div., USA
CAPRON D	Glass-filled PA 6	AlliedSignal Inc., Engineering Plastics Div., USA
CAPROLAN	Thermoplastic PUR	Elastogran Polyurethane AG, Germany
CARBAICAR	UF resins and compounds	Aicar SA, Spain
CARBOFOL	Plastics sheeting (for geomembranes and roofing membranes)	HT Troplast AG, Germany

TRADE NAME	NATURE	PRODUCER OR SUPPLIER
CARBONEX	PC sheeting	ICI Acrylics
CARBOTEX	PC; (PC + PBT) blends	Kotetsu & Co., Japan (Distributors: Civiera & Silver International Inc., (CSI), USA
CARBOTHANE	TPE (PC + PUR) blend (medical grade)	Thermedics Inc., USA
CARIFLEX	Styrenic block copolymers (SIS and SBS types)	Shell Chemicals
CARIL	Expandable beads of (PS + PPO) blends	Shell Chemicals
CARILON	Aliphatic polyketones: poly(olefin ketone) polymers	Shell Chemicals
CARINA	PVC resins	Shell Chemicals
CARINEX	PS-HI	Shell Chemicals
CARLONA	Polyolefins	Shell Chemicals
CARODEL	Polyarylates	ICI
CATAPYRRIC	Flame-retardant polyolefin compounds	AEI Compounds, Division of General Electric Co., UK
CEBIAN	SAN	Daical Chemical Industries,Japan
CECOR	Recyclable honeycomb reinforcement for plastics composites	Hexcel Corp., USA
CEFOR	PP	Shell Chemicals
CELANESE N	PA 6 and 66	Hoechst; Hoechst Celanese Corp., USA
CELANEX	PBT polymers and compounds	Hoechst; Hoechst Celanese Corp., USA
CELAZOLE	Polybenzimidazole resins	PBI Performance Parts, USA (a division of Hoechst)
CELCON	Acetal polymers	Hoechst; Hoechst Celanese Corp., USA
CELLIDOR	CA; CAB; CP	Albis
CELLOBOND	Phenolic resins and compounds	BP Chemicals
CELLTREK	Polyols for PUR foam systems	Dow
CELSTRAN	PA; PC; (PC + ABS) blends; PP compounds; PUR elastomers; PBT; PPS	Polymer Composites Inc., USA
CENTREX	ASA and (ASA + AES) blends	Monsanto (being transferred to Bayer in 1996)
CERTRAN	High-modulus PE fibres	Hoechst Celanese Corp., USA
CEVIAN	ABS (flame retardant)	Daicel Polymers Ltd., UK;Hoechst Celanese Corp., USA
CEVIAN B	(ABS + PBT) blend	Daicel Polymers Ltd., UK;Hoechst Celanese Corp., USA
CEVIAN N	SAN	Daicel Polymers Ltd., UK;Hoechst Celanese Corp., USA
CEVIAN V	ABS	Daicel Polymers Ltd., UK;Hoechst Celanese Corp., USA
CEVOLIT	Fibre-reinforced thermosetting resins	Cellpack AG, Switzerland
C-FLEX	TPE (styrenic type)	Consolidated Polymer Technologies Inc., Div. of Linvatec Corp., USA
CHARPRO	Flame-retardant polyolefin compounds (incorporating alumina trihydrate)	Dow
CHEMFLUOR	PVDF	Norton Performance Plastics Corp., USA
CHEMIGUM	TPE (olefinic type); nitrile rubber (NBR)	Goodyear
CHEMPOL	RIM systems: UP resins and PUR	Cook Composites & Polymers Co., USA
CLARADEX	ABS	Shin-A Corp., USA
CLARENE	EVA; EVOH	Solvay
CLARITECH	Silicone rubber compounds	Silicone Products & Technology Inc., USA
CLEARFLEX	PE-LLD; PE-VLD (very low density)	EniChem
CLEARTUF	Poly (ethylene naphthalate) (PEN)	Goodyear
CLOCEL	PUR structural foams	Baxenden Chemicals Ltd., UK

TRADE NAME	NATURE	PRODUCER OR SUPPLIER
CLYSAR	Shrink films	Du Pont
COMMAX	Silicone resins and compounds	Tecknit Inc., USA
COMPODIC	PA 6 and 12; EVA; PPS	DIC Trading (USA) Inc., USA
COMTUF	PA 6, 66 and 612; PBT; PET	ComAlloy International Corp., USA
CONAPOXY	Epoxy resins and compounds	Conap Inc., USA
CONATHANE	PUR casting resins	Conap Inc., USA
CONDUCTOMER	Range of antistatic polymer compounds	Synthetic Rubber Technologies, USA
CONDUX	PVC-P; (ABS + PVC) blends	Advanced Dynamics Corp., Specialty Compounds Div., USA
CONEX	Meta-aramid fibre (based on *m*-phenylene isophthalamide)	Teijin Ltd., Japan
COPO	Styrene- butadiene rubber (SBR) impact modifiers for styrenics	DSM
CORFAM	Composite leather-like synthetic material	Du Pont
CORIAN	Acrylic-based cultured marble material	Du Pont
CORO-FOAM	PUR foam systems	Cook Composites & Polymers Co., USA
CORTEM ALLOYS	Compounds of PEEK, and of liquid-crystal polymers, with special low-softening glass reinforcement	Corning Inc., USA
CORTON	PP compounds	PolyPacific Pty. Ltd., Australia
CORVIC	PVC resins	European Vinyls Corp. (EVC); ICI; and associates worldwide.
COSMIC	Epoxy resins; diallyl phthalate and isophthalate compounds	Cosmic Plastics Inc., USA
CRASTIN	PBT	Du Pont
CRASTON	PPS	Ciba-Geigy
CREANIT	PMMA sheet (stone texture)	Röhm GmbH, Germany
CRISPER	Synthetic paper produced from PS (claimed superior permability over PP-based paper)	Toyobo Co., Japan
CRISTAMID	Semi-aromatic PA	Elf Atochem
CRYLAC	(PVC + NBR) blends	Silac SA, France
CRYSTIC	UP resins and compounds	Scott Bader Co. Ltd., UK
CYANAPRENE	PUR systems (for elastomers and foams)	Air Products & Chemicals Inc., USA
CYCOGEL	ABS	Nova Polymers Inc., USA (distributors for Proquigel, Brazil)
CYCOLAC	ABS	GE Plastics
CYCOLIN	(ABS + PBT) blends	GE Plastics
CYCOLOY	(PC + ABS) blends	GE Plastics
CYCOVIN	(PVC + ABS) blends	GE Plastics
CYMEL	MF moulding compounds	Cytec Industries Inc., USA (business unit of American Cyanamid Co.)
CYREX	(PC + acrylic) blends	CYRO Industries, USA
CYROLITE	MBS copolymers; acrylics	Röhm GmbH, Germany
DACRON	PET fibres	Du Pont
DAI-EL	TPE compounds (fluoropolymer type)	Daikin America Inc., USA
DAIFLON	PCTFE	Daikin Kogyo Co., Japan
DAISO	Diallyl phthalate compounds	Daikin Kogyo Co., Japan
DAPLEN	PE and PP materials, including filled compounds and blends	PCD Polymere GesmbH, Austria
DAPLEN TC	"Technical compound" series of reactor-produced PP compounds and blends	PCD Polymere GesmbH, Austria

TRADE NAME	NATURE	PRODUCER OR SUPPLIER
DECELITH	PVC compounds and extruded products	Eilenburger Chemie Werke AG, Germany
DEGALAN	PMMA	Degussa
DELOXEN	PPO (PPE)	VAMP Technology, Italy
DELRIN	Acetal homopolymers, including DELRIN P - an easy moulding, high-heat-stability grade (replacing the former DELRIN II)	Du Pont
DEMNUM	Range of perfluoropolyethers	Daikin Kogyo Co., Japan
DENIBLEND	Polymer blends	VAMP Technology, Italy
DENICARB	PC	VAMP Technology, Italy
DENILEN	PP compounds	VAMP Technology, Italy
DENISAB	ABS	VAMP Technology, Italy
DENISAN	SAN	VAMP Technology, Italy
DENISTYR	PS compounds	VAMP Technology, Italy
DENITER	PBT	VAMP Technology, Italy
DENKA VINYL	Impact-modifiers for PVC	Denki Kagaku Kogyo, Japan
DENYL	PA 6 and 66	VAMP Technology, Italy
DERAKANE	Liquid vinyl ester resins containing styrene monomer	Dow
DESMOFLEX	PUR elastomers (cast, cold-cured)	Bayer; Miles Inc., USA
DESMOPAN	Thermoplastic elastomeric PUR for injection moulding, extrusion and calendering	Bayer; Miles Inc., USA
DESMOPHEN	PUR structural foams	Bayer
DEVCON	Epoxy resins	ITW Devcon Corp., USA
DEXCO	Hot-melt resins	Exxon Chemical
DEXEL	CA moulding compounds	Courtaulds Chemicals, UK
DEXFLEX	TPE (olefinic type)	D & S Plastics International, USA
DIAKON	PMMA	ICI Acrylics
DIALAC	ASA	Mitsubishi
DIAMOND	ABS; (ABS + PC) blends	Diamond Polymers, USA
DIEGLAS	UP moulding compounds	Glastic Corp., USA
DIELECTRITE	UP moulding compounds (BMC)	Industrial Dielectrics Inc., USA
DIMENSION	(PA + PPO) blends	AlliedSignal Inc., USA
DION	Range of chemical-resistant UP and vinyl ester resins for GRP applications	Reichhold Chemicals Inc., USA
DIPRANE	PUR systems	Kemira Polymers, UK and The Netherlands
DISTITRON	UP resins	Lonza
DIXIE SPAR	Water-soluble sodium salt of an acrylic acid/acrylamide polymer	Dixie Chemical Co. Inc., USA
DOBECKOT	Epoxy resins and compounds	BASF
DOWLEX	PE-LD; PE-HD	Dow
DOWLEX IP	"Improved processing" DOWLEX grades	Dow
DOWLEX NG	"New generation" DOWLEX grades (improved toughness)	Dow
DOWSAN	SAN	Dow
DQDA	EVA	Union Carbide
DUFLEX	PVC	Vinyl Ltd., UK
DULARIT	Epoxy resins and compounds	Henkel
DUMILAN	EVA	Mitsui Sekka, Japan
DUOGAL	PP sheeting	Polygal, Israel
DURA CAP	Weather-resistant capstock for PVC extrusions	Geon Co., USA
DURACON	Acetal copolymer (closely similar to HOSTAFORM)	DC Polyplastics Co., Japan
DURAFLEX	Polybutylene	Shell Chemicals
DURAL	Rigid PVC; (PVC + ABS) blend	Alpha Chemical & Plastics Corp., USA
DURALEX	(PVC + PUR) blends	Alpha Chemical & Plastics Corp., USA

TRADE NAME	NATURE	PRODUCER OR SUPPLIER
DURAPOL	UP resins	Isola Werke GmbH, Germany
DURASTRENGTH	Acrylic impact modifiers for PVC	CECA (Elf Atochem)
DURATECH 3300	Silicone rubber compounds	Silicone Products & Technology Inc., USA
DURAX	Phenolic compounds	Isola Werke GmbH, Germany
DUREL	Polyarylate	Hoechst, Hoechst Celanese Corp., USA
DURETHAN (Formerly NYDUR)	PA polymers and compounds	Bayer; Albis; Miles Inc. USA
DURETHAN A	PA 66 polymers and compounds	Bayer; Albis; Miles Inc. USA
DURETHAN B	PA 6 polymers and compounds	Bayer; Albis; Miles Inc. USA
DURETHAN C	Copolyamide-based compounds	Bayer; Albis; Miles Inc. USA
DUREZ	Phenolic resins and compounds; diallyl phthalate compounds; alkyd resin compounds; UP resin compounds	Occidental Chemical Corp., Durez Div., USA
DURODET	Thermosetting GRP compounds (including SMC)	Mitras Kunststoffe GmbH, Germany
DUROID	See RT/DUROID	
DUROLITE	GRP syntactic foam	Mitras Kunststoffe GmbH, Germany
DUROLON	PC	Policarbonatos do Brazil, Brazil
DUTRAL TP	Modified PP	Montell Italia, Italy; Himont USA
DYFLOR	PVDF	Huels
DYLARK	SMA copolymers; (SMA + PBT) blends	Arco Chemical Co., USA
DYLENE	PS	Arco Chemical Co., USA
DYNAFLEX	TPE (SEBS-based) - some grades modified with EXACT polyolefins	GLS Corp. Thermoplastic Elastomers Div., USA
DYNAMAR PPA	Fluoroelastomer-based processing aids for PE	3M Co. Specialty Fluoropolymers Dept., USA
DYNEEMA	PE fibres of very high modulus and strength (produced by a gel-extrusion process)	DSM High Performance Fibres, The Netherlands
DYTRON	TPE (olefinic type) - range includes grades crosslinkable by radiation (DYTRON XL)	Monsanto
EASTAR	PETG copolyester	Eastman Chemical
EBAC	Copolymers of ethylene with n-butyl acrylate (used as tie-layers in coextrusion)	Chevron Chemical Co., USA
ECCOBOND	Thermosetting (light-curable) adhesives	Grace Specialty Polymers, USA
ECDEL	TPE (copolyester type)	Eastman Chemical
ECOBOARD	Plastics wood substitute produced from reclaimed materials (PE-LD and PE-HD)	American Ecoboard Inc., USA
ECORYL	Acrylic emulsions	Elf Atochem
ECOMATY AX	Water-soluble vinyl alcohol copolymer (for film production and moulding)	The Nippon Synthetic Chemical Co., Japan
ECOTHENE	PE-HD	Quantum Chemical Co., USA
EGEPLAST	Rigid PVC	Vinyl Ltd., UK
EKADUR	PBT	Kunststoff-Bearbeitungs GmbH, Germany
EKALON	PC	Kunststoff- Bearbeitungs GmbH, Germany
EKALOY	Polymer blends	Kunststoff- Bearbeitungs GmbH, Germany
EKATAL C	Acetal copolymer	Kunststoff- Bearbeitungs GmbH, Germany
EKATAL H	Acetal homopolymer	Kunststoff- Bearbeitungs GmbH, Germany
EKAVYL	PVC	Elf Atochem

TRADE NAME	NATURE	PRODUCER OR SUPPLIER
EKTAR	Thermoplastic polyesters (including PCT, PCTA, PCTG, PETG) ; TPE (Olifinic type); (PC + PBT) blends	Eastman Chemical
ELA	TPE (polyester type)	Teijin Ltd., Plastics Technology and production Dept., Japan
ELASLEN	PE-C	Showa Denko K.K. Japan
ELASTAMINE	Reactive polyether amines	Huntsman Chemical Corp.
ELASTOLLAN	TPE (PUR type)	BASF; Elastogran GmbH, Germany
ELASTOPAN	PUR casting systems	Elastogran GmbH, Germany
ELASTOPAN AQUA	PUR system (polyether type) for water blown foam production	Elastogran GmbH, Germany
ELASTOSIL	Silicone rubbers	Wacker
ELASTOTEC	TPE (ether/ester random block copolymer)	BASF
ELASTUF	TPE (copolyester type)	Goodyear, Polyester Div., USA
ELECTRAFIL	Antistatic moulding compounds filled with carbon or metal fibres (based on ABS; PA; PC)	DSM
ELECTROGLAS	PMMA	Glasflex, Div. of Polycast Technology Corp., USA
ELIX	Polymeric (SMA-type) modifiers for PVC (previously in the CADON range)	Monsanto
ELTEX	PE-HD	Solvay
ELTEX P	PP (highly isotactic)	Solvay
ELVALOY	EVA terpolymers used as polymeric modifiers in PVC	Du Pont
ELVANOL	EVOH	Du Pont
ELVAX	EVA	Du Pont
EMAC	Ethylene/methyl acrylate copolymer	Chevron Chemical Co., USA
ERMAREX	(PPO + PA) blends	MRC Polymers Inc., USA
EMBLET	Polyester film	Nyltech Italia, Italy
EMBRYON	Polyolefin resins for carpet-yarn production (claimed modified for resistance to soiling)	Neste Chemicals, Finland
EMICLEAR	ABS compounds; modified PPO (PPE) compounds	Toshiba International, USA
EMPEE	Polyolefin compounds	Monmouth Plastics Inc., USA
ENATHENE	Ethylene/butyl acrylate copolymer	Quantum Chemical Co., USA
ENCORE	Range of recycled engineering thermoplastics (polyester; liquid-crystal; polyacetal; PA; PPS)	Hoechst Celanese Corp., USA
ENGAGE	TPE (olefinic type), including grades for clingfilm production (e.g. ENGAGE KC 8852)	Dow
ENPLEX	(PVC + ABS) blends	Kanegafuchi Chemical Industry Co., Japan
ENPNITE	Compounds based on PA 6 and PP	Chisso America Inc., USA
ENVEX	Polyimide materials (including formable sheet and machinable products)	Rogers Corp., USA
EPALEX	PP resins and compounds	PolyPacific Pty. Ltd., Australia
EPIC	Thermosetting resins based on combinations of epoxy resins with isocyanurates	Bayer
EPIKOTE	Epoxy resins	Shell Chemicals

TRADE NAME	NATURE	PRODUCER OR SUPPLIER
EPISOL	PVC resins (emulsion type)	Eletrochemical Industries, Israel; Longfield Chemicals Ltd., UK
EPIVYL	PVC resins (suspension type)	Eletrochemical Industries, Israel; Longfield Chemicals Ltd., UK
EPOCAST	Epoxy and polyimide resins	Furane Products, Ciba-Geigy Corp., USA
EPODITE	Epoxy resins and compounds	Showa Denko K. K., Japan
EPON	Epoxy resins	Shell Chemicals
EPONITE	Epoxy-resin-based compounds	Shell Chemicals
EPOSAND	Epoxy resin compound	Shell Chemicals
EPOPHEN	Epoxy resins and compounds	Borden Chemicals & Plastics, USA
EPROSIN	Epoxy resins and compounds	Chemapol, Czech Republic
ERACLEAR	PE-LLD	EniChem
ERACLENE	PE-HD	EniChem
ERCOM	Recycled SMC	Mitras Kunststoffe GmbH, Germany
ERTALON	PA 6 (including machinable materials)	Polymer Corp., USA
ERTALYTE	PET (including machinable materials)	Polymer Corp., USA
ESBRID	PA 6 and 66 compounds	Thermofil Inc., USA
ESBRITE	PS	Sumimoto Chemical Co. Ltd., USA
ESCOMER	Polyethylene waxes	Exxon Chemical
ESCOR	Ethylene/acrylic acid copolymers	Exxon Chemical
ESCORENE	PE (various grades); PP compounds	Exxon Chemical
ESCORENE ULTRA	EVA; EMA; EAA	Exxon Chemical
ESCOREZ	Hydrocarbon resins	Exxon Chemical
ESDASH	PP	Thermofil, UK
ESLON	PET, and other engineering thermoplastics	Cheil Synthetics Inc., South Korea; Plastribution Engineering Polymers Ltd., UK
ESPLADUR	Phenolic moulding compounds	Raschig AG, Germany
ESPRENE	Modified PP	Sumimoto Chemical Co., Japan
ESTALLOY	TPE (styrenic type)	GLS Corp., USA
ESTALOC	TPE (PUR type) compounds (including reinforced grades)	BF Goodrich
ESTAMID	TPE (polyester/amide type, based on PA 12)	Upjohn Corp., USA
ESTANE	TPE (PUR type)	BF Goodrich
ESTEFORM	UP resins and compounds	Chromos-Ro Polimeri, Yugoslavia
ESTYRENE	Styrene/methyl methacrylate	Petroplas Ltd., UK
ETAR	PET	Eastman Chemical
ETC	PP compounds	Plastribution Engineering Polymers Ltd., UK
ET-POLYMER	Modified PE	Borealis
ET-SEMICON	Conductive version of ET-POLYMER	Borealis
EUROPRENE N	Nitrile rubbers (NBR)	EniChem Elastomers, Italy and UK
EUROPRENE N OZO	(PVC + NBR) blends	EniChem Elastomers, Italy and UK
EVAL	EVOH	Eval Co. of America, USA
EVALCA	EVOH	Eval Co. of America, USA
EVATANE	EVA, and other ethylene copolymers (some used as tie-layers in multilayer extrusion)	Elf Atochem
EVICOM	PVC compounds	European Vinyls Corp. (EVC); EVC Compounds (UK) Ltd., UK
EVIPOL	PVC resins	European Vinyls Corp. (EVC)

TRADE NAME	NATURE	PRODUCER OR SUPPLIER
EVOPRENE	TPE: styrenic (SBS and SEBS) and olefinic types	Evode Plastics Ltd., UK
EXACT	PE-LLD and PE-ULD (produced with metallocene catalysts)	Exxon Chemical
EXATRON	Epoxy resins and moulding compounds	EniChem
EXCEED	PE-LLD	Exxon Chemical
EXOLITE	PMMA	CYRO Industries, USA
EXTIR	Expandable PS	EniChem SpA, Styrenics and Engineering Plastics Div., Italy
EXTRON	PP polymers and compounds	PolyPacific Pty. Ltd., Australia
EXXELOR	Functionalised polymers	Exxon Chemical
EXXTRAL	TPE (olefinic type)	Exxon Chemical
EYPEL-A	Foamed polyphosphazene (used as flame-resistant insulation)	Albermarle Corp., USA
EYPEL-F	Polyphosphazene elastomers	Albermarle Corp., USA
FABENYL	PA 6 and 66	Tubize Plastics, Belgium (subsidiary of Nyltech)
FABELTAN	TPE (PUR type)	Tubize Plastics, Belgium (subsidiary of Nyltech)
FAIRPRENE	Elastomer-coated fabrics	Du Pont
FARBOSET	Epoxy resins	Farboil Co., USA
FARADEX	Range of conductive compounds (containing about 1% of steel fibres)	DSM
FE	Magnetisable polymers	Lati Engineering Thermoplastics SpA, Italy; Lati UK Ltd.
FENOCHEM	Phenolic resins and compounds	Chemiplastica SpA, Italy
FENOFORM	Phenolic resins and compounds	Chromos-Ro Polimeri, Yugoslavia
FERRENE	PE-HD	Ferro
FERROCLINE	Modified PP	Ferro
FERROFLO	PS	Ferro
FERROLENE	PP	Ferro
FERROPAK	Polyolefin compounds	Ferro
FERROTRON	PTFE	Polymer Corp., USA
FERROZELL	Silicone resins and compounds	Ferrozell, Zachs & Co., Germany
FIBERFIL	Various thermoplastic polymers	DSM
FIBERFIL J	PSU compounds	DSM
FIBERITE	Various thermosetting resins and compounds (including MF, PF, UP, vinyl ester, epoxy, silicone, polyimide)	ICI Composites
FIBERLOC	Glass-reinforced PVC compounds (with special adhesion promoter on the glass)	Geon Co., USA; Hydro Geon, UK
FIBERSTRAN	Various thermoplastic compounds (including PA, PC, PS, SAN)	DSM
FIBREDUX	Epoxy resin prepregs	Ciba-Geigy
FILMON	Cast PA films	Nyltech Italia, Italy
FILTRAM	Geotextiles	Exxon Chemical
FINA	Polyolefins; PS	Fina Oil & Chemical Co., USA; Fina Chemicals, Petrofina SA., Belgium
FINACLEAR	TPE (SBS type)	Fina Oil & Chemical Co., USA; Fina Chemicals, Petrofina SA., Belgium
FINAPRENE	TPE (styrenic type)	Fina Oil & Chemical Co., USA; Fina Chemicals, Petrofina SA., Belgium
FINAPRO	PP resins and compounds	Fina Oil & Chemical Co., USA; Fina Chemicals, Petrofina SA., Belgium
FINATHENE	PE-HD	Fina Oil & Chemical Co., USA; Fina Chemicals, Petrofina SA., Belgium
FLEXCLEAR	Flexible PVC sheeting	O' Sullivan Corp., USA
FLEXEL	TPE (vinyl alloy type)	BF Goodrich
FLEXIRENE	PE-LD	EniChem
FLEXOMER	Range of ethylene copolymers of ultra-low density and low stiffness	Union Carbide
FLORAFLON	PVDF	Elf Atochem
FLOWMAT	UP compounds (SMC)	Cook Composites & Polymers Co., USA

TRADE NAME	NATURE	PRODUCER OR SUPPLIER
FLUON	Fluoropolymer compounds. (Original ICI trade name for PTFE resins)	LNP Engineering Plastics
FLUOREL	TPE (fluoropolymer type)	3M Co., USA
FLUOROCOMP	Fluoropolymer compounds with various reinforcements, fillers and lubricants	LNP Engineering Plastics
FLUOROSINT	PTFE stock shapes	Polymer Corp., USA
FLUOTOP T	Two-part coating system for PVC (a ter-polymer keying layer and PVDC top coat)	Ferrari SA, France
FOREX	Expanded rigid plastics sheeting (grade "EPC" based on PVC)	Airex AG, Switzerland
FORMCAST	Epoxy resin systems for casting and impregnation	Formulated Resins Ltd., UK
FORTIFLEX	PE-HD	Solvay
FORTILENE	PP	Solvay
FORTRON	PPS resins and compounds	Hoechst; Hoechst Celanese Corp., USA
FP	Fluoropolymer materials and compounds	LNP Engineering Plastics
FREEFLOW	UP resins	Cook Composites & Polymers Co., USA; Freeman Distribution, UK
FREEMIX	UP moulding compounds	Cook Composites & Polymers Co., USA; Freeman Distribution, UK
FUDOWLITE	Diallyl phthalate moulding compounds	Fudow Chemical Co., Japan
FULLMID	PA 6 and 66	Sebi SpA, Italy
FUSABOND	Compatibilisers (including grafts of maleic anhydride on polyolefins) for polymer blending and recycling	Du Pont
GAFITE	PBT	Hoechst
GAFLEX	TPE (copolyester type)	Hoechst
GAFTUF	PBT	Hoechst
GAPEX	PP compounds (glass-reinforced; impact-modified)	Ferro Corp., Filled & Reinforced Plastics Div., USA
GAPEX ALLOY	(PP + PA) blends	Ferro Corp., Filled & Reinforced Plastics Div., USA
GARAPRENE	PVC; thermoplastic blends	Evode Plastics Ltd., UK
GARDLITE	PMMA	Southern Plastics Co., USA
GARY PVC	PVC cable-covering compounds	Evode Plastics Ltd., UK
GEDEXCEL	Expanded PS	Elf Atochem
GELON A 100	Amorphous PA (used as layers in packing films and blow-mouldings)	GE Plastics
GELOY	(PVC + ASA) blends	GE Plastics
GEMAX	(PPO + PPT) blends	GE Plastics
GENAL	Phenolic resins and compounds	GE Plastics
GEOLAST	TPE: (PP + NBR) blend	Monsanto
GEON	PVC resins, compounds and blends	Geon Co., USA; Hydro Geon, UK
GEON HTX	"High heat" PVC compounds; (PVC + ABS) blends	Geon Co., USA; Hydro Geon, UK
GLASCOL	Acrylic emulsion polymers	Allied Colloids, UK
GLASKYD	Diallyl phthalate moulding compounds	Cytec Industries Inc., USA (Business unit of American Cyanamid Co.)
GLO-CLEAR	Flexible PVC sheeting	O' Sullivan Corp., USA
GORETEX	PTFE compounds	W. L. Gore & Associates Inc., USA
GRANLAR	Liquid-crystal polymers	Graumont Inc., USA
GRANULAR	MF compounds	Plastic Manufacturing Co., USA
GREENFLEX	EVA	EniChem

TRADE NAME	NATURE	PRODUCER OR SUPPLIER
GRILAMID	PA 12	EMS-Chemie AG, Switzerland; EMS Grilon, UK and USA
GRILAMID EL	TPE (PA block copolymer type)	
GRILAMID TR	Amorphous (transparent) PA (with aliphatic and cycloaliphatic repeating units in the molecular chain)	
GRILBOND	Adhesion promoters	
GRILESTA	Copolyester powder coatings	
GRILON	PA 6; TPE (PA type)	
GRILON C	PA 612	
GRILON EL	TPE (PA block copolymer type)	
GRILON T	PA 66	
GRILONIT	Epoxy resins	
GRILPET	PBT	
GRILTEX	PA and polyester adhesives	
GRIVORY	Partly aromatic, amorphous PA	
GRIVORY GV	Glass-reinforced PA	
GRIVORY HTV	PA copolymers	
HALAR	CTFE ; ECTFE	Ausimont
HALOFLEX	Vinyl/acrylic polymer latices used in surface coatings	ICI
HALON ET	ETFE	Ausimont
HALON TFE	PTFE	
HANALAC	ABS	Miwon USA
HAREX	UP moulding compounds	Resopal Werke, Germany
HAYSITE	UP moulding compounds	Haysite Reinforced Plastics, USA
HDPX	Self-crosslinkable PE	Borealis
HELM	PVC resins (emulsion type)	Vinyl Ltd., UK
HERAMID	PA 6	Franz Rauscher AG, Germany; Plastribution Engineering Polymers Ltd., UK
HERCUPRENE	TPE (styrenic type)	J-Von, LP, USA
HEREX	Rigid PVC foam (crosslinked, closed-cell)	Airex AG, Switzerland
HETRON	UP resins and compounds	Ashland Chemical Inc., USA
HICOND	PE-HD polymers and compounds	United Composites Inc., USA
HID	PE-HD	Chevron Chemical Co., USA
HIFAX	PE; TPE (olefinic type)	Montell Italia, Italy; Himont USA
HI-FLEX	UP moulding compounds	Budd Co., Plastics Div., USA
HIGLAS	Glass-reinforced PP compounds	Montell Italia, Italy; Himont USA
HILENE	Polyolefins	Vinmar Inc., USA and The Netherlands
HILOY	PA 6 and 66; PBT; PET	ComAlloy International Corp., USA
HIPERTUF	Poly(ethylene naphthalate)	Shell Chemical
HIPS	PS-HI	Lucky Styrenics, South Korea (distributor: Franklin Polymers Inc., USA)
HISTAT	PE-LLD; PE-HD	United Composites Inc., USA
HITANOL	PS	Hitachi Chemical Co., Japan
HI-VAL	Polyolefins; PS-HI; acetal polymers	General Polymers Inc., USA

TRADE NAME	NATURE	PRODUCER OR SUPPLIER
HIVALLOY	Reactor-made olefin copolymers and blends	
HIVALLOY G	Styrene/propylene copolymers	Montell North America, USA
HIVALLOY T	Styrene/propylene/maleic anhydride terpolymers	
HIVALLOY W	Propylene/acrylic copolymers	
HOSTACOM	Reinforced PP compounds	
HOSTADUR	PBT	
HOSTAFLON ET	ETFE	
HOSTAFLON FEP	FEP	Hoechst; Hoechst Celanese Corp., USA
HOSTAFLON TF	PTFE	
HOSTAFLON TFA	PFA	
HOSTAFLON TFB	VDF/TFE/HFP terpolymer	
HOSTAFORM	Acetal copolymers	
HOSTAGLAS	PET sheeting	Hoechst
HOSTALEN	PE-HD; PP	
HOSTALEN GUR	PE-UHMW	
HOSTALIT	PVC resins and blends	
HOSTALIT H	Polymeric impact modifiers (EVA-based)	Hoechst; Hoechst Celanese Corp., USA
HOSTALLOY	PE blends	
HOSTAPREN	PVC-C	
HOSTATEC	PEK	
HOSTYREN XS	AES	
HUNTSMAN	PS (general-purpose); PS-HI	Huntsman Chemical Corp.
HX	Liquid-crystal polymers and compounds	Du Pont
HYDLAR	Formable and machinable PA 66 and TPE materials	A. L. Hyde Co., USA
HYFLON	PFA; MFA (methylfluoroalkoxy ethylene polymer)	Ausimont
HYLAR	PVDF	
HYLITE	Deep-drawable sheet (PP core between aluminium skins) for car body panels	Hoogovens Hylite BV, The Netherlands
HYPALON	Chlorosulfonated PE	Du Pont
HYPERLAST	PUR systems	Kemira Polymers, UK and The Netherlands
HYPERLITE	PET compounds	Kaneka Corp., Japan
HYPOL	TDI/polyether prepolymer for PUR	W. R. Grace & Co., USA
HYSOL	Epoxy resins and compounds	Dexter Corp., USA
HYTEMP	Polymeric impact modifiers (acrylic)	Zeon Chemicals Inc., USA
HYTREL	TPE (copolyester type)	Du Pont
HY-VIN	PVC	Hydro Polymers Ltd., UK
HYVIS	Polyisobutylene	BP Chemicals
IDEMITSU	PC	Idemitsu Petrochemical Co., Japan; Shuman Plastics Inc., USA; Plastribution Engineering Polymers Ltd., UK
ILLANDUR	UP resins and compounds	Dr. Illing, Germany
IMPET	PET compounds	Hoechst, Hoechst Celanese Corp., USA
IMRON	PUR enamel	Du Pont
INDOPOL	Polybutylene	Amoco
INNOVEX	PE-LD	BP Chemicals
INSPIRE	High-temperature-resistant PC copolymers	Dow
INSULTRUC	UP moulding compounds	Industrial Dielectrics Inc., USA
IOTEK	Ionomers	Exxon Chemical
IPC	Glass-reinforced PPS compounds	International Polymer Corp., USA

TRADE NAME	NATURE	PRODUCER OR SUPPLIER
IPETHENE	PE-LD	Carmel Olefins, Israel and UK
ISOCHEMBOND	Epoxy resins and formulations	Furane Products, Ciba-Geigy Corp., USA
ISOFOAM	Liquid systems for PUR foams	Baxenden Chemicals Ltd., UK
ISOMIN	MF resins and compounds	Perstorp Ferguson Ltd., UK
ISONOL	PUR systems for rigid foams	Dow
ISOPLAS	Crosslinkable PE	Micropol Ltd., UK
ISOPLAST	Rigid thermoplastic PUR materials	Dow
ISOSTAC	PUR systems	EniChem
ISPLEN	PP	Repsol Quimica SA, Spain
IUPIACE	Modified PPO (PPE)	Mitsubishi
IUPILON	PC	
IUPITAL	Acetal copolymer	
IXAN	PVC and copolymers	Solvay
IXEF	Polyarylamide, reinforced compounds	
JACKDAW	PA 6 and 66	Jackdaw Polymers Ltd., UK
JEFFAMINE	Reactive polyether amines	Huntsman Chemical Corp.
JET-FLEX	AES terpolymer materials	Multibase Inc., USA
JGP	ABS; PP; PA	JGP Ltd., UK
JONYLON	PA 6 and 66 compounds	BIP Ltd., UK
J-PLAST	TPE (styrenic type)	J-Von, LP, USA
JUVINIL	PVC	Jugovinil, Yugoslavia
KADEL	Poly(arylether ketone) (PAEK)	Amoco Chemical
KALADAR	Poly(ethylene napthalate) resin	ICI
KALADEX	Poly(ethylene naphthalate) film	ICI Films
KALREZ	Perfluoroelastomer parts	Du Pont
KAMAX T	Acrylic/imide copolymers	AtoHaas
KANE ACE	ABS, MBS, and acrylic copolymers used as modifiers and process aids (especially in PVC)	Kanegafuchi Chemical Industry Co., Japan; Kaneka Texas Corp., USA
KANELITE	PS	
KAPEX	Closed-cell plastics foam	Airex AG, Germany
KAPRONET	Purging compound	GE Plastics
KAPTON	Polyimide film	Du Pont
KARLEX	PC	Ferro
KASA-KARBON	Thermoplastic compounds	Eilenburger Chemie Werke AG, Germany
KAY-FAX	Polyimide	Toa Gosei Chemical Industry Co., Japan
KELBURON	TPE (olefinic type)	DSM
KELDAX	Sound-barrier resin	Du Pont
KEL-F	CTFE; fluoropolymer compounds	3M Co., USA
KELON A	PA 66 compounds (mineral-filled)	Lati Engineering Thermoplastics SpA, Italy; Lati UK Ltd
KELON B	PA 6 compounds (mineral-filled)	
KELON C	PA 6/66 compounds (mineral-filled)	
KELRINAL	PE-C	DSM
KELTAN	EPDM	
KELTAN TP	Rubber-modified PP compounds based on reclaimed material	
KEMATAL	Acetal copolymers	Hoechst; Hoechst Celanese Corp., USA
KEPITAL	Acetal polymers	Korea Engineering Plastics Co., Ltd., South Korea

TRADE NAME	NATURE	PRODUCER OR SUPPLIER
KERIMID	Polyimide	Nyltech, France, Italy and UK
KERIPOL	UP resins	Phoenix, Germany
KERMEL	A meta-aramid fibre made from trimellitic anhydride and 4,4-diisocyanatodi-phenylmethane	Rhône-Poulenc, France
KEVLAR	A para-aramid fibre, based on poly(*p*-phenylene terephthalamide)	Du Pont
KF	PVF film	Kureha Chemical Industry Co., Japan
KIBISAN	SAN	Chi Mei Corp., Taiwan
KINEL	Polyimide moulding compounds	Ciba-Geigy
KOBLEND	(PVC + ABS) and (PC + AES) blends	EniChem
KODABOND	Copolyester	Eastman Chemical
KODAPAK	Thermoplastic polyesters (for packaging applications)	Eastman Chemical
KODAR	Thermoplastic polyester compounds	Eastman Chemical
KODAR A 150	PECTA	Eastman Chemical
KODAR 5445	PCTG	Eastman Chemical
KODAR PETG	PETG	Eastman Chemical
KOHINOR	PVC compounds (rigid and flexible)	Pantasote Polymers Inc., USA
KOPA	PA 66, polymers and compounds	Kolon Industries, South Korea
KOPET	PET	Plastribution Engineering Polymers Ltd., UK
KOSTIL	SAN	EniChem
KRASTEN	PS	Chemapol, Czech Republic
KRATON	TPE (styrenic type)	Shell Chemical
KRATON D	SBS	Shell Chemical
KRATON G	SEBS	Shell Chemical
K-RESIN	Styrene/butadiene copolymers	Phillips Chemicals
KRYNAC	Butyl rubber	DSM
KRYTOX	Perfluoropolyethers	Du Pont
KYDENE	(PVC + acrylic) blends	AtoHaas
KYDEX	Acrylic-modified PVC formable sheet	Kleerdex Co., USA
KYNAR	PVDF	Elf Atochem
LACBLOC	TPE (SEBS type)	Silac, France
LACQRENE	PS; PS-HI; and styrenic compounds	Elf Atochem
LACQTENE	PE-LD; PE-LLD; PE-HD	Elf Atochem
LACQTENE P	PP	Elf Atochem
LACQVYL	PVC resins	Elf Atochem
LADENE	Polyolefins and other polymers	SABIC (Saudi Basic Industries Corp.)
LAMITEX	Silicone resins and compounds	Franklin Polymers Inc., USA
LAPEX	PES	Lati Engineering Thermoplastics SpA, Italy; Lati UK Ltd.
LARFLEX	Elastomer-modified PP	Lati Engineering Thermoplastics SpA, Italy; Lati UK Ltd.
LARIL	Modified PPO (PPE)	Lati Engineering Thermoplastics SpA, Italy; Lati UK Ltd.
LAROFLEX	Blends of PVC with modifying polymers	BASF
LARTON	PPS polymer and compounds	Lati Engineering Thermoplastics SpA, Italy; Lati UK Ltd
LASTANE	TPE (PUR type)	Lati Engineering Thermoplastics SpA, Italy; Lati UK Ltd
LASTIFLEX	(PVC + terpolymer) blends	Lati Engineering Thermoplastics SpA, Italy; Lati UK Ltd
LASTIL	SAN compounds; modified PPO (PPE)	Lati Engineering Thermoplastics SpA, Italy; Lati UK Ltd
LASTILAC	ABS compounds	Lati Engineering Thermoplastics SpA, Italy; Lati UK Ltd
LASTILAC 10	(PC + ABS) blends	Lati Engineering Thermoplastics SpA, Italy; Lati UK Ltd
LASTIROL	PS; PS-HI compounds	Lati Engineering Thermoplastics SpA, Italy; Lati UK Ltd
LASUF	PSU polymers and compounds	Lati Engineering Thermoplastics SpA, Italy; Lati UK Ltd
LATAMID 6	PA 6 polymers and compounds	Lati Engineering Thermoplastics SpA, Italy; Lati UK Ltd
LATAMID 66	PA 66 polymers and compounds	Lati Engineering Thermoplastics SpA, Italy; Lati UK Ltd
LATAMID 68	PA 6/66 copolymers and compounds	Lati Engineering Thermoplastics SpA, Italy; Lati UK Ltd

TRADE NAME	NATURE	PRODUCER OR SUPPLIER
LATAMID 12	PA 12 polymers and compounds	Lati Engineering Thermoplastics SpA, Italy; Lati UK Ltd
LATAN	Acetal copolymers and compounds	
LATENE	PP homopolymers and compounds	
LATENE EP	Ethylene/propylene copolymers	
LATENE HD	PE-HD	
LATER	PBT polymers and compounds	
LATIBLEND	(PBT + ABS) and (PBT + PC) blends	
LATILON	PC polymers and compounds	
LATILUB	Internally lubricated compounds	
LATISHIELD	Compounds for EMI-shielding products	
LATISTAT	Antistatic compounds	
LECUTHERM	Epoxy resins and compounds	Bayer
LECUPREN	UP resins and compounds	
LEGUVAL	UP resins and compounds	
LEVAFLEX	Modified PP	
LEVAPREN	EVA (wide range of VA contents)	
LEVASINT	EVOH copolymers	
LEXAN	PC polymers, compounds and products	GE Plastics
LEXAN GR	PC suitable for sterilisation by gamma radiation	
LEXAN HP	PC film with coating resistant to UV and abrasion	
LEXAN XT	PC/silicone block copolymer	
LISA	PC	Bayer; Miles Inc., USA
LITEN	PE-HD	Longfield Chemicals Ltd., UK
LOCALEN	EVA	BASF
LOMOD	TPE (copolyester type)	GE Plastics
LOPOX	Epoxy resins and moulding compounds	Orkem, France
LOTADER	Various terpolymers	Elf Atochem
LOTRENE	PE-LD	Orkem, France
LOTREX	PE-LLD	
LOTRYL	Ethylene copolymers and terpolymers (including EVA)	Elf Atochem
LUBRICOM	PA 6	ComAlloy International Corp., USA
LUBRICOMP	Internally lubricated compounds (various polymers)	LNP Engineering Plastics
LUBRILOY	Internally lubricated PA 66 compounds	
LUBRIZOL 7441	Hydrogenated butadiene/styrene copolymer (used as viscosity-improving additive in mineral oils)	Lubrizol Corp., USA
LUCALEN	Ethylene copolymers	BASF
LUCALOR	Chlorinated PVC (PVC-C)	Elf Atochem
LUCEL	Polyacetal	Lucky Ltd., South Korea; Arrowcrest Industrial Ltd, UK
LUCKY	ABS; PE-HD; PMMA; PS; SAN	
LUCOBIT	Bituminised ethylene copolymers	BASF
LUCOLENE	PVC compounds	Elf Atochem
LUCOREX	PVC compounds	
LUCOVYL	PVC resins	
LUCRYL	PMMA; acrylic materials (including capping compounds for PVC)	BASF
LUFLEXEN	PE-LLD, produced with metallocene catalysts	
LUMIFLON	Soluble fluoropolymers (for application as barrier and protective coatings)	ICI (originally Asahi Glass Ltd., Japan)
LUPAN	SAN compounds	Lucky Ltd.,South Korea; Arrowcrest Industrial Ltd., UK
LUPIC	Biaxially oriented plastics films	Mitsui & Co., UK and USA

TRADE NAME	NATURE	PRODUCER OR SUPPLIER
LUPOL	Polyolefins	Lucky Ltd.,South Korea; Arrowcrest Industrial Ltd., UK
LUPOLEN	PE-LD; PE-HD; EVA	BASF
LUPOLEN UHMW	PE-UHMW	BASF
LUPOS	ABS compounds	Lucky Ltd.,South Korea; Arrowcrest Industrial Ltd., UK
LUPOX	PBT	Lucky Ltd.,South Korea; Arrowcrest Industrial Ltd., UK
LUPOY	(PC + ABS) blends	Lucky Ltd.,South Korea; Arrowcrest Industrial Ltd., UK
LUPRANATE	MDI (polymeric and simple) for PUR	BASF
LURAN	PS; SAN	BASF
LURAN S	Acrylonitrile/stryrene/acrylate (ASA); ABS/methacrylate graft copolymers	BASF
LURANYL	Styrenic blends; PPO (PPE) blends with PA	BASF
LUSTRAN	ABS; (PVC + ABS) blends; PS	Monsanto (transfer to Bayer planned for 1996)
LUSTRAN SAN	SAN	Monsanto (transfer to Bayer planned for 1996)
LUSTRAN HH	Acrylonitrile/styrene/methacrylate	Monsanto (transfer to Bayer planned for 1996)
LUSTREX	PS; PS-HI	Monsanto (transfer to Bayer planned for 1996)
LUSTRO	Formable plastics sheets (CA; CAB;PETG)	Lustro Plastics Co., USA
LUTOFAN	PVC resins and blends	BASF
LUVICAN	Poly(vinyl carbazole)	BASF
LUXATE	Isocyanate prepolymers	Olin Corp., USA
LUXIS	PA 66 polymers and compounds	Westover Color & Chemical Corp., USA
LYOCELL	Generic name for solvent-spun cellulosic fibres	Various producers: e.g. Lenzing, Austria; Courtaulds, UK
LYTEX	Epoxy SMC	Quantum Composites Inc., USA
MAGNUM	ABS	Dow
MACROBLEND	(PC + PET) blends, including reinforced and filled compounds	Bayer; Miles Inc., USA
MAKROFOL	PC films	Bayer
MAKROLON	PC polymers and compounds	Bayer; Miles Inc., USA; Albis
MELECCA	Styrene copolymers (including styrene/maleimide and its blends with PA), and compounds	Denki Kagaku Kogyo, Japan and USA
MALON	PET	M A Industries Inc., Polymers Div., USA
MAMAX	PET	M A Industries Inc., Polymers Div., USA
MAPLEN	PE-LD; PE-HD	Montell Italia, Italy
MARANYL	PA 66 and 6/66	LNP Engineering Plastics
MARLEX	PE-HD; PP	Phillips Chemical Co., USA
MARVYLEX	PVC compounds (including some based on ultra-high molecular weight resin)	L.V.M. NV., Belgium
MAXCELS	PUR foaming systems	Foam Enterprises, USA
MAXFLEX	PUR foaming systems	Foam Enterprises, USA
MAXPAK	PUR foaming systems	Foam Enterprises, USA
MEGOLON	Cable-covering compounds	Lindsay & Williams Ltd., UK
MELDIN	Polyimide: formable sheet; injection-moulding resins	Furon Co., USA
MELINAR PET	PET	ICI Polyester Intermediates and Polymers, UK
MELINEX	PET films	ICI Films
MELMEX	MF moulding compounds	BIP Plastics Ltd., UK
MELOPAS	MF moulding compounds	Raschig AG, Germany
MERLON	PC	Bayer; Miles Inc., USA
MENZOLIT	UP resins and compounds	Menzolit Werke GmbH, Germany
METABLEN C	Polymeric modifiers (MBS type)	Elf Atochem

TRADE NAME	NATURE	PRODUCER OR SUPPLIER
METHYLON	Phenolic resins	Occidental Chemical Corp., USA
MICALISS	Synthetic paper based on PET and PAN	Schoeller Hoesch, Germany
MICAMID	Synthetic paper based on polyimide	Schoeller Hoesch, Germany
MICROTHENE	PE (LD, LLD and HD) and PP powders, mainly for rotational moulding and powder coating	Quantum Chemical Co., USI Div., USA
MILASTOMER	TPE (olefinic type)	Mitsui Plastics Inc., USA
MINDEL	PSU compounds	Amoco Performance Products Inc., USA
MINDEL A	(PSU + ABS) blends	Amoco Performance Products Inc., USA
MINDEL B	PSU blend compounds	Amoco Performance Products Inc., USA
MINDEL S	PSU blend compounds	Amoco Performance Products Inc., USA
MINLON	PA 66 compounds (mineral-filled)	Du Pont
MIPOLAM	PVC flooring materials	HT Troplast AG, Germany
MIPOPLAST	PVC films	HT Troplast AG, Germany
MIRROPHANE	Metallised films and sheeting	Gomar Manufacturing Co., Inc., USA
MISTABOND	PUR foaming systems	MarChem Corp., USA
MISTIQUE	PVC bottle-blowing compounds	Geon Co., USA
MODAR	Low-viscosity methacrylate resins (curable to acrylic thermosets)	ICI Acrylics
MOLDESITE	Phenolic resins and compounds	SPREA, Italy
MOLTOPREN	Rigid PUR foam systems	Bayer; Miles inc., USA
MODUR	MDI/polyester prepolymer	Bayer; Miles inc., USA
MONTELL	PE-LD; PP	Montell Italia, Italy
MOPLEN	PP	Montell Italia, Italy
MORTHANE	TPE (PUR type)	Morton International Inc., USA
MOSTEN	PP	Longfield Chemicals Ltd., UK
MULTIBASE	ABS	Multibase SA, France Multibase Inc., USA
MULTIBASE G	TPE (SEBS type)	Multibase SA, France Multibase Inc., USA
MULTI-FLAM	PP compounds	Multibase SA, France Multibase Inc., USA
MULTI-FLEX	TPE (olefinic and styrenic types)	Multibase SA, France Multibase Inc., USA
MULTI-PRO	PP compounds	Multibase SA, France Multibase Inc., USA
MULTI-SAN	SAN	Multibase SA, France Multibase Inc., USA
MULTRATHENE	MDI/polyester prepolymer	Bayer; Miles Inc., USA
MYLAR	PET films	Du Pont
MYOFLEX	Epoxy resins and compounds	Schweizerische Isola Werke, Switzerland
NAKAN	PVC	Elf Atochem
NAXEL	PC	MRC Polymers Inc., USA
NCPE	General letter code for grades of PE and some ethylene copolymers. E.g. NCPE 22: PE-UHMW. NCPE 5028 and 5119: EVA. NCPE 5860: EEA copolymer (previously marketed by BP Chemicals as NOVEX LE 1860)	Borealis
NEALID	PP moulding compounds	Neste OY, Finland
NEOFLON	Fluoropolymers	Daikin America Inc., USA Daikin Kogyo Co., Japan
NEOFLON EP	ETFE	Daikin America Inc., USA Daikin Kogyo Co., Japan
NEOFLON PFA	PFA	Daikin America Inc., USA Daikin Kogyo Co., Japan
NEONIT	DAP compounds	Ciba-Geigy
NEOPOL	Long-fibre-reinforced PA, PP and polyester compounds	Neste OY, Finland
NEOPOLEN	Polyolefin foam (cushioning material)	BASF
NEOPRENE	Polychloroprene	Du Pont
NERAFEN	PPO (PPE)	Spolane, Czech Republic
NIKALET	Epoxy resins and compounds	Nippon Carbide Industry Co., Japan
NIPOFLEX	EVA	Toyo Soda Co., Japan

TRADE NAME	NATURE	PRODUCER OR SUPPLIER
NIPOL	Polymeric impact modifier (NBR type)	Zeon Chemicals Inc., USA
NIPOLIT	PVC resins	Chisso Corp., Japan
NIPOLON	PE-LD; PE-HD	Toyo Soda Co., Japan
NOBLEN	PP	Sumimoto Plastics America Inc., USA
NO-HEAT	Plastics tubing and mouldings, solvent-expanded, shrinkable by air-drying	Insulation Systems Inc., USA
NOMEX	Meta-aramid fibre based on poly(*m*-phenylene isophthalamide)	Du Pont
NORASOL	Water-soluble polycarboxylate	Elf Atochem
NORCHEM	PE polymers and compounds	Quantum Chemical Co., USI Div., USA
NORDEL	Hydrocarbon rubbers (modified PP)	Du Pont
NORFLEX	PS	Norddeutsche Seekabelwerke AG, Germany
NORSOMIX	UP resins and compounds	Orkem, France
NORSOREX	Polynorbornene elastomers	Elf Atochem
NORTUFF	Ethylene copolymers; PP polymers and compounds	Quantum Chemical Co., USI Div., USA
NORVINEX	PVC compounds (including silane-crosslinkable grades)	Hydro Plast AB, Sweden; Hydro Polymers Ltd., UK
NORVINYL	PVC suspension polymers	Hydro Plast AB, Sweden; Hydro Polymers Ltd., UK
NORYL	PPO (PPE)	GE Plastics
NORYL GTX	(PPO + PA) blends	GE Plastics
NOVABLEND	PVC compounds	Novatec Plastics & Chemicals Co. Inc., USA
NOVACOR	PS	Novacor Chemicals Inc., USA
NOVACYCLE	Recycled PVC compounds	Novatec Plastics & Chemicals Co. Inc., USA
NOVADUR	PBT	Mitsubishi Engineering Plastics, USA
NAVALAR	Polymeric impact modifiers (ABS type)	Nova Polymers Inc., USA
NOVALAST	TPE (olefinic type)	Nova Polymers Inc., USA
NOVALENE	TPE (EPDM)	Nova Polymers Inc., USA
NOVALLOY A	(ABS + PA) blends	Daicel Polymers Ltd., UK
NOVALLOY B	(ABS + PBT) blends	Daicel Polymers Ltd., UK
NOVALLOY E	Antistatic ABS compounds	Daicel Polymers Ltd., UK
NOVALLOY S	(ABS + PVC) blends	Daicel Polymers Ltd., UK
NOVALOY	ABS compounds, including (PVC + ABS) blends	Novatec Plastics & Chemicals Co. Inc., USA
NOVAMID	PA 6	Mitsubishi Engineering Plastics, USA
NOVAPOL	PE-LD; PE-LLD; PE-HD	Novacor Chemicals Inc., USA
NOVAREX	PC	Mitsubishi Gas Chemical Co., Inc., Japan
NOVATEC	Polyolefins	Mitsubishi Gas Chemical Co., Inc., Japan
NOVATEX	Polyolefins	Mitsubishi Gas Chemical Co., Inc., Japan
NOVEX	PE-LD (originally also some ethylene copolymers, e.g. EEA - see under NCPE)	BP Chemicals
NOVODUR	ABS, AES	Bayer; Miles Inc., USA
NOVOLEN	PP homopolymers, copolymers and specialist grades	BASF
NOVOLEX	See NCPE 5860 under NCPE	
NOVON	Biodegradable (starch-filled) thermoplastic materials	Novon Products Group, USA
NOXON	Polymer claimed reactive with ozone (offered for ozone-removal devices)	Hoechst
NOYAREX	High-clarity polymers for optical applications	Mitsui & Co., USA and UK
NUCREL	Ethylene/methacrylic acid copolymer	Du Pont

TRADE NAME	NATURE	PRODUCER OR SUPPLIER
NUEVOWOOD	Plastics wood substitute, based on reclaimed materials (PE-HD filled with peanut hulls)	Environmental Solutions Inc., USA
NYCOA	PA 6; 66 and 6/66	Nylon Corp. of America, USA
NYDUR	Now DURETHAN (q.v.)	
NYLAMID	PA 6; 66 and 12	Polymer Service Corp., USA
NYLATRON	PA compounds	DSM
NYLASTAT	Antistatic cast PA compounds	Nylcast Ltd., Nylontechnics Div., UK
NYLUBE	Lubricated cast PA compounds`	Nylcast Ltd., Nylontechnics Div., UK
NYPEL	PA 6	AlliedSignal Inc., USA
NYREF	PA (N-MXD6)	Solvay
NYRIM	PA 6 for RIM	DSM
OILON	Cast PA compounds oil-lubricated	Nylcast Ltd., Nylontechnics Div., UK
OLDAPAL	UP resins	Buesing & Fasch GmbH, Germany
OLDODUR	Integral-skin rigid foams	Buesing & Fasch GmbH, Germany
OLDOFILL	Foams for packing applications	Buesing & Fasch GmbH, Germany
OLDOFLEX	Integral-skin flexible foam	Buesing & Fasch GmbH, Germany
OLDOFLEX S	Microcellular footwear soling compositions	Buesing & Fasch GmbH, Germany
OLDOPREN	TPE (PUR type)	Buesing & Fasch GmbH, Germany
OLDOTHERM	Rigid foams (mainly for thermal insulation)	Buesing & Fasch GmbH, Germany
OLEHARD	PP compounds	Chisso America Inc., USA
ONGROVIL	PVC resins	Borsodi Vegyi, Hungary
OPPANOL	Polyisobutylene	BASF
OPTEMA	Ethylene/methyl acrylate copolymers	Exxon Chemical
OPTENE EVA	EVA (range of VA contents)	Borealis
OPTENE ECB	Compound of EBA with bitumen (material of sheeting for roofing applications)	Borealis
OPTICITE	PS	Dow
OPTIX	Acrylic polymers, compounds and sheeting	Plaskolite Inc., USA
OPTUM	Olefinic blends	Ferro
OREVAC	Olefin - based terpolymers	Elf Atochem
ORGABLEND	(PP + PA) blends produced from recycled polymers (properties claimed comparable with the corresponding virgin ORGALLOY blends)	Elf Atochem
ORGAFLEX	Polyphosphazene	Elf Atochem
ORGALAN	PC	Elf Atochem
ORGALLOY	(PP + PA) blends	Elf Atochem
ORGAMIDE	PA 6	Elf Atochem
	Ultra-fine plastics powders	Elf Atochem
	PBT compounds	Elf Atochem
ORGAVYL	PVC resins	Elf Atochem
ORMOCER	Extra-hard surface coating for PC sheeting (ORganic MOdified CERamic) claimed to approach glass in abrasion resistance	Bayer
OROGLAS	PMMA	AtoHaas
OROMID	Recycled PA 6 and 66	Nyltech, France, Italy and UK
OXYBLEND	PVC resins and PVC-U compounds	Occidental Chemical Corp., Vinyls Div., USA
OXYCLEAR	PVC resins and PVC-U compounds	Occidental Chemical Corp., Vinyls Div., USA
OXYFLEX	Flexible PVC compounds	Occidental Chemical Corp., Vinyls Div., USA
PALAPREG	Resins and resin systems for BMC and SMC	BASF
PALATAL	UP resins	BASF

TRADE NAME	NATURE	PRODUCER OR SUPPLIER
PALCLEAR	PVC-U sheeting	Palram Ltd./Paltough Ltd., Israel
PALDOOR	PVC door panels	
PALIGHT	Cellular PVC sheets	
PALOPAQUE	Industrial PVC sheeting	
PALRUF	Corrugated PVC sheeting	
PALSUN	PC sheeting (UV-protected)	
PANLITE	PC compounds	Teijin Chemicals Ltd., Japan and USA
PANTALAST	(PVC + EVA) blends	Pantasote Polymers Inc., USA
PANTOXYL	Bonding agents (MF-based)	Condea Chemie GmbH, Germany
PARAGLAS	PMMA	Degussa
PARALOID	Acrylic resins used as modifiers in PVC (initially also known as ACRYLOID in Europe and UK)	Rohm & Haas Co., USA
PARAPET	PMMA for laser-read discs	Mitsui & Co., USA and UK
PARAPLEX	Epoxy resins and compounds	Rohm & Haas Co., USA
PARAPOL	Polybutylene	Exxon Chemical
PAXON	PE-HD	Paxon Polymer Co., USA
PEBAX	TPE (polyether block amide)	Elf Atochem
PEKEVIC	PVC	Neste OY, Finland
PELLETHANE	TPE (PUR type)	Dow
PELPRENE P	TPE (polyether-ester type)	Toyobo Co., Ltd., Japan
PELPRENE S	TPE (polyester type)	
PEMEX	PE-HD	Petroleos Mexicanos (PEMEX), Mexico; Plastribution Engineering Polymers Ltd., UK
PENTASTAT TZ	Static-dissapative PVC sheeting	Klöckner Pentaplast Ltd., UK
PERBUNAN	Nitrile rubber (NBR)	Bayer
PERMASTAT	Antistatic compounds: ABS; (PC + ABS) and (PC + polyester) blends	RTP Co., USA
PERRIPRENE	Range of TPE, including olefinic and styrenic types	Perrite Rubber Compounds, UK
PERSPEX	Acrylic polymers and sheeting	ICI Acrylics
PETKIM	PVC resins (suspension type)	Vinyl Ltd., UK
PETLON	PBT	Bayer; Miles Inc., USA
PETRA	PET	AlliedSignal Inc., USA
PETROTHENE	PE (most types, including crosslinkable grades); ethylene copolymers (including EVA); PP	Quantum Chemical Co., USI Div., USA
PEVIKON	PVC (including paste-forming resins)	Hydro Plast AV, Sweden; Hydro Polymers Ltd., UK
PIBIFLEX	PBT copolymers	EniChem
PIBITER	PBT copolymers and hot-melt compounds	
PIOLEN	Packaging films (single and multi-layer)	Spohn Verpackungswerke, Germany
PLASLUBE	Range of lubricated plastics compounds	DSM
PLASPOL	Plastics powders	Ashland Plastics (UK) Ltd., UK
PLASTECH	Thermoplastic compounds	Cabot Plastics International, Belgium; Cabot Corp., USA
PLASTILIT	PVC compounds	Solvay
PLASTOTHANE	TPE (PUR type)	Morton International Inc., USA
PLATABOND	Co-polyamide hot-melt compounds	Elf Atochem
PLATHERM		
PLATILON	PP polymers	

TRADE NAME	NATURE	PRODUCER OR SUPPLIER
PLENCO	PET; phenolic, MF, and UP resin compounds	Plastics Engineering Co., USA
PLENEX	TPE (styrenic type)	Vista Chemical Co., USA
PLEXALLOY HZ	PC: sheeting produced from recycled material	Röhm GmbH, Germany
PLEXAR	PE-LLD; PE-HD; EVA	Quantum Chemical Corp., USI Div., USA
PLEXAR 909	Purging compound for PA and EVOH	Quantum Chemical Corp., USI Div., USA
PLEXIDUR	PMMA	AtoHaas
PLEXIGLAS	Acrylic polymers; PMMA sheeting	AtoHaas
PLEXIMID	Poly(methyl methacrylimide) (PMMI)	AtoHaas
PLIOCORD VP	Vinylpyridine copolymer	Goodyear Chemicals Europe
PLIOVIC	PVC resins (paste-forming type)	Goodyear Chemicals Europe
PLUSTEC	PA	Polyram Ram On Industries, Israel
PLUTON	Aromatic ladder polymers	3M Co., USA
PLYOPHEN	Phenolic resins	Occidental Chemical Corp., USA
POCAN	PBT and blends	Bayer; Miles Inc., USA
POLIDUX	ABS; SAN	Petroplas Ltd., UK
POLIFIL	PP	Polifil, USA
POLLOPAS	UF resins and compounds	Perstorp
POLY BD	Polybutadiene (hydroxyl-terminated)	Elf Atochem
POLYBOND	Compatibilisers for polymer alloying and recycling. Range includes grafts of maleic anhydride on polypropylene (POLYBOND 350), used also as coupling agents for glass reinforcement or mineral fillers for polyolefins	Uniroyal Chemical Co., USA
POLYCAL	PVC compounds	Elf Atochem
POLYCHEM	DAP compounds	Budd Co., Plastics Div., USA
POLYCLEAR	PET (bottle grade)	Hoechst; Hoechst Celanese Corp., USA
POLYCLIP	PC glazing panels	Polygal, Israel
POLYCOMP	Range of various thermoplastic compounds	LNP Engineering Plastics
POLYCURE	Crosslinkable PE-HD	Borealis
POLY-DAP	DAP compounds	Cosmic Plastics Inc., USA
POLYFINE	PP	Tokuyama Soda Co. Ltd., Japan
POLYFLAM	Flame-retarded compounds: PP; PS; PE-LLD	A. Schulman Inc., USA
POLYFLON	PTFE	Diakin Kogyo Co., Japan; Diakin America Inc., USA
POLYFOL	Oriented PS sheeting for thermoforming	Senoplast UK Ltd., UK
POLYFORT	PP polymers and compounds	A. Schulman Inc., USA
POLYGAS	PE-HD	Solvay
POLYLAC	ABS	Chi Mei Corp., Taiwan
POLYMAN	PS; (PVC + ABS) blends	A. Schulman Inc., USA
POLYMIST	PTFE	Ausimont
POLYNIL	PA polymers and compounds	Nilit Ltd., Israel
POLYOX	Water-soluble polyoxyethylene	Union Carbide
POLYPUR	TPE (PUR type)	A. Schulman Inc., USA
POLYREX	PS homopolymers	Chi Mei Corp., Taiwan
POLYRITE	UP resin systems	Polyply Inc., USA
POLYSET	Epoxy resin compounds	Morton International Inc., USA
POLY STAR	PS	Idemitsu Petrochemical Co., Japan; Shuman Plastics Inc., USA; Plastribution Engineering Polymers Ltd., UK

TRADE NAME	NATURE	PRODUCER OR SUPPLIER
POLY-STONE	Granite-effect masterbatches	Plastics Color-Chip, Div. of PMC Inc., USA
POLYSTYROL	PS (general purpose); PS-HI; SB	BASF
POLYTIMBRE	Plastics wood substitute based on reclaimed PE-HD	Custom-Pac Extrusions Inc., USA
POLYTRON	Rigid PVC compounds (including antistatic grades)	Geon Co., USA; Hydro Geon, UK
POLYTROPE	PP compounds; TPE (olefinic type)	A. Schulman Inc., USA
POLY-X	PC structural glazing sheets	Polygal, Israel
PORENE	ABS	Thai Petrochemical Industry Co., Thailand
PREMI-GLAS	Epoxy resin SMC; UP resin compounds	Premix Inc., Moulding Compound Div., USA
PREMI-JECT	UP resin compounds	Premix Inc., Moulding Compound Div., USA
PREVAIL	(ABS + PUR) blends	Dow
PREVEX	Modified PPO (PPE)	GE Plastics
PRIMACOR	Ethylene/acrylic acid copolymer (used as adhesive tie-layers in coextrusion)	Dow
PRIMEF	PPS compounds	Solvay
PRISM	Liquid PUR systems for RIM	Bayer; Miles Inc., USA
PROCOM	PP compounds	BASF
PROCOND	PP compounds	United Composites Inc., USA
PRODORIT	Epoxy resins and compounds	T.I.B. Chemie (Shell), Germany
PRO-FAX	PP	Montell Italia, Italy; Himont USA
PROGILITE	Phenolic resins and compounds	Nyltech, France, Italy and UK
PROLASTIC	TPE (olefinic and styrenic types)	Discas Inc., USA
PROLOY	(PC + ABS) blends	GE Plastics
PROPADEX	Recycled PP	Mainetti Recycling Ltd., UK
PROPAFILM	Oriented PP films	ICI Films
PROPAK	PP compounds	PolyPacific Pty. Ltd., Australia
PROPATHENE	PP	ICI
PROPYFLEX	Multi-layer PP film for medical-grade bags and bottles	Kobusch/Sengewald Verpackungen GmbH, Germany
PULSE	Blends of PC (CALIBRE) with ABS (MAGNUM)	Dow
PVC-SEMICON	Conductive PVC compounds	Borealis
PYRONYL	Flame-retarded PA compounds	Synlon Ltd., UK
QUESTRON	High-performance PP	Montell Italia, Italy
QUARITE	Acrylic sheet with a "granite" look	Aristech Chemical Corp., USA
QUATREX	Epoxy resins and compounds	Dow
QUELFLAM	Rigid PUR foams	Baxenden Chemicals Ltd., UK
QUESTRA	Metallocene-catalysed crystalline syndiotactic polystyrene	Dow
RADEL A	Poly(ether sulfone) compounds	Amoco Chemicals (Europe), Switzerland; Amoco Performance Products Inc., USA
RADEL R	Poly(phenyl sulfone) compounds	Amoco Chemicals (Europe), Switzerland; Amoco Performance Products Inc., USA
RADIBLEND	Blends: (PC + ABS); (PC + PET); (PP + PA)	Radicinovacips SpA, Italy
RADIFLAM	PA 6 and 66 (flame-retarded)	Radicinovacips SpA, Italy
RADILON A	PA 66	Radicinovacips SpA, Italy
RADILON S	PA 6	Radicinovacips SpA, Italy
RADILUX	PC	Radicinovacips SpA, Italy
RADITER	PBT; PET	Radicinovacips SpA, Italy
RALUPOL	UP resins	Raschig AG, Germany
RAVEPOX	Epoxy resins and compounds	EniChem
RAVIKRAL	ABS	EniChem

TRADE NAME	NATURE	PRODUCER OR SUPPLIER
RAVINIL	PVC resins	European Vinyls Corp. (EVC); ICI; and associates world-wide
RECOTECH 2300	Silicone rubber compounds	Silicone Products Technology Inc., USA
RECYPER	Range of compatibilisers for polymer blends	Mitsubishi Gas Chemical Co., Polyolefins Div., Japan
REFAX	Recycled PP	Montell Italia, Italy; Himont USA
REGALTECH	Medical-grade sheeting and films (PVC; PP)	O' Sullivan Corp., USA
RENAISSANCE	Recyclable PVC sheeting	O' Sullivan Corp., USA
RENY	Aromatic copolymer PA compounds	Mitsubishi Engineering Plastics, USA
RENYL	PA 6	Nyltech, France, Italy and UK
REPLAY	Recycled PS	Huntsman Chemical Corp.
REPLUS	Reformulated PA	Polyram Ram On Industries, Israel
REPOLEM	Acrylic emulsions	Elf Atochem
REPREAN	Ethylene copolymers	Discas Recycled Products Corp., USA
RESARIT	PMMA	Resart GmbH, Germany; Plastics Plus Ltd., UK
RESART	MF resins and compounds	Resart GmbH, Germany; Plastics Plus Ltd., UK
RESARTHERM	UP resin moulding compounds	Resart GmbH, Germany; Plastics Plus Ltd., UK
RESIFOAM	PUR foam systems based on RESINOL polyols and URESTYL isocyanates	Resina Chemie BV, The Netherlands
RESIMENE	MF compounds	Monsanto
RESINOID	Phenolic resins and compounds	Resinoid Engineering Corp., Materials Div., USA
RESINOL	Thermosetting resin materials (including polyols)	Raschig AG, Germany
RESIPOL	UP resins and systems	Raschig AG, Germany
RESOTHERM	UP resins	Resart GmbH, Germany; Plastics Plus Ltd., UK
RESTIL	SAN	Montell Italia, Italy
RESTIRAN	ABS	Montell Italia, Italy
RETAIN	PE-LD; PE-HD	Dow
RETPOL	PP compounds	PolyPacific Pty. Ltd., Australia
REVINEX	Latex for use in cementitious compositions	Doverstrand Ltd., UK
REXENE	PE-LLD; PP	Rexene Products Co., USA
REXFLEX	PP	Rexene Products Co., USA
REXFLEX FPO	Low-density propylene homopolymers and copolymers (FPO = "flexible polyolefin")	Rexene Products Co., USA
RFX	PC prismatic sheet	Polygal, Israel
RHENALYD	Alkyd resins	Condea Chemie GmbH, Germany
RHEOVIS	Acrylic rheology modifiers	Allied Colloids, UK
RHODESTER CL	Liquid-crystal polymers	Nyltech, France, Italy and UK
RHODOPAS	SB	Nyltech, France, Italy and UK
RIBLENE	PE-LD	EniChem
RICON	Polymeric impact modifiers (polybutadiene and SB types)	Ricon Resins Inc., USA
RIGIDEX	PE-HD	BP Chemicals
RIGIPORE	Expandable PS	BP Chemicals
RIGOLAC	UP resins	Showa High Polymer Co., Japan

TRADE NAME	NATURE	PRODUCER OR SUPPLIER
RILSAMID T	PA with low water absorption	Elf Atochem
RILSAN	PA 11	Elf Atochem
RILSAN A	PA 12	Elf Atochem
RILSAN ESY	PA 11 powder for electrostatic spraying	Elf Atochem
RIMLINE	PUR RIM systems	ICI
RIMPLAST	PA 12 ; silicones; TPE (polyester and PUR types)	Huels
RITEFLEX	TPE (polyester type)	Hoechst; Hoechst Celanese Corp., USA
RODRUN	Liquid-crystal polymer	Unitika Co. Ltd., Japan
RONFALIN	ABS	DSM
RONFALOY	ABS-type polymeric modifiers ; (ABS + PBT) alloy	DSM
ROPOL	PE-LD; PE-HD	Chem. Komb, Borzesti, Romania
ROPOTEN	PE-LD	Neftochim, Belgium
ROSEVIL	PVC resins	Chem. Komb, Borzesti, Romania
ROSITE	UP moulding compounds	Rostone Corp., USA
ROTAMID	PA 6 for rotational casting	Nylacast Ltd., Nylontechnics Div., UK
ROTOTRON	PP rotational moulding compounds	Rototron Corp., USA
ROYALENE	EPDM	Uniroyal Chemical Co., USA
ROYALTUFF	Grafted-EPDM modifiers for engineering thermoplastics (e.g. ROYALTUFF X-330: SAN/EPDM graft)	Uniroyal Chemical Co., USA
RT/DUROID	Fluoropolymer materials for machining and die-cutting	Rogers Corp., USA
RTP	Various polymeric compounds	RTP Co., USA
RUMBER	Plastic wood substitute based on reclaimed materials (PE-HD and rubber tyres)	Rumber Materials Inc., USA
RUTAMID	PA 6 and 66	Bakelite AG, Germany; Bakelite Polymers Ltd., UK
RUTAPHEN	Phenolic compounds	Bakelite AG, Germany; Bakelite Polymers Ltd., UK
RUTAPOX	Epoxy resins and compounds	Bakelite AG, Germany; Bakelite Polymers Ltd., UK
RYNITE	PBT; PET	Du Pont
RYTON	PPS polymers and compounds. (Also the PPS polymer material of SULFAR fibres)	Phillips Chemical Co., USA
RYULEX	(PC + ABS) blends	Dianippon Ink and Chemicals Inc., Japan
SABRE	(PC + polyester) blends	Dow
SADUR	Acrylonitrile/methyl methacrylate (A/MMA) copolymers	Resart GmbH, Germany; Plastics Plus Ltd., UK
SALCARE	Acrylic plastics for personal-care products	Allied Colloids, UK
SANREX	SAN	Mitsubishi Monsanto Chemical Co., Japan
SANTOCLEAR	PS	Mitsubishi Monsanto Chemical Co., Japan
SANTOLITE	Polyaryl ketones	Monsanto
SANTOPRENE	TPE (olefinic TPV type)	Advanced Elastomer Systems (joint venture by Monsanto and Exxon)
SARAN	Vinyl chloride/vinylidene chloride/acrylonitrile	Dow
SARLINK	TPE (olefinic and styrenic types)	DSM
SATINFLEX	Flexible PVC compounds	Alpha Chemical & Plastics Corp., USA
SAXEROL	Plastics sheeting: ABS; PS	Eilenburger Chemie Werke AG, Germany
SAXOLEN	Plastics sheeting: PE; PP	Eilenburger Chemie Werke AG, Germany
SAXOREC	Recycled PE and PP polymers and sheeting	Eilenburger Chemie Werke AG, Germany

TRADE NAME	NATURE	PRODUCER OR SUPPLIER
SBD	Syndiotactic polybutadiene: a vulcanisable TPE	Japan Synthetic Rubber Co., Japan
SCARAB	UF moulding powders	BIP Ltd., UK; BIP Chemicals Inc., USA
SCARNOL	EVOH	Nippon Gohsei, Japan (distributor Morton International Inc., USA)
SCHU-LINK	Crosslinkable PE-HD	A. Schulman Inc., USA
SCLAIR	PE-LLD; PE-HD	Du Pont Canada
SCLAIRLINK	Crosslinkable PE	Du Pont Canada
SCOLEFIN	PE-HD	Buna AG, Germany
SCONAPOR	Expandable polystyrene	Buna AG, Germany
SCONATER	ABS	Buna AG, Germany
SCOPACRYL	Acrylic polymers	Buna AG, Germany
SCOPYROL	PS	Buna AG, Germany
SCOTCHCAST	Epoxy resins and compounds	3M Co., USA
SCOVINAT	PVA	Buna AG, Germany
SCOVINYL	PVC resins (including graft copolymers: e.g. SCOVINYL SC 6569 HV - graft of vinyl chloride onto an elastomeric acrylate)	Buna AG, Germany
SEBILEN	PP compounds	Sebi SpA, Italy
SEBIMID	PA 6 and 66	Sebi SpA, Italy
SEBLEND A	(PC + ABS) blends	Sebi SpA, Italy
SEBS-COMPOUND	TPE (SEBS) compounds	Borealis
SEBS-SEMICON	Conductive SEBS compounds	Borealis
SECURIFLEX	Anti-lacerative windscreen system: PVB-interleaved glass backed with PUR film	Saint-Gobain Vitrage, France
SELAR	Polymers for barrier applications	Du Pont
SELAR PA	Amorphous PA	Du Pont
SELAR PT	PBT (with a comonomer)	Du Pont
SELAR RB	A hydrocarbon copolymer	Du Pont
SELAR OH	EVOH	Du Pont
SELCHIM	PVC resins	Solvay
SELECTOGAL	PC prismatic sheet	Polygal, Israel
SELECTRON	UP resins	PPG Industries, Polymer Products, USA
SELON	PVC compounds	Sebi SrA, Italy
SETAPOL	UP resins	Synthese BV, The Netherlands
SETILITHE	CA	Plastribution Engineering Polymers Ltd., UK
SEVAMID	PA 6	Seva Nylon SpL, Italy; Petroplas Ltd., UK
SHIMOCO	UP resins and compounds	DSM
SHINKOLAC	ABS	Mitsubishi Engineering Plastics Inc., USA
SHINKOLITE	PMMA	Mitsubishi Engineering Plastics Inc., USA
SHUMAN	(PVC + ABS) blends	Shuman Plastics Inc., USA
SICOAMIDE	PA 6 and 66	Amberley Thermoplastics Ltd., UK
SICRON	PVC compounds	European Vinyls Corp., (EVC)
SILASTIC	Silicone resins	Dow Corning Corp., USA
SI-LINK	Crosslinkable PE compounds (based on copolymers of ethylene and vinyltrimethoxy silane)	Union Carbide
SILIPACT	Silicone resins and compounds	Lonza
SILTEM	Polyetherimide/silicone block copolymers	GE Plastics
SILTEMP	Silicone resins and compounds	GE Silicones, USA
SINKRAL	ABS	EniChem
SINTIMID	Polyimides	Lenzing, Austria
SINVET	PC	EniChem

TRADE NAME	NATURE	PRODUCER OR SUPPLIER
SIRCEL	PS polymers	Societa Italiana Resine SpA, Italy
SIRESTER	UP resins	
SIRFEN	Phenolic compounds	
SIRITLE	UF resins and compounds	
SKYGREEN	Biodegradable aliphatic polyester (polyethylene succinate)	Sunkyong Industries, South Korea
SMOKEGUARD HF	Flame-retarded, smoke-suppressed compounds (halogen-free)	Evode Plastics Ltd., UK
SNIALENE	PP	Nyltech, France, Italy and UK; Nylon Corp. of America, USA
SNIAMID	Pa 6 and 66, polymers and compounds	
SNIASAN	ABS	
SNIATAL	Acetal polymers	
SNIATER	PBT	
SOARBLEN	EVA copolymers	Nichimen America Inc., USA
SOARLEX		
SOARNOL	EVOH copolymers (main use as barrier layers in multi-layer extrusions)	Elf Atochem; Nichimen America Inc., USA
SOFTLEX	EVA copolymers	Nippon Petrochemicals Co., Japan
SOLARENE	PS	Plastribution Engineering Polymers Ltd., UK
SOLEF	PVDF	Solvay
SOLPRENE	SB thermoplastic rubbers (used as modifiers for plastics and bitumen)	Phillips Petroleum Co., USA
SOLVIC	PVC resins	Solvay
SPARLUX	PC	
SPECTAR	PETG copolyester	Eastman Chemical
SPECTRIM	PUR; RIM systems	Dow
SPESIN	PBT compounds	Kolon Industries, South Korea
STAFLENE	PE-HD	Nisseki Plastic Chemical Co. Ltd., Japan
STAMYLAN HD	PE-HD	DSM
STAMYLAN LD	PE-LD	
STAMYLAN P	PP	
STAMYLAN UH	PE-UHMW	
STAMYLEX	PE-LLD	
STAMYROID	Modified amorphous PP	
STAMYTEC	Highly crystalline, low-density polyolefin (production based on Chisso technology)	
STANULOY	PET; (PC + PET) blends	MRC Polymers Inc., USA
STANYL	PA 46 polymers and compounds	DSM
STAPRON	SMA copolymers	
STAPRON C	(PC + ABS) blends	
STAPRON E	(PC + PET) blends	
STAPRON S	Elastomer-modified SMA (High MA content)	
STARFLEX	SBS block copolymers	Star Plastics Co., USA
STARLITE	Complex, allegedly polymer-based material of undisclosed composition, claimed by the originator (Maurice Ward, UK) to be resistant to laser and nuclear-blast heat. Reputedly tested by British and USA military establishments.	In development stage (in 1996)
STAT-KON	Various antistatic and conductive compounds	LNP Engineering Plastics
STATRITE	TPE (polyether type) alloying additives for static-dispersive compounds of polyolefins, PVC, and styrenics	BF Goodrich
STEPANFOAM	PUR foaming systems	Stepan Co., USA
STEREON	SB block copolymers	Firestone Synthetic Rubber and Latex Co., USA

TRADE NAME	NATURE	PRODUCER OR SUPPLIER
STERNITE	ABS polymers	Elf Atochem
STYCAST	Epoxy resins and compounds	Emerson & Cuming Inc., USA
STYCOND	Styrenic copolymers and blends	United Composites Inc., USA
STYLAC	ABS polymers	Asahi Dow Chemical Co. Ltd., Japan
STYLEX	PS	Mitsubishi Petrochemical Co. Ltd., Japan
STYRADEX	Reprocessed PS	Mainetti Recycling Ltd., UK
STYRAFIL	Glass-filled SMA compounds	DSM
STYROBLEND	SB compounds, including flame-retarded (STYROBLEND FR)	BASF
STYROCELL	Expandable PS	Shell Chemicals
STYRODUR	PS extrusion materials	BASF
STYROL	Styrene copolymers	Denki Kagaku Kogyo, Japan and USA
STYROLUX	PS-HI (clear SB copolymers)	BASF
STYRON	PS	Dow
STYROPLUS	SB compounds	BASF
STYROPOR	Expandable PS	BASF
SULLIVAC	(PVC + ABS) blend materials for thermoforming	O' Sullivan Corp., USA
SUMIKATHENE	PE-HD; PE-LLD; EVA	Sumimoto Chemical Co. Ltd., Japan
SUMILITE FST	Poly(aryl sulfone)	Sumimoto Chemical Co. Ltd., Japan
SUMIPEX	PMMA	Sumimoto Chemical Co. Ltd., Japan
SUNPRENE	PVC resins and compounds	Elf Atochem
SUNTEC	PE-HD; PE-LD	Asahi Chemical Industry Co. Ltd., Japan
SUNTOP	Foamed, corrugated PC sheeting	Palram Ltd./Paltough Ltd., Israel
SUNTRA	PPS compounds	Sunkyong Industries, South Korea
SUNTUFF	Corrugated PC sheeting	Palram Ltd./Paltough Ltd., Israel
SUPEC	PPS compounds, including (PPS + PEI) blends and glass-filled grades	GE Plastics
SUPERCLEEN	Flexible PVC compounds	Alpha Chemical & Plastics Corp., USA
SUPLEN	Modified PP (high rigidity and clarity)	Polybrasil, Brazil
SUPRAMIN	UF and MF resins	Societa Italiana Resine Spa, Italy
SUPRAPLAST	Phenolic compounds	Süd-West Chemie AG, Germany
SUPRASEC	Isocyanate prepolymers	ICI Polyurethanes, Belgium
SUPREL	(PVC + ABS) blends	Vista Chemical Co., USA
SURLYN	Ionomer resins	Du Pont
SWEDCAST	PMMA	Swedlow Inc., USA
SYCAR	Silicon-carbon thermosetting resins and resin systems	Hercules
SYMALIT	PP/glass mat composite	GE Plastics
SYNLON	PA 6 and 66 compounds	Synlon Ltd., UK
SYNOLIT	UP resins	DSM
TAFFEN	Thermoplastic composites (including thermoformable PP sheeting)	Exxon Chemical
TAITALAC	ABS	Bolcof Plastic Materials, USA
TARNOFORM	PPO (PPE)	Nitrogen Works, Tarnow, Poland

TRADE NAME	NATURE	PRODUCER OR SUPPLIER
TARPEE	Laminated PE sheeting	Hagihara Industries Inc., Japan
TEAMEX	PE-VLD (for packaging films and geomembranes)	DSM
TECHLON	PA 6 and 66	Vamp Technology, Italy
TECHNORA	A copolymeric aramid fibre based on *p*-phenylene and 3,4-oxydiphenylene terephthalamide	Teijin Chemicals Ltd., Japan and USA
TECHNYL	PA 6 and 66	Nyltech, France, Italy and UK
TECHSTER E	PET	
TECHSTER T	PBT	
TECHNOFLON	Fluoropolymers	Ausimont
TECHNOPRENE	Glass-fibre-reinforced PP	EniChem
TECOFLEX	TPE (PUR type)	Thermedics Inc., USA
TECOLITE	Phenolic compounds	Toshiba Chemical Products Co., Japan
TECOPHILIC	TPE (PUR type)	Thermedics Inc., USA
TECOPLAST	TPE (PUR type)	
TEDIFLEX	PUR systems	EniChem
TEDILAST		
TEDIPUR		
TEDIRIM		
TEDITHERM		
TEDLAR	PVF	Du Pont
TEDUR	PPS	Albis
TEFLON	Fluoropolymers: FEP; PFA; PTFE	Du Pont
TEFZEL	ETFE	
TEKMILON	Fibres and filaments of very high-modulus, high-crystallinity PE (similar to DYNEEMA)	Mitsui Petrochemical Industries Co. Ltd., Japan
TELCAR	TPE (olefinic type)	Teknor Apex Co., Plastics Div., USA
TELENE	Range of RIM systems	BF Goodrich
TEMPRITE	PVC-C resins and compounds	
TENAC	Acetal polymers and blends	Asahi Denka Kogyo, Japan; Asahi America Inc., USA
TENCEL	A solvent-spun cellulosic fibre	Courtaulds, UK
TENITE	CA; CAB; CAP; PP; PET	Eastman Chemicals
TENITE PE	PE-LD; PE-LLD	
TENSOL	Solvent cements for bonding plastics products	ICI
TEPEX	PA 66/carbon fibre composite sheet	Du Pont
TERBLEND	(PC + ASA) and (PC + ABS) blends	BASF
TERLAC	TPE compounds, based on SBS	Silac, France
TERLURAN	ABS	BASF
TERLUX	ABS (clear resins)	
TERLUX CLEAR	MABS	
TERRAM	Geotextiles	Exxon Chemical
TEXALON	PA 6; 66; and 610	Texapol Corp., USA
TEXAPOL	PBT; acetal polymers	
TEXIN	TPE (PUR type); thermoplastic PUR; and (PUR + PC) blends	Bayer; Miles Inc., USA
TEXTOLITE	Silicone compounds	GE Silicones, USA
THERBAN	Butadiene/acrylonitrile rubbers, regular and hydrogenated (NBR and HNBR) used as impact modifiers	Bayer
THERMAFLO	PVC footwear soling compounds	Evode Plastics Ltd., UK
THERMATECH	Silicone rubber compounds(4300 series)	Silicone Products and Technology Inc., USA
THERMEX	PBT	ComAlloy International Corp., USA
THERMOCOMP	Various thermoplastic compounds	LNP Engineering Plastics

TRADE NAME	NATURE	PRODUCER OR SUPPLIER
THERMODET	Extruded thermoplastic films and sheeting	Mitras Kunststoffe GmbH, Germany
THERMOFLEX	TPE (SEPS type)	Plastics Technology Service Ltd., UK
THERMOTRON	Fibres containing zirconium carbide, claimed to re-emit heat when absorbing visible light	Unitica Co. Ltd., Japan
THERMX	Copolyester	Eastman Chemical
THUNDERON SS-N	Conductive, antistatic acrylic fibres	Nippon Sanmo Co., Japan
TIMBRON	PS-based plastics substitute for wood	Glynwed Ltd., UK
TIPOLEN	PE-LD; PE-MD	Tiszai Vegyi Kombinat, Hungary
TIPOX	Epoxy resins and compounds	
TISCO	Various thermoplastic polymers and compounds	Tisco Srl, Italy
TIVAR	PE-UHMW	Poly-Hi Solidur, Div. of Menasha Corp., USA
TONEN	PP	TCA Plastics Inc., USA (Div. of Tonen Sekiyu Kagaku, Japan)
TOPALLOY	PP compounds	
TOPAS	Transparent copolymers of ethylene and cyclic olefins	Hoechst; Hoechst Technical Polymers, USA
TOPLEX	(PC + ABS) blends	Multibase Inc., USA
TOPOREX	PS (homopolymers)	Mitsui Toatsu Chemicals Inc., Japan
TORAY PBT	PBT polymers	Toray Industries Co. Ltd., Japan; Toray Plastics America Inc., USA
TORAYCA	PA 6 and 66	
TORAYFAN	Biaxially stretched PP films	Toray Industries Co. Ltd., Japan; Toray Plastics America Inc., USA
TORLON	Polyamide-imide resins and compounds	Amoco Performance Products Inc., USA
TORNAC	Hydrogenated nitrile rubber (HNBR)	DSM
TOYOLAC	ABS; (PC + ABS) blends	Toray Industries Co. Ltd., Japan; Toray Plastics America Inc., USA
TPX	Poly(4-methyl-pentene-1)	Mitsui Petrochemical Industries Ltd., Japan; Mitsui Plastics Inc., USA
TRAMID	PA 6	Sam Yang Co. Ltd., South Korea
TREFSIN	TPE: blends based on butyl rubber	Exxon Chemical
TRETOCRETE	SB latex for bonding cementitious compositions	Doverstrand Ltd., UK
TRIAX	Blends: (PA + ABS); (PC + ABS); (PA + polyolefin)	Monsanto (being transferred to Bayer in 1996)
TRIBIT	PBT polymers and compounds	Sam Yang Co. Ltd., South Korea
TRIBOLON XT	Polyimide moulding compounds	Rogers Inc., USA
TRILEN	PP compounds	Sam Yang Co. Ltd., South Korea
TRIMAX BOARD	Plastics board (wood substitute) based on reclaimed materials (PE-HD and others) with glass-fibre reinforcement	Trimax Inc., USA
TRIPET	PET polymers and compounds	Sam Yang Co. Ltd., South Korea
TRIREX	PC polymers and compounds	
TRIVIN	Rigid PVC blends	Vi-Chem Corp., USA
TROCAL	PVC roofing membranes and profiles	HT Troplast AG, Germany
TROCELLEN	Crosslinked PE foam products	

TRADE NAME	NATURE	PRODUCER OR SUPPLIER
TROGAMID T	Amorphous (transparent) PA	Huels
TROLITAX	Epoxy compounds	HT Troplast AG, Germany
TROLON	Phenolic compounds	Huels
TROSIFOL	Poly(vinyl butyral) film	HT Troplast AG, Germany
TROSIPLAST	PVC blends	Huels
TROVIDUR	PVC	Huels
TUFLIN	PE-LLD	Union Carbide
TUF-STIF	Rigid PVC	Georgia Gulf Corp., PVC Div., USA
TUFTANE	PUR film and sheeting	Lord Corp., USA
TWARON	An aramid fibre, similar to KEVLAR	Akzo Fibres and Polymers, The Netherlands
TYPLAX	EVA; PP	Typlax Products, USA
TYRIL	PS-HI; SAN	Dow
TYRIN	PE-C	Dow
UBE	PA 6; 66; and 12	UBE Industries, Japan and USA
UCAR	PVC solution resins; PE-LD; PE-LLD	Union Carbide
UCARDEL	(PSU + SAN) blends	Union Carbide
UDEL	PSU	Amoco Chemical
UHMW	PE-UHMW	Motell Italia, Italy Himont USA
ULTEM	PEI	GE Plastics
ULTEM LTX	(PEI + PC) blends	GE Plastics
ULTRABLEND	Blends based on PBT	BASF
ULTRAC	PE-UHMW	AlliedSignal Inc., USA
ULTRA CAP	PVDF-based capstock for PVC extrusions, ABS, steel, aluminium and wood (claimed superior to DURA CAP)	Geon Co., USA
ULTRADUR	PBT polymers and compounds	BASF
ULTRADUR S	(PBT + ASA +PET) glass-filled blends	BASF
ULTRFORM	Acetal copolymers	BASF
ULTRALASTIC	PUR sheeting	Poly-Hi Solidur, Div. of Menasha Corp., USA
ULTRAMID	PA 6; 66; 610 and copolymers	BASF
ULTRANYL	(PA + PPO) blends	BASF
ULTRAPEK	Poly (arylether ketones) :PEK; PEEK	BASF
ULTRPLAS	MF resins and compounds	Perstorp
ULTRASON E	PES compounds	BASF
ULTRASON S	PSU compounds	BASF
ULTRASTYR	ABS	EniChem
ULTRATHENE	EVA	Quantum Chemical Co., USI Div., USA
ULTRA TUFLIN	PE-LLD (produced by "Unipol II" Technology)	Union Carbide
ULTRAX	Liquid-crystal polymer	BASF
UNICHEM	PVC compounds	Union Carbide
UNIPOL PP	Random copolymers of PP and an alpha-olefin (applicable as heat-sealing materials)	Union Carbide
UNIPURGE	PE purging compound	Union Carbide
UNITE	Compatibilisers and coupling agents (grafts of maleic anhydride on polyolefins)	Aristech Chemical Corp., USA
UNIVAL	PE-HD (for blow moulding)	Union Carbide
UPILEX	Formable polyimide sheet	UBE Industries, Japan and USA
U-POLYMER	Polyarylate (similar to ARDEL)	Unitica Co. Ltd., Japan
UREOL	PUR resin systems	Ciba-Geigy
UREPAN	TPE (vulcanisable PUR type)	Bayer
URESTYL	PUR - rigid foam components and systems	DSM
UROCHEM	UF resins and compounds	Chemiplastica SpA, Italy

TRADE NAME	NATURE	PRODUCER OR SUPPLIER
UROMIX	UP compounds	Ubbik Nederland BV, The Netherlands
URTAL	ABS polymers	EniChem
UTHANE	TPE (PUR type)	Urethanes India, India
VALITE	Phenolic resins and compounds	Lockport Thermosets Inc., USA
VALOX	Compounds based on thermoplastic polyesters (including PBT; PBT blends; PET; and PCT)	GE Plastics
VALTEC	PP homopolymers (including spherical-particle, additive-coated form)	Montell Italia, Italy; Himont USA
VALTRA	Styrenic polymers	Chevron Chemical Co., USA
VAMAC	TPE (ethylene/acrylic copolymer)	Du Pont
VAMPALLOY	Flame-retarded blends	Vamp Technology, Italy
VAMPAMID 6	Flame-retarded PA 6 compounds	Vamp Technology, Italy
VAMPAMID 66	Flame-retarded PA 66 compounds	Vamp Technology, Italy
VAMPCARB	Flame-retarded PC compounds	Vamp Technology, Italy
VAMPLEN	Flame-retarded PP compounds	Vamp Technology, Italy
VAMPORAN	Flame-retarded PPO (PPE) compounds	Vamp Technology, Italy
VAMPSAB	Flame-retarded ABS compounds	Vamp Technology, Italy
VAMPSON	Flame-retarded PSU compounds	Vamp Technology, Italy
VAMPSTYR	Flame-retarded PS compounds	Vamp Technology, Italy
VAMPTER	Flame-retarded PBT compounds	Vamp Technology, Italy
VANDAR	PBT (high-impact) compounds	Hoechst; Hoechst Celanese Corp., USA
VARCUM	Phenolic resins	Occidental Chemical Corp., USA
VARIMAT	UP and vinyl ester resin compounds	Interplastic Corp., USA
VARLAN	PVC resins	DSM
VECTOR	Styrenic block copolymers	Dexco Polymers, USA
VECTRA	Liquid-crystal polymers and compounds	Hoechst; Hoechst Celanese Corp., USA
VECTRA V140	(LCP + PPS) blends, glass-reinforced	Hoechst; Hoechst Celanese Corp., USA
VEDRIL	PMMA	Montell Italia, Italy
VEGARD PET	Glass-reinforced PET sheet (stamp-formable)	Kurary Sumimoto Metal Industries, Japan
VELITE	Acrylic resin	Elf Atochem
VENPET	PET	Longfield Chemicals Ltd., UK
VERSALON	Fatty-acid amides (used in hot-melt adhesives, and in epoxy resins)	Cray Valley, UK
VERSAMID	Fatty-acid amides - curing agents for epoxy resins	Cray Valley, UK
VERTON	PA compounds (long-fibre-reinforced)	LNP Engineering Plastics
VESPEL	Range of non-melting polyimide resins for direct forming of sheet and parts	Du Pont
VESTAMID	PA 6 and 66; PEBA	Huels
VESTIFORM	PMMA	Huels
VESTOBLEND	(PPO + PA) blends	Huels
VESTODUR	PBT	Huels
VESTOLEN A	PE-HD	Huels
VESTOLEN BT	Polybutylene	Huels
VESTOLEN EM	PP elastomer-modified	Huels
VESTOLEN P	PP	Huels
VESTOLIT	PVC resins	Huels
VESTOPAL	UP resins	Huels
VESTOPLAST	Amorphous olefin copolymers	Huels
VESTOPREN	Polyolefin impact modifier	Huels
VESTORAN	PPO (PPE)	Huels
VESTYPOR	Expandable PS	Huels
VESTYRON	PS; PS-HI	Huels
VHMW	PE-UHMW	Montell Italia, Italy; Himont USA

TRADE NAME	NATURE	PRODUCER OR SUPPLIER
VIBRATHANE	PUR coating resins	Uniroyal Chemical Co., Urethane Chemical Div., USA
VICLON	PVC	Kureha Chemical Industry Co., Japan
VICOM	PVC compounds	EVC Compounds, Italy and UK
VICTREX	PEEK; PEK; PES; and compounds	Victrex USA Inc., Victrex Sales Ltd., UK
VIDAR	PVDF	Solvay
VILIT	Vinyl copolymers	Huels
VINAKON	PVC-U	Hydro Geon, UK
VINEX	PVAL	Air Products & Chemicals Inc., USA
VINIDUR	PVC resins (including VC/acrylate copolymers)	BASF
VINLOC	Range of heat-shrinkable plastics tubing	Canussa, UK
VINNAPAS	PVAC-EVA	Wacker Chemie
VINNOL	PVC resins	Wacker Chemie
VINOFLEX	PVC resins	BASF
VINURAN	Polymeric modifiers for PVC	BASF
VINYCHLON	PVC resins	Mitsui Toatsu Chemical Inc., Japan
VIPLA	PVC resins	European Vinyls Corp. (EVC)
VIPLAVIL	Vinyl copolymer resins	European Vinyls Corp. (EVC)
VIPPROP	Recycled PP	Victor International Plastics Ltd., UK
VISICO	Systems for chemically crosslinkable PE	Borealis
VISTA	Rigid PVC compounds	Vista Chemical Co., USA
VISTAFLEX	TPE (olefinic type)	Exxon Chemical
VISTALON	Ethylene-propylene rubber (EPM)	Exxon Chemical
VISTANEX	Polyisobutylene	Exxon Chemical
VISTEL	Rigid PVC compounds	Vista Chemical Co., USA
VITON	Fluoroelastomers	Du Pont
VITON FREE FLOW	Fluoropolymer processing aids for polyolefins	Du Pont
VITRON	PVC compounds (blow-moulding grades)	European Vinyls Corp. (EVC)
VIVYPAK	PET (bottle grades)	EniChem
VOLOY	PA 6; 66; 612; PBT; PET	ComAlloy International Corp., USA
VOLTALEF	PCTFE	Elf Atochem
VORALAST	Fast-cure PUR (polyether) system for RIM (especially of shoe soles)	Dow
VORANOL	Polyols for PUR foams	Dow
VULKOLLAN	PUR elastomers (cast, heat-cured)	Bayer
VYDYNE	PA 66, polymers and compounds	Monsanto
VYNCOLIT	Phenolic compounds	Vyncolit NV, Belgium
VYNITE	(PVC + nitrile rubber) blends	Alpha Chemical & Plastics Corp., USA
VYRAM	TPE (olefinic type)	Advanced Elastomer Systems, USA
VYTHENE	(PVC + PUR) blends	Alpha Chemical & Plastics Corp., USA
WELITE	PBT	Wellman Inc., Plastics Div., USA
WELLAMID	PA 6 and 66	Wellman Inc., Plastics Div., USA
WELPET	PET	Wellman Inc., Plastics Div., USA

TRADE NAME	NATURE	PRODUCER OR SUPPLIER
WELVIC	PVC compounds	European Vinyls Corp. (EVC); ICI and associates world-wide
WHITCON TL	Fluoropolymers used as lubricant additives	ICI
WINGTACK	Polyterpene resin	Goodyear Chemicals
XANTAR	PC	DSM
XC	PP	Borealis
XENOY	(PC + PBT) blends	GE Plastics
XT-POLYMER	Acrylic polymers	CYRO Industries, USA
XTRABOND	PA 6, adhesive grade	AlliedSignal Inc., USA
XYCON	Polyester/PUR hybrid thermosetting resins	Amoco Chemical
XYDAR	Liquid-crystal polymer	Amoco Chemical
XYRON	PPO (PPE) materials	Asahi Chemical Industry Co., Japan
YPAREX	Extrudable adhesive polymer	DSM
ZEMID	PE-HD polymers and compounds	Du Pont
ZENITE	Liquid-crystal polymers (based on aromatic copolymers)	Du Pont
ZEONEX	Optical grade resins (amorphous cycloolefins with ethylene as the main comonomer)	Zeon Chemicals Inc., USA; Nippon Zeon Co., Japan
ZETPOL	Polymeric impact modifier (HNBR)	Zeon Chemicals Inc., USA; Nippon Zeon Co., Japan
ZYEX	A PEEK fibre developed by ICI mainly for tennis-racquet strings	Zyex Ltd., UK
ZYLAR	Acrylic terpolymers (S/B/MMA)	Novacor Chemicals Inc., USA
ZYTEL	PA 6; 66; 612; 11; and PA compounds	Du Pont

PART 3

SECTION 2

TRADE NAMES OF PLASTICISERS

The trade names of a number of commercial plasticisers and/or plasticiser ranges are listed in column 3 of table 1 below. The listing is according to chemical type as identified by its generic name in colmn 1.

In addition, column 2 of the table gives the type-reference number which appears in the table of abbreviated terms(letter symbols) for plasticisers in Section 3 of Part 2.

Table 2 below lists alphabetically suppliers of the plasticisers named in table 1.

TABLE **1** Trade names of some commercial plasticisers

CHEMICAL TYPE	REFERENCE NO. CITED IN SECTION 3 OF PART 2	TRADE NAMES OF COMMERCIAL PRODUCTS AND/OR PRODUCT RANGES, WITH (IN BRACKETS) REFERENCES TO THE RELEVANT SUPPLIERS LISTED IN TABLE 2		
PHTHALATES	1	BISOFLEX(B.6)	KODAFLEX(E.1)	REPROXAL (D.3)
		BUFA(A.3)	KP(F.2)	SANSOCIZER (N.2)
		CHEMIAL(C.6)	KYOWA(K.2)	SANTICIZER (M.5)
		CP-CIZER(C.5)	LINEVOL ESTER (S.5)	SCANDINOL (S.2)
		CRESTAPOL(S.3)	LINPLAST(C.11)	SEKISUI(S.4)
		DELETE(U.4)	MIRFLEX(R.3)	SICOL(E.4)
		DIACIZER(M.4)	MOLLAN(C.7)	SISAS(S.6)
		DIPLAST(D.2)	MONOCIZER (D.1)	STAFLEX(R.1)
		EDENOL(H.4)	MONSANTO (M.5)	STARFOL(H.7)
		EVIPLAST(C.8)	MORFLEX(M.6)	STANICIZER (S.7)
		EXTERPLAST (C.10)	NATROFLEX (H.12)	SYNOLATE(S.9)
		FLIXIMEL(M.2)	NESTE(N.1)	UCB(U.5)
		GARBEFLEX(R.2)	NUOPLAZ(H.7)	UNIFLEX(U.3)
		GENOMOLL(H.6)	PALATINOL (B.1)	UNIMOLL(B.2)
		HALLCO(H.1)	PLASTHALL (H.1)	URAPLAST(D.5)
		HATCO(H.3)	PLASTIFCANTE (A.2)	VESTINOL(H.7)
		HATCOL(H.3)	PLASTIFIANT K(U.1)	WACKER(W.1)
		HERCOFLEX(H.5)	POLYCIZER (D.1) (H.2)	WITAMOL(H.7)
		HEXAPLAS(I.1)	PX(A.10)	WOLFLEX(V.2)
		INDUFLEX(C.4)	REICHHOLD (R.1)	
		JAYFLEX(E.5)	REOMOL(C.9)	
PHOSPHATES	2	AMGARD(A.5)	HALLCO(H.1)	PLIABRAC(A.5)
		ANTIBLAZE(A.5)	KP(F.2)	REOFOS(C.9)
		DISFLAMOLL (B.2)	KRONITEX(F.2)	SANTICIZER (M.5)
		FLEXOL(U.4)	LINDOL(A.4)	TOLGARD(A.5)
		FYROL(A.4)	MONSANTO(M.5)	TOLPAZ(A.5)
		GARBEFOS(R.2)	NATROFLEX (H.2)	VIRCOL(A.5)
		GENOMOLL(H.6)	PHOSFLEX(A.4)	

CHEMICAL TYPE	REFERENCE NO. CITED IN SECTION 3 OF PART 2	TRADE NAMES OF COMMERCIAL PRODUCTS AND/OR PRODUCT RANGES, WITH (IN BRACKETS) REFERENCES TO THE RELEVANT SUPPLIERS LISTED IN TABLE 2		
TRIMELLITATES	3	BISOFLEX(B.6)	JAYFLEX(E.5)	PLASTHALL (H.1)
		CRESTAPOL(S.3)	LINPLAST(C.11)	PLASTIFICANTE (A.2)
		DIPLAST(A.6)	MONOCIZER (D.1)	PX(A.10)
		HALLCO(H.1)	MORFLEX(M.6)	RUCOFLEX(R.5)
		HATCO(H.3)	NATROFLEX (H.2)	STAFLEX(R.1)
		HEXAPLAS(I.1)	NUOPLAZ(H.7)	UNIFLEX(U.3)
ALIPHATIC DIESTERS (Esters of aliphatic dicarboxylic acids)	4	ADIMOLL(B.2)	INDUFLEX(C.4)	POLYCIZER(D.1) (H.2)
		ADIPATE(U.5)	JAYFLEX(E.5)	PRIPLAST(U.2)
		ADIPOL(F.2)	KODAFLEX(E.1)	PX(A.10)
		BENDOL(I.2)	LEXOLUBE(I.2)	REICHHOLD(R.1)
		BISOFLEX(B.6)	LINPLAST(C.11)	REOMOL(C.9)
		CP-CIZER(C.5)	MIRFLEX(R.3)	SANSOCIZER (N.2)
		CRYOFLEX(S.1)	MOLLAN(C.7)	SICOL(E.4)
		DELETE(U.4)	MONOPLEX (H.1)	SCANDINOL(S.2)
		DIACIZER(M.4)	MONSANTO (M.5)	STAFLEX(R.1)
		DIPLAST(D.2)	MORFLEX(M.6)	STANICIZER(S.7)
		DOLPLAS(M.1)	NATROFLEX (H.2)	UCB(U.5)
		EDENOL(H.4)	NUOPLAZ(H.7)	UNIFLEX(U.3)
		FLIXIMEL(M.2)	PALATINOL (B.1)	URAPLAST(D.5)
		HALLCO(H.1)	PLASTHALL (H.1)	VESTINOL(H.7)
		HATCO(H.3)	PLASTIFICANE (A.2)	WACKER(W.1)
		HATCOL(H.3)	PLASTOLEIN (E.3)	WAREFLEX (S.1)
		HEXAPLAS(I.1)	PLASTOMOLL (B.1)	WITAMOLL (H.7)
POLYMERIC (POLYESTER) PLASTICISERS	5	ADEKACIZER (A.1)	HERCOFLEX (H.5)	REOPLEX(C.9)
		ADILEN(T)	LEXOLUBE(I.2)	RESOFLEX(C.1)
		ARBEFLEX(R.4)	MORFLEX(M.6)	SANTICIZER (M.5)
		CRESTAPOL(S.3)	PALAMOLL (B.1)	STARFOL(H.7)
		DELETE(U.4)	PARAPLEX(H.1)	TP(M.7)
		DRAPEX(W.2.1)	PLASTHALL (H.1)	ULTRAMOLL (B.2)
		EDENOL(H.4)	PLASTIFICANTE (A.2)	UNIFLEX(U.3)
		FLIXIMEL(M.2)	PLASTOLEIN (E.3)	URAPLAST(D.5)
		GARBEFLEX(R.2)	POLYCIZER(D.1)	VESTINOL(E.2)
		GLYPLAS(L)	POLYMIX(A.6)	WEPLAST(C.10)
		HALLCO(H.1)	PRIPLAST(U.2)	WITAMOL(H.7)
		HALOREZ(W.2.2)	REICHHOLD(R.1)	WOLFLEX(V.2)

CHEMICAL TYPE	REFERENCE NO. CITED IN SECTION 3 OF PART 2	TRADE NAMES OF COMMERCIAL PRODUCTS AND/OR PRODUCT RANGES, WITH (IN BRACKETS) REFERENCES TO THE RELEVANT SUPPLIERS LISTED IN TABLE 2		
EPOXY PLASTICISERS	**6**	ADILEN(T)	FLEXOL(U.4)	PLASTOFLEX (A.4)
6.1 - Epoxidised soya bean oil		ADMEX(H.7)	HALLCO(H.1)	PLASTOLEIN (E.3)
6.2 - Epoxidised linseed oil		DIEPOX(D.2)	KRONOX(F.2)	PRIPLAST(U.2)
		DRAPEX(W.2.1)	LANKROFLEX (A.3)	REOPLAST(C.9)
6.3 - Epoxy-alkyd esters		ECEPOX(E.2)		
		EDENOL(H.4)	MONOPLEX(H.1)	SOA(U.4)
		EPOXICIZER(D.1)	FARAPLEX(H.1)	STANICIZER (S.7)
		EPOXOL(A.7)	PEROXIDOL(R.1)	VIKOFLEX(E.2)
		ESTABEX(A.4)	PLAS-CHECK(F.1)	
		ESTYNOX(C.2)	PLASTHALL(H.1)	
ESTERS OF ALIPHATIC MONOCARBOXYLIC ACIDS	**7**			
Esters of higher aliphatic acids	7.1	ABRACOL(B.4)	FLIXIMEL(M.2)	NATROCHEM (H.2)
		ADILEN(T)	GLYCOLUBE(L)	PARAPLEX(H.1)
		ALDO(L)	HALLCO(H.1)	PARICIN(C.2)
		BISOFLEX(B.6)	HERCOFLEX(H.5)	PRIPLAST(U.2)
		CONTRASTAT (R.1)	KAOWAX(K.2)	RICATYL(R.1)
		DEPLASTOL(H.4)	KEMSTER(H.8)	STAFLEX(R.1)
		EDENOL(H.4)	KESSCOFLEX(S.8)	TEGMER(H.1)
		EM 28(R.1)	KP(F.2)	UNIFLEX(U.3)
		EMERY(E.3)(U.2)	LIPINOL(H.7)	VINYLUBE(L)
		ETHOSPERSE(L)	MOLLIT(B.2)	
		FLEXRICIN(C.2)	MORFLEX(M.6)	
Esters of lower aliphatic acids and deivatives: Acetates,	7.2			
propionates,butyrates	7.2.1	DIACETIN(B.2)	MOLLIT(B.2)	TRIACETIN (B.2)
		FLEXOL(U.4)	PRIPLAST(U.2)	TRIBUTYRIN (E.1)
		KENPLAST(K.1)	SAIB(E.1)	TRIPROPIONIN (E.1)
		KESSCOFLEX(S.8)	TEGDA(B.2)	
		KODAFLEX(E.1)	TEXANOL(E.1)	
Glycolates	7.2.2	DIACIZER(M.4)	HALLCOMID (H.1)	KODAFLEX (E.1)
		DIOLPATE(M.1)	KP(F.2)	
CITRATES	**8**	CITROFLEX(M.6)	ESTAFLEX(A.4)	
		CITROFOL(B.3)	FLIXIMEL(M.2)	
ESTERS OF AROMATIC OR CYCLIC CARBOXYLIC ACIDS	**9**			
Benzoates	9.1	BENZOFLEX(V.1)	DELETE(U.4)	MOLLIT(B.2)
		BISOMER(B.6)	KENPLAST(K.1)	NYCOFLEX(N.3)
		CETAMOLL(B.1)	KODAFLEX(E.1)	NUOPLAZ(H.7)

CHEMICAL TYPE	REFERENCE NO. CITED IN SECTION 3 OF PART 2	TRADE NAMES OF COMMERCIAL PRODUCTS AND/OR PRODUCT RANGES, WITH (IN BRACKETS) REFERENCES TO THE RELEVANT SUPPLIERS LISTED IN TABLE 2		
Abietates	9.2	ABALYN(H.5)	HERCOLYN(H.5)	PICCOLASTIC (H.5)
		FLEXALYN(H.5)	METALYN(H.5)	PICCOVAR(H.5)
SULFONIC ACID DERIVATIVES (Sulfonates; sulfonamides)	**10**	DELLATOL BBS(B.2)	MESAMOLL(B.2)	SANTICIZER (M.5)
		ISAPLAST(B.5)	QUADROL(B.1)	UCB(U.5)
		KETJENFLEX (A.4)	REPROXAL(D.3)	
CHLORINATED PARAFFINS AND OLEFINS	**11**	ALAIFLEX(P.1)	CPF(W.2.2)	MONOCIZER (D.1)
		CERECLOR(I.1)	BIABLO(O)	NATROCHEM (H.2)
		CLORAFIN(H.5)	ELETROFINE (E.2)	PAROIL(D.4)
		CHLOREZ(D.4)	FLEXCHLOR(W.2.1)	PLASTHALL (H.1)
		CHLOROFLO (D.4)	FYARESTOR (W.2.1)	PLASTICHLOR (H.2)
		CHLOROWAX(O)	HALLCOMID (H.1)	SUPER-CHLOR(D.4)
		CHLORPARAFFIN(H.6)	MIRFLEX(R.3)	WITACLOR (H.7)
ETHERS, POLY-HYDRIC ALCO-HOLS, KETONES	**12**	ABITOL(H.5)	ESTYNOX(C.2)	POLYCIZER (H.2)
		CARBOWAX(U.4)	KARION(M.3)	SANTICIZER (M.5)
REACTIVE AND POLYMERISABLE PLASTICISERS	**13**	BISOMER(B.6)	SARTOMER(S.1)	
		SANTOSET(M.5)	TAC(E.2)	
HYDROCARBONS (liquid and solid)	**14**	DRAKEOL(P.2)	HB(H.5)	SONNEBORN (W.2.3)
		DUTREX(S.5)	KENFLEX(K.1)	VIPLEX(C.13)
		ENERFLEX(B.6)	KENPLAST(K.1)	WITCO(W.2.2)
		ESCOFLEX(P.3)	LIPINOL(H.7)	
MISCELLANEOUS PLASTICISERS	**15**	MOLLIT(B.2)	URESIN(H.6)	

NOTE - Trade names of polymers incorporated in other polymers as "solid plasticisers" (e.g. NBR in PVC) and, generally, as polymeric modifiers of various kinds and functions, are listed in Section 3 of Part 3.

PART 3

SECTION 2

TABLE **2**

REFERENCES CITED AFTER THE TRADE NAMES IN TABLE 1	PLASTICISER SUPPLIERS
A.1	Adeca Argus Chemical Co., Japan
A.2	Adisa Iberica SA, Spain
A.3	Akcros Chemicals
A.4	Akzo Nobel
A.5	Albright and Wilson
A.6	Alusuisse Italia SpA, Italy
A.7	American Chemical Service, USA
A.8	Amoco Chemical
A.9	Argus Chemicals, Belgium; Argus Div., Witco Corp., Polymer Additives Group, USA
A.10	Aristech Chemical Corp., USA
B.1	BASF
B.2	Bayer; Miles Inc., USA
B.3	J. A. Benckiser, Germany
B.4	A. Boake, Roberts & Co., UK
B.5	Dr. Th. Böhme KG, Germany
B.6	BP Chemicals
C.1	Cambridge Industries Co., USA
C.2	CasChem Inc., USA
C.3	CdF Chemie, France
C.4	Cespa SA, Spain
C.5	Chang Chem Petrochemical Co., Japan
C.6	Chemial SpA, Italy
C.7	Chemie Linz GmbH, Austria
C.8	Chemol Impex, Hungary
C.9	Ciba-Geigy
C.10	Coim SpA, Italy
C.11	Condea Chemie GmbH, Germany
C.12	Condesia SA, Spain
C.13	Crowley Chemical Co. Inc., USA
D.1	Dainippon Ink and Chemicals Inc., Japan
D.2	Destillerie Italiane SpA, Italy
D.3	Deutche Texaco AG, Germany (now Huntsman Chemical Corp. subsidiary)
D.4	Dover Chemical Corp., USA
D.5	DSM
D.6	Du Pont
E.1	Eastman Chemical
E.2	Elf Atochem
E.3	Emery Group (Div. of Henkel Corp., USA)
E.4	EniChem
E.5	Exxon Chemical
F.1	Ferro Corp., Bedford Chemical Div., USA; Ferro (Holland) BV
F.2	FMC Corp., Process Additives Div., USA
H.1	C.P. Hall Co., USA
H.2	Harwick Chemical Corp., USA
H.3	Hatco Corp., USA

PLASTICISER SUPPLIERS

REFERENCES CITED AFTER THE TRADE NAMES IN TABLE 1	
H.4	Henkel
H.5	Hercules Inc., USA
H.6	Hoechst; Hoechst Celanese Corp., Polymer Additives, USA
H.7	Huels; Huls America Inc., USA
H.8	Humco Chemical Div., USA (subsidiary of Witco Corp., Polymer Additives Group)
I.1	ICI Petrochemicals and Plastics Div., UK; ICI Polymer Additives, USA
I.2	Inolex Chemical Co., USA
K.1	Kenrich Petrochemicals Inc., USA
K.2	Kyowa Hakko Kogyo Co., Japan
L	Lonza G & T Ltd, Switzerland (member of the Alusuisse-Lonza Group)
M.1	Macpherson B & T Polymers, UK
M.2	Melle SA, France
M.3	E. Merck, Germany
M.4	Mitsubishi Petrochemical Co. Ltd., Japan
M.5	Monsanto
M.6	Morflex Inc., USA
M.7	Morton Intrnational Inc., Industrial Chemicals and Additives, USA
N.1	Neste Chemicals NV, Belgium
N.2	New Japan Chemical Co., Japan
N.3	Nyco SA, France
O	Occidental Chemical Corp., USA
P.1	Pechiney St. Gobain, France
P.2	Pennzoil Products Co., Penreco Div., USA
P.3	Plastics and Chemicals Inc., USA (distributors for Kum Yang Co., Ltd)
R.1	Reichhold Chemie AG, Switzerland; Reichhold Chemicals Inc., USA
R.2	Rhône Poulenc, France
R.3	Rio Rodano SA, Spain
R.4	Robinson Bros. Ltd., UK
R.5	Ruco Polymer Corp., USA
S.1	Sartomer Co., USA
S.2	Scandiflex, Denmark
S.3	Scott Bader Ltd., UK
S.4	Sekisui Chemical Co. Ltd., Japan
S.5	Shell Chemicals
S.6	Sisas SpA, Italy
S.7	Stanchem Inc., USA
S.8	Stepan Co. PVO Dept., USA
S.9	Synres BV, The Netherlands
T	Traquisa SA, Spain
U.1	Ugine Kuhlman (Produits Chemie), France
U.2	Unichema International, The Netherlands
U.3	Union Camp Corp., Chemical Products Div., USA
U.4	Union Carbide Corp., Specialty Chemicals Div., USA
U.5	Union Chimique Belge, Belgium
V.1	Velsicol Chemical Corp., USA
V.2	Victor Wolf Ltd., UK
W.1	Wacker Chemie, Germany

REFERENCES CITED AFTER THE TRADE NAMES IN TABLE 1	PLASTICISER SUPPLIERS
W.2.1	Witco Corp., USA, Polymer Additives Group
W.2.2	Witco Corp., USA, Golden Bear Div.
W.2.3	Witco Corp., USA, Petroleum Specialties Group

PART 3

SECTION 3

TRADE NAMES OF MISCELLANEOUS COMPONENTS OF PLASTICS MATERIALS

TRADE NAME	NATURE AND/OR APPLICATION	PRODUCER OR SUPPLIER	REMARKS
ABCURE	Peroxide initiators for UP resins	Abco Industries Ltd., USA	
ABRIL WAX	Lubricants (ester and amide types)	Abril Industrial Waxes, UK	
ACCUREL	Porous particles (powder or granules) of various polymers, for use as carriers for liquid additives (absorbed into the pores)	Akzo Nobel Faser, Germany	
ACRAWAX C	Lubricants and antiblocking and dispersion agents	Lonza	
ACRYLOID	See PARALOID		
ACTIMER FR	Reactive (monomer) bromine-containing flame retardants	Dead Sea Bromine Group, Israel; Eurobrom BV, The Netherlands	E.g. FR 1025 M: pentabromo-benzyl acrylate
ACTIVEX	Chemical blowing agent (endothermic, carbonate type)	J. M. Huber Corp., Chemicals Div., USA	
ACTONE	Dispersing agents for colorants	AlliedSignal Inc., Performance Additives Div., USA	
ADINE	Brominated flame retardants	CECA (Elf Atochem)	
ADK STAB	Heat stabilisers for plastics	Asahi Denka Kogyo, Japan	E.g. ADK STAB CPS: cadmium-free stabiliser for PVC
ADVAPAK	Range of multifunctional additives (mainly for PVC)	Morton International Inc., Industrial Chemicals and Additives, USA	
ADVASTAB	Range of tin-based heat stabilisers for PVC	Morton International Inc., Industrial Chemicals and Additives, USA	
ADVAWAX	Lubricants, slip agents, antiblocking agents	Morton International Inc., Industrial Chemicals and Additives, USA	
AEROSIL	Range of fumed silicas for use as thixotropic, anti-plate-out and anticaking agents	Degussa	
AFFLAIR	Range of mica-based pearlescent and metallic-lustre pigments	EM Industries Inc., Pigments Div., USA	
AGERITE	Antioxidants (amine and thiocarbamate types)	R.T. Vanderbilt Co. Inc., USA	
AJICURE	Latent curing agents for epoxy resins	Ajinomoto USA, Inc.	
AKTISIL	Silane-treated silica fillers (of the SILLITIN range)	Franz Hoffmann & Sohne KG, Germany	
AKROCHEM	Lubricants (various types)	Akrochem Corp., USA	

TRADE NAME	NATURE AND/OR APPLICATION	PRODUCER OR SUPPLIER	REMARKS
ALCOPOL	Range of surface-active agents	Allied Colloids, UK	
ALDOSPERSE	Antifogging agents	Lonza	
ALKALUBE	Lubricants (glycerol ester type)	Rhône-Poulenc, France	
ALUGLASS	Aluminium-coated glass fibre for use in antistatic and conductive compounds	Palindent, Israel	
ALPHA TALC	White talc fillers	Alpha Calcit Fuellstoff GmbH & Co. KG, Germany	
ALVINOX	Antioxidants (high-MW hindered-phenol type)	3V Sigma, Italy	
AMAPLAST	Range of colorants for plastics	Color-Chem International Corp., USA	
AMGARD	Range of phosphorus-based flame retardants (including phosphate-ester flame-retardant plasticisers)	Albright and Wilson	
AMICAL	Calcium carbonate fillers	Franklin Industrial Minerals, USA	
ANCAMIDE	Curing agents (amide type) for epoxy resins		
ANCAMINE	Curing agents (amine type) for epoxy resins	Anchor Chemicals Co. Ltd., UK	
ANCAREZ	Modifiers for epoxy resins		
ANCOMER	Modifying additives for plastics	Ancomer Ltd., UK	E.g. ANCOMER 11 (TMPTM - q.v.) a chemical crosslinking agent for PVC
ANTIBLAZE	Range of flame retardants for PUR systems	Albright and Wilson	
ANTISTATICO	Antistatic agents	3V Sigma, Italy	
APCO	Amine curing agents for epoxy resins	Hexcel Corp., Resins Group, USA	
AQUALINE	Water-borne mould-release agents	Dexter Corp., USA	
AQUAMIDE	Water-based curing agents (amide type) for epoxy resins	Anchor Chemical Co., UK	
AQUAMINE	Water-based curing agents (amine type) for epoxy resins		
ARCOL	Polyols	Arco Chemicals Co., USA	Used mainly in PUR flexible foam systems
ARMID	Antislip agents	Akzo Nobel	
ARMOCURE	Range of aliphatic polyamine curing agents for epoxy resins	Akzo Nobel	
ARMOSLIP	Internal mould-release agents (fatty-acid amide type)	Akcros Chemicals; Akzo Nobel	
ARMOSPHERES	Silicate microspheres	Georg M Langner & Co., Germany	
ARMOSTAT	Antistatic agents	Akzo Nobel	E.g. ARMOSTAT 1000 - glycerol monostearate
ASP	Clay (hydrous kaolin) fillers	Engelhard Corp., USA	

TRADE NAME	NATURE AND/OR APPLICATION	PRODUCER OR SUPPLIER	REMARKS
ATEPAS	Viscosity depressant	Dr. Th. Böhme KG,	
ATMER	Internal lubricants; antistatic agents; and other surface-effect additives	ICI Polymer Additives, USA; ICI Surfactants, Belgium	
ATMOS	Mould-release, antistatic, and antifogging agents for PVC and polyefins	Witco Corp. Polymer Additives Group, USA	Food-contact-grade additives
ATMUL			
ATOMITE	Ground calcium carbonate fillers	ECC International	
AURACOTE	Colorant dispersions in plasticisers for PVC	Hilton Davis Chemical Co., USA	
AVRELLE	Liquid dispersion of silica filler in methyl methacrylate	ICI Acrylics	For production, by casting, of sinks and work surfaces
AXTIL	Nucleating agents	AlliedSignal Inc., Performance Additives Div., USA	
AZOBIS	Chemical blowing agents (ADC and modified ADC types)	Toyo Hydrazine Industry Co., Japan	
AZOBUL	Chemical blowing agent (AZO type)	CECA (Elf Atochem)	
AZTEC	Range of peroxide initiators for UP resins	Aztec Catalyst Co., USA	
BA	Chemical blowing agents (custom-formulated) for thermoplastics	Polyvel Inc., USA	E.g. BA 20T for EVA and PE-LD
BAKELITE	Range of curing agents for epoxy resins	Union Carbide Chemicals and Plastics Co., USA	
BAROLUB	Range of lubricants (most types)	Baerlocher	
BARORAPID	Processing aids (acrylic types)		
BAROSTAB	Heat stabilisers (various types)		
BAROSTAT	Antistatic agents		
BAYERTITAN	Titanium dioxide pigments for plastics and surface coatings	Bayer	
BAYFERROX	Synthetic iron oxide pigments	Bayer; Miles Inc., USA	
BAYMOD	Impact modifiers for PVC ("A" series - ABS type: "L" series - EVA type)		Former trade name: LEVAPREN
BAYPLAST	Pigments (organic type)		
BAYTEC TS	Isocyanates (TDI) for PUR systems		
BEE	Liquid colorants	Morton International Inc., USA	
BEKI-SHIELD	Very fine stainless steel fibres for use in EMI and RFI shielding and antistatic compounds	Bekaert NV SA, Belgium; Bekaert Steel Wire Corp., USA	BEKI-SHIELD GR - chopped, sized fibres; BEKI-SHIELD VK-continuos filaments (tow)
BIOMET	Biocides (e.g. diphenylstibine - 2 ethyl hexanoate)	Elf Atochem	

TRADE NAME	NATURE AND/OR APPLICATION	PRODUCER OR SUPPLIER	REMARKS
BLACK PEARLS	Pelletised furnace carbon black	Cabot Corp., Special Blacks Div., USA	
BLANC FIXE MICRO	Fine-particle barium sulfate	Sachtleben Chemie GmbH, Germany	
BLANKOPHOR	Optical brightening agent	Bayer; Miles Inc., USA	
BLENDEX	Impact modifiers (ABS type) mainly for PVC (also for PC and polyesters)	GE Specialty Chemicals Inc., USA	
BOLIDEN MICA	Mica fillers	Boliden AG, Sweden	
BONDTINT	Reactive polymeric colorants for PUR systems	Milliken Chemical Co., USA	
BOROGARD ZB	Biocide additive	US Borax Inc., USA	
BRITOMYA	Calcium carbonate fillers	Omya Croxton & Garry Ltd., UK	
BROMKLOR	Bromine- and chlorine-substituted aliphatic compounds for use as flame-retardant plasticisers	Ferro Corp., Keil Chemical Div., USA	
BROMO XP	Brominated copolymers of isobutylene and *p*-methyl styrene, used as elastomeric impact modifiers in thermoplastics and some epoxy resins, and as alloy compatibilisers	Exxon Chemical	
BUBBLEKUP	Range of hollow ceramic microspheres (fillers for liquid resin systems)	PQ Corp., USA	
BUCA	Range of clay (hydrous kaolin) fillers	Engelhard Corp., Specialty Minerals and Colors Group, USA	
BULAB	Catalysts (tertiary-amine type) for epoxy resins and PUR systems	Buckman Laboratories Inc., USA	
BURGESS	Range of clay fillers	Burgess Pigment Co., USA	
BUSAN 11 - M1	Flame retardant, smoke and glow suppressant, based on magnesium metaborate	Buckman Laboratories Inc., USA	Some fungicidal and UV stabilising action claimed
BUSPERSE 47	Lubricant and pigment dispersant (fatty-acid amide type)	Buckman Laboratories Inc., USA	
CAB-O-SIL	Range of fumed silicas (thixotropic and viscosity-control agents; anti-plate-out agents for PVC; anticaking agents)	Cabot	
CACHALOT	Processing aid for flexible PVC	M. Michel & Co., Inc., USA	
CADMOLITH	Cadmium-based colorants (red and yellow)	SCM Chemicals	
CADON	Impact modifiers (SMA type) for PVC	Monsanto	Also encountered under the trade name "ELIX"
CALBLACK	Carbon black	Repsol Quimica SA, Spain	
CALCIDAR	Calcium carbonate fillers	Pluess Stauffer (and Omya sales offices)	
CALCIPLAST	Fine-particle, coated calcium carbonate fillers	Alpha Calcit Fuellstoff GmbH, Germany	
CALIBRITE	Calcium carbonate fillers	Pluess Stauffer (and Omya sales offices)	

TRADE NAME	NATURE AND/OR APPLICATION	PRODUCER OR SUPPLIER	REMARKS
CAMEL	Range of ground calcium carbonate fillers	Genstar Stone Products Co., USA	E.g. CAMEL WHITE (ultra-fine grade); CAMEL TEX (for BMC)
CAO	Range of phenol-based antioxidants	PMC Specialties Group Div., PMC Corp., USA	
CARADATE	Isocyanates	Shell Chemicals	
CARADOL	Polyols	Shell Chemicals	
CARAPOR	Additive for PUR systems	Shell Chemicals	
CARB-O-FIL	Carbon filler (ground coal) for PS, PUR, UP resins and phenolic resins	Shamokin Filler Co., Inc., USA	
CARBOKUP	Surface-treated calcium carbonate fillers	Malvern Minerals Co., USA	
CARBOREX	Calcium carbonate fillers	Pluess Stauffer (and Omya sales offices)	
CARDINAL CC	Organotin heat stabilsers for PVC	Cardinal Stabilisers Inc., USA	
CARDURA	Chemical intermediates	Shell Chemicals	
CARSTAB	Antioxidants (thioester synergist type)	Morton International Inc., Morton Performance Chemicals, USA	
CAS	Calcined, anhydrous calcium sulfate	US Gypsum Co., USA	
CASCADE	Talc fillers	Chemag GmbH, Germany	
CASIFLAX	Wollastonite fillers	Quarzwerke GmbH, Germany	
CATA-CHEK	Range of tin-based catalysts for PUR systems, and heat stabilisers for PVC	Ferro Corp., Bedford Chemical Div., USA	
CATALPO	Clay (hydrous kaolin) fillers	Engelhard Corp., Specialty Minerals and Colors Group, USA	
CELITE	Diatomaceous silica (flatting and antiblocking agent)	Celite Corp., USA	
CELLCOM	Range of chemical blowing agents for most plastics	Kum Yang Co. Ltd., South Korea	Distributor: Plastics and Chemicals Inc. USA
CELLTREK	Polyols for PUR foam systems	Dow	
CELOGEN	Range of chemical blowing agents for PVC, polyolefins, and some other plastics	Uniroyal Chemical Co. Inc., USA	
CELUKAVIT	Processing aids	CECA (Elf Atochem)	
CERAMCEL	Range of hollow ceramic microspheres	Microcell Technology Inc., USA	
CERECLOR	Chlorinated paraffins: flame retardant plasticiser extenders	ICI	
CERES	Soluble colorants (dyes)	Bayer; Miles Inc., USA	
CHAMELEON	Range of photochromic and thermochromic colorant masterbatches for polyolefins and styrenics	Victor International Plastics, UK	

TRADE NAME	NATURE AND/OR APPLICATION	PRODUCER OR SUPPLIER	REMARKS
CHEMOCEL	Chemical blowing agent concentrates for thermoplastics	Clark-Cooper Corp., USA	
CHEMSTAT	Mould-release, slip and antiblocking agents	Chemax Inc., USA	
CHIMASSORB	UV stabilisers (various types)	Ciba-Geigy	
CHLOREZ	Solid chlorinated paraffins of high chlorine content: flame retardant additives for various polymers	Dover Chemical Corp., USA	E.g. CHLOREZ 760: high melting, 74% chlorine
CHLOROWAX	Liquid chlorinated paraffins and alpha olefins: flame retardants, and plasticiser extenders for PVC	Occidental Chemical Corp., USA	
CHROMA-TEK	Range of liquid colorant dispersions for thermosetting resins	Chroma-Tek, USA	Subsidiary of Alpha Corp., USA
CINQUASIA	Range of pigments (gold; red; blue; green)	Du Pont	
CLEARTINT	Transparent-colorant concentrates for polyolefins	Milliken Chemical Co., USA	
COCURE	Catalysts for castable PUR elastomer systems	Caschem Inc., USA	
COLORCON HT	Range of pigments (aluminium-lake type)	Colorcon. Berwind Pharmaceutical Services Inc., USA	HT = "high tint"
COLOR CRYSTALS	Range of colour concentrates for thermoplstics	Teknor Color Co., USA	
COLORLINE	Range of liquid colorants	Colortek Vertriebs GmbH, Germany	
COLOR-MAX	Colorant dispersions in solvents and plasticisers	W.J. Ruscoe Co., USA	
COLORPAC	Pigments for UP resins	Burns & Russel Co., Specialty Chemical Div., USA	
COLOR-SPHERES	Colorant concentrates in fine-pellet form	Progressive Polymers Inc., USA	
COMBOLOOB	Lubricants (combined wax types)	Astor Wax Corp., USA	
COMISTAB	Heat stabilisers (various types) for PVC	Comiel SpA, Italy	
CONACURE EA	Range of hardeners for epoxy resins	Conap Inc., USA	
CONACURE AH	Range of curing agents, and polyols for PUR systems	Conap Inc., USA	
CONAP DS	Inorganic-colour pastes for epoxy and PUR systems	Conap Inc., USA	
CONDUCTEX	Carbon powders for antistatic and conductive plastics compounds	Columbian Chemicals Co., USA	
CONREC	Range of composite stabiliser masterbatches (including ones for recycled thermoplastics)	Constab Polymer-Chemie GmbH, Germany	
COPAC	Promoter for UP resin systems (Co/K type)	Huels AG, Germany; Huls America Inc., USA	
COPO	Elastomeric (SBR type) impact modifiers for styrenics	DSM	

TRADE NAME	NATURE AND/OR APPLICATION	PRODUCER OR SUPPLIER	REMARKS
CORAX	Carbon powders for antistatic and conductive plastics compounds	Degussa	
COSMIC BLACK	Bone-black pigments	Ebonex Corp., USA	Some grades FDA-approved for food-contact applications
COTINS	Organotin curing agents for PUR systems (elastomers and foams)	Caschem Inc., USA	
CRISTAL	Titanium dioxide pigments for plastics and surface coatings	Cristal - The National Titanium Dioxide Co. Ltd., Saudi Arabia	
CRODACID	Lubricants (fatty-acid type) for thermoplastics	Croda Universal	
CRODAMIDE	Lubricants (fatty-acid amide type) for thermoplastics	Croda Universal	
CROFON	Fibre optics (PMMA core, fluoropolymer skin)	Du Pont	Used for dashboard illumination in motorcars
CROMOPHTAL	Pigments (organic type)	Ciba-Geigy	Many FDA-approved for food-contact applications
CT 264	Antistatic agents	Boehringer Ingelheim KG, Germany	
CUNILATE	Biocide: copper-bis(8-hyroxyquinoline)	Morton International Inc., Industrial Chemicals and Additives, USA	
CURENE	Range of curing agents for PUR systems	Anderson Development Co., USA	
CURITHANE	Range of curing agents for PUR and PIR foam systems	Dow Plastics, USA	
CYANOX	Range of antioxidants (various types) and antioxidant blends	Cytec Industries Inc., USA	Trading unit of American Cyanamid Co.
CYASORB	Range of UV stabilisers (various types)	Cytec Industries Inc., USA	
CYASTAB	Heat stabilisers (lead-based) for PVC	Cytec Industries Inc., USA	
CYASTAT	Antistatic agents	Cytec Industries Inc., USA	
CYCOM NCG	Nickel-coated carbon powder for antistatic and conductive plastics compounds	Cytec Industries Inc., USA	
CYLACELL	Inorganic physical blowing agent with nucleating action in polyolefins. Some applications also in rubbers (including EPDM) and UP resins	Sylatec Corp., USA	Particulate sodium silicate blown by water of hydration into closed microspheres (at temperatures above 150°C)
DABCO	Range of catalysts for PUR systems (some also for PIR foam systems)	Air Products and Chemicals Inc., USA	DABCO T : tin-based. DABCO NEM, NCM, NMM: morpholine-based
DAXAD	Dispersing agents for additives in polymer emulsions	W. R. Grace & Co, Organic Chemicals Div., USA	

TRADE NAME	NATURE AND/OR APPLICATION	PRODUCER OR SUPPLIER	REMARKS
DAY-GLO	Range of luminescent colorants	Day-Glo Color Corp., USA	
DECHLORANE PLUS	Flame retardant (chlorinated, non-reactive, non-plasticising) for thermoplastics and thermosets	Occidental Chemical Corp., USA	
DEGALAN	Processing aids (acrylic type) for PVC	Degussa	
DEGAMIN	Isophorone diamine hardener for epoxy resins	Degussa	
D.E.H.	Range of amine hardeners for epoxy resins	Dow	
DEHYDAT	Antistatic agents	Henkel	
DEVOLITE	China clay (kaolin) fillers	ECC International	
DI-AQUA	Aqueous alkaline cleaner for removal of silicone residues (including release agents)	Merck	
DICALITE	Diatomaceous silica fillers and antiblocking agents; pearlite fillers	Grefco Inc., USA	
DICAPERL	Hollow glass microspheres	Grefco Inc., USA	
DI-CUP	Dicumyl peroxide	Hercules	
DIGESIL	Range of digestants for removal of silicones, fluorosilicones, polysulfides, and PUR from surfaces	Merck	Claimed effective also with some inks and surface coatings
DILEXO	Emulsifiers for polymer dispersions	Condea Chemie GmbH, Germany	
DIME	Dimethylcyclohexylamine (DMCHA) - catalyst for PUR systems	Shell Chemicals	
DIMUL S	Mould release; antistatic and antifogging agent for polyolefins and PVC	Witco Corp., Polymer Additives Group, USA	
DISFLAMOLL	Flame-retardant plasticisers (phosphate ester type) for PVC	Bayer; Miles Inc., USA	
DIS-RES	Colorant dispersion for UP and epoxy resins, and PUR systems	Advance Coatings Co., USA	
DRAKEOL	Lubricants; dispersing agents; and mould-release agents (mineral oil type)	Penreco Div., Pennzoil Products Co., USA	
DREWPLAST	Range of lubricants for PVC (glycerol ester and polyoxyethylene types)	Stepan Co., USA	
DRIKALITE	Calcium carbonate fillers	ECC International	
DUALITE	Hollow microspheres (of PVDC and some other polymers) externally coated with a filler or pigment	Pierce & Stevens Corp., USA	
DURAMITE	Calcium carbonate fillers	ECC International	
DURASTRENTH	Acrylic impact modifier for PVC	CECA (Elf Atochem)	Weathering-resistant

TRADE NAME	NATURE AND/OR APPLICATION	PRODUCER OR SUPPLIER	REMARKS
DURCAL	Calcium carbonate fillers	Plues-Staufer (and Omya sales offices)	
DYHARD	Range of hardeners (DCDA type) for epoxy resins	SKW Chemicals Inc., USA	
DYNAMAR	Processing aids (fluoropolymer type) for polyolefins, EVA, PET, and PVC	3M Co., Specialty Fluoropolymers Dept., USA	
DYNAPURGE ELITE	A one-step purging compound for extrusion and injection moulding equipment	Euro Polymers BV, The Netherlands	
DYNASYLAN	Crosslinking agents	Huels AG, Germany	
DYOBARIT	Barium sulfate fillers	Worlee-Chemie GmbH, Germany	
EASTMAN INHIBITOR	Antioxidants (e.g. OABH - metal deactivator for polyolefins) and UV stabilisers (e.g. RMB for cellulosics; HHBP for PVC and styrenics)	Eastman Chemical	
EASTOBRITE OB	Optical brightening agent	Eastman Chemical	
EASTOFLEX **EASTOTAC**	Tack additives for polyolefins and rubber	Eastman Chemical	
ECCOSPHERES	Hollow glass microspheres (fillers for weight reduction and syntactic foams)	Emerson and Cuming Inc., USA	
ECEPOX	Epoxidised oils	CECA (Elf Atochem)	
ECON-GLASS	Solid glass beads for use as fillers	D. J. Enterprises Inc., USA	
ECONO-BLAK	Carbon-black concentrates for polyolefins and polystyrene	Rondy & Co., USA	
ECOSTAR	Hydrophobic starch (offered as matting agent for PVC plastisols, and as biodegradability promoter in polyolefins)	Ecostar International, USA	
ECR GLAS	Corrosion-resistant glass fibres	Owens-Corning Fiberglas Corp., USA	
EDENOL	Stabilisers for PVC (Ca/Zn, and metal-free types)	Henkel	
ELASTAMINE	Reactive polyether amines	Huntsman Chemical Corp.	Formerly Texaco Chemical Co.
ELASTOCARB	Magnesium carbonate (flame-retardant and smoke-suppressant grades)	Morton International Inc., USA	
ELFTEX	Furnace carbon black	Cabot Corp., Special Blacks Div., USA	
ELIX	Polymeric modifiers (SMA type) for PVC	Monsanto	Previously in the CADON range
ELTESOL	Catalysts	Albright & Wilson	
ELVALOY	EVA-based terpolymers (used as impact modifiers in PVC)	Du Pont	
EMAFIL	Colorant masterbatches	Clariant Corp.,	Formerly Sandoz Chemicals

TRADE NAME	NATURE AND/OR APPLICATION	PRODUCER OR SUPPLIER	REMARKS
EMCOR	Reinforcing fibres (short whiskers) for phenolic materials (used as asbestos replacement)	Engelhard Corp., Specialty Minerals and Colors Group, USA	Used inter alia in phenolic brake linings
ENDUROLENE	Perylene maroon colorants	Cookson Pigments Inc., USA	
ENDURONONE	Red colorants (anthraquinone and naphtol types)	Cookson Pigments Inc., USA	
ENDUROPHTAL	Blue and green colorants (phtalocyanine type)	Cookson Pigments Inc., USA	
ENDUROQUIN	Red, violet, and magenta colorants (quinacridone type)	Cookson Pigments Inc., USA	
ENVIROSTRAND	Antimony oxide concentrates in strand form (some grades include other flame retardants)	Anzon Ltd., UK; Anzon Inc., USA	
EPAMINE	Curing agents (aliphatic amine type) for epoxy resins	Epic Resins, Div. Of General Fiberglass Supply Inc., USA	Offered as replacements for MDA hardeners
EPICELL	Chemical blowing agents (exothermic type) for PE, EVA, and plastics processed at up to 150°C	EPI Environmental Products Inc., USA	
EPICOR	Chemical blowing agents (endothermic type) for various plastics	EPI Environmental Products Inc., USA	E.g. EPICOR 753 - concentrate for use in PS and PPO alloys
EPIKURE	Curing agents (various types) for epoxy resins	Shell Chemicals	
EPISTATIC	Lead stabilisers for PVC	Eagle-Picher Industries Inc., Chemicals Div., USA	
EPODIC	Epoxy resins diluent	Anchor Chemical Co., UK	
EPOLENE	Lubricant and mould-release waxes (polyolefin and glycerol monostearate types); compatabilisers (maleic-graft type)	Eastman Chemical	
EPOLITE	Curing agents (various types) for epoxy resins	Hexcel Corp., Resins Group, USA	
EPOTUF	Curing agents (various types) for epoxy resins	Reichhold Chemicals Inc., Reactive Polymers Div., USA	
EPOXICAL CORE FILL	Expanded shale aggregate (a light-weight filler for glass-reinforced epoxy resins)	P.S.H. Industries Inc., USA	
ESCOMER	Lubricants (synthetic wax type)	Exxon Chemical	
ESPEROX	Range of peroxyester and peroxyketal initiators and curing agents for UP resins	Witco Corp., Polymer Additive Group, USA	
ESTABEX	1. Heat stabilisers (epoxy type) for PVC 2. Some biocides (e.g. ESTABEX ABF - an organoarsenic compound)	Akzo Nobel	
ESTOFIL	Colorant masterbatches	Clariant Corp.	Formerly Sandoz Chemicals
ESTOFIL S	Polymer-soluble dyes	Clariant Corp.	
ESTOL	Lubricants (fatty-acid ester type)	Unichema International, The Netherlands Unichema North America, USA	

TRADE NAME	NATURE AND/OR APPLICATION	PRODUCER OR SUPPLIER	REMARKS
ETHACURE	Curing agents (amine type) for epoxy resins	Albemarle Corp	Formerly Ethyl Corp., Chemicals Group
ETHANOX	Antioxidants (phenolic, and some other types - e.g. ETHANOX 398 - a fluorophosphonite)	Albemarle Corp	Formerly Ethyl Corp., Chemicals Group
EVANSTAB	Antioxidants (thioester type) for polyoloefins, PS, and ABS	Hampshire Chemical Corp., Evans Chemetics Div., USA	
EVERSORB	UV stabilisers (various types)	Everlight Chemical Industrial Corp., Taiwan	
EXOCEROL	Chemical blowing agents (exothermic, and exothermic/endothermic blends)	Boehringer Ingelheim KG, Germany; BI Chemicals, Henley Div., USA	
EXPANCEL	Thermoplastic microspheres containing encapsulated gas which expands on heating. Used as a physical blowing agent.	Expancel-Nobel Industries, Div. of Expancel Inc., USA	E.g. EXPANCEL 091: acrylic-based, gas-filled microspheres
EXPANDEX	Chemical blowing agents (exothermic, substituted-tetrazole type)	Uniroyal Chemical Co., Inc., USA	Used in high-melting thermoplastics
EXXELOR	Polyolefin-based impact modifiers, functionalised (by comonomer grafting) for enhanced compatibility with various thermoplastics	Exxon Chemical	
EZn-CHEK	Heat stabilisers for PVC	Ferro Corp., Bedford Chemical Div., USA	
FASCAT	Catalysts for UP resins	CECA (Elf Atochem)	
FERROSIL	Ferro-aluminium silicate filler	Kaopolite Inc., USA	
FIBERGLAS	Glass fibres	Owens-Corning Fiberglas Corp., USA	
FIBERFRAX	Mineral (alumina-silica) fibres used as reinforcement in composites, and in thermal insulation applications	The Carborundum Co., Fibers Div., USA	
FIBERKAL	Fibrous clay (calcined kaolin) fillers	Kaopolite Inc., USA	
FICEL	Chemical blowing agents (exothermic type)	Witco Corp., Polymer Additives Group, USA	
FILESTER	Pigments (organic type) and soluble colorants (dyes)	Ciba-Geigy	
FINNTITAN	Titanium dioxide pigments	Kemira OY, Finland	
FIREBRAKE	Range of flame-retardant, smoke-suppressant additives (zinc-borate based)	Borax Consolidated, UK; US Borax Inc., USA	
FIREMASTER	Range of flame retardants	Great Lakes Chemical Corp., USA	
FIRESHIELD	Flame retardants (based on antimony oxide)	Laurel Industries Inc., USA	
FIREX	Flame retardants	Dr. Th. Böhme, Germany	
FLAKEGLAS	Glass flakes	Owens-Corning Fiberglas Corp., USA	

TRADE NAME	NATURE AND/OR APPLICATION	PRODUCER OR SUPPLIER	REMARKS
FLAMTARD	Inorganic flame retardant additives for plastics	Alcan Aluminium Corp., Alcan Chemicals Div., USA	E.g. FLAMTARD Z - zinc borate; FLAMTARD M7 - magnesium hydroxide
FLECTOL	Antioxidants (amine type)	Monsanto	
FLEXCHLOR	Chlorinated paraffins: flame-retardant plasticiser extenders for PVC	Witco Corp., Polymer Additives Group, USA	
FLUOLITE	Optical brightening agent	ICI	
FM	Range of chemical blowing agents in pellet concentrates	Quantum Chemical Corp., USA	Range includes exothermic, endothermic and reactive types
FOAM-KON	Range of chemical blowing agent concentrates for thermoplastics	LNP Engineering Plastics	
FOMREZ	Range of catalyst for PUR elastomers and foam systems	Witco Corp., Polymer Additives Group, USA	
FORANE	Range of blowing agents	Elf Atochem	
FORDCAL	Calcium carbonate fillers	ECC International	
FORMACEL	Physical blowing agents (HCFC type) for polystyrene foams	Du Pont	FORMACEL S claimed "ozone friendly"
FR	Series of brominated flame retardant additives for various plastics	Dead Sea Bromine Group, Israel; Ameribrom Inc., USA	Range contains, inter alia, TBS and PBBA (q.v.)
FRANKLIN FIBER	Grades of calcium sulfate microfibre for use in thermosetting resins	US Gypsum Co., USA	Suggested, inter alia, as asbestos replacement
FREKOTE	Mould-release agents	Dexter Corp., USA	
FUSABOND	Coupling agents (anhydride-grafted polyolefin type) for reinforcements in resins	Du Pont	
FYARESTOR	Brominated/chlorinated paraffins: flame-retardant additives (mainly for PVC, PUR, and epoxy resins)	Witco Corp., Polymer Additives Group, USA	
FYRAWAY	Range of flame-retardant additives for various plastics	Harwick Chemical Corp., USA	
FYREBLOC	Antimony oxide concentrates in pellet form (some grades include other flame retardants)	Anzon Ltd., UK; Anzon Inc., USA	
FYROL	Range of flame retardants (halogenated phosphate-ester type) for PUR systems	Akzo Nobel	
GARBEFIX	Antioxidants (phenolic and phosphite types)	Great Lakes Chemical Corp., USA	
GASIL	Synthetic silicas (thickeners, antiblocking, anti-caking and anti-plate-out agents)	Crosfield Ltd., UK	

TRADE NAME	NATURE AND/OR APPLICATION	PRODUCER OR SUPPLIER	REMARKS
GENIPLEX A	Accelerator for sulfur-curing rubbers	Schering AG, Germany	
GENISET MD	Nucleating agent (dimethylbenzylidene sorbitol) for PP and polyolefins	Schering AG, Germany	
GENITRON	Range of chemical blowing agents	Schering AG, Germany	
GIVSORB UV	UV stabilisers (phenylformamide type)	Givaudan-Roure Corp., USA	
GLENDION	Polyols for PUR systems	EniChem	
GLOMAX	Calcined kaolin fillers	Dry Branch Kaolin Co., USA	
GLYCOLUBE	Range of lubricants, and antistatic, antifogging, antiblocking, and slip agents)	Lonza	
GLYCOMUL	Range of antistatic, antifogging and dispersing agents	Lonza	
GLYCOSPERSE	Antifogging and dispersing agents	Lonza	
GLYCOSTAT	Antistatic agents	Lonza	
GRAFIL	Carbon fillers (various grades) and fibre products (tow, tape) for use as reinforcements	Courtaulds, UK	
GRAF-LITE	Low-density graphite filler for use in EMI and RFI shielding plastics materials	IMI Ltd.. USA	
GRANUPLAST	Range of "granite-look" additives and masterbatches for thermoplastics	Safas Corp., USA	
GRAPHTOL	Pigments (organic type)	Clariant Corp.	Formerly Sandoz Chemicals
HAL-LUB	Lubricants (lead-based) for PVC	Halstab Div., Hammond Lead Products Inc., USA	
HALSTAB	Heat stabilisers (lead-based) for PVC	Halstab Div., Hammond Lead Products Inc., USA	
HARSHAW	Colorants, including inorganic (HARSHAW METEOR) and organic pigments	Engelhard Corp., Specialty Minerals and Colors Div., USA	
HELIO FAST	Pigments (organic type)	Bayer; Miles Inc., USA	
HELIOGEN	Pigments (organic type)	BASF	
HEXAFIL	Clay fillers	ECC International	
HIMOD	Range of muscovite mica fillers	KMG Minerals Inc., USA	
HI-PFLEX	Surface-treated calcium carbonate fillers	Specialty Minerals Inc., USA	
HI-POINT	Ketone peroxide initiators for UP resins	Witco Corp., Polymer Additives Group, USA	
HI-SIL	Fine-particle, precipitated silica fillers	PPG Industries, Fiber Glass Products, USA	
HITOX	Composite titanium dioxide pigments	Hitox Corp., USA	
HOLCOBATCH	Pigment concentrates in heat-fusible carrier	Holland Colors Americas Inc., USA	Suitability for most plastics claimed

TRADE NAME	NATURE AND/OR APPLICATION	PRODUCER OR SUPPLIER	REMARKS
HOLON KILLER	Masterbatches of antimicrobial additives	Nikko Co. Ltd., Japan	
HOLOSPHERES	Hollow ceramic microspheres	IMI Ltd., USA	
HOMBITAN	Titanium dioxide pigments	Sachtleben Chemie GmbH, Germany	
HONEY WAX	Mould-release waxes	Specialty Chemicals Co., USA	
HORNA	Pigments (inorganic type)	Ciba-Geigy	E.g. HORNACHROM - chrome yellow
HOSTAFLAM	Range of flame retardants	Hoechst; Hoechst Celanese Corp., USA	
HOSTALIT H	Impact modifiers (EVA type) for PVC	Hoechst; Hoechst Celanese Corp., USA	
HOSTALUB	Range of lubricants (various types)	Hoechst; Hoechst Celanese Corp., USA	
HOSTALUX	Optical brightening agents	Hoechst; Hoechst Celanese Corp., USA	
HOSTANOX	Range of antioxidants (thioether, phenolic, and organophosphite types)	Hoechst; Hoechst Celanese Corp., USA	
HOSTAPERM	Pigments (organic type)	Hoechst; Hoechst Celanese Corp., USA	
HOSTASOL	Soluble colorants (dyes)	Hoechst; Hoechst Celanese Corp., USA	
HOSTASTAB	Heat stabilisers (various types) for PVC	Hoechst; Hoechst Celanese Corp., USA	
HOSTASTAT	Antistatic agents	Hoechst; Hoechst Celanese Corp., USA	
HOSTATRON	Range of chemical blowing agents	Hoechst; Hoechst Celanese Corp., USA	
HOSTAVIN	Range of UV stabilisers	Hoechst; Hoechst Celanese Corp., USA	
HY-DENSE	Lubricants (metal stearate type)	Mallinckrodt Specialty Chemicals Co., USA	
HYDRAL	Alumina trihydrate flame-retardant powders, including ultrafine grades	Aluminum Co. of America (ALCOA), USA	
HYRDAMAX	Range of alumina trihydrate and magnesium hydroxide fine-particle flame retardants	Climax Performance Materials Corp., USA	Subsidiary of Amax Inc.
HYDRIN	Epichlorohydrin rubber (ECO) impact modifier	Zeon Chemicals Inc., USA	
HYDRITE	Range of kaolin fillers	Dry Branch Kaolin Co., USA	
HYDROCARB	Calcium carbonate fillers	Pluess-Staufer (and Omya sales offices)	
HYDROCEROL	Chemical blowing agents (mainly endothermic, of acid/carbonate type), some with nucleating action	Boehringer Ingelheim KG, Germany; BI Chemicals, Henley Div., USA	E.g. HYRDOCEROL BIH - a combination of citric acid and sodium bicarbonate
HYSTRENE	Lubricants and mould-release agents (fatty-acid type) for polystyrene	Witco Corp. Polymer Additives Group, USA	
HYTEMP	Impact modifiers (polyacrylate type)	Zeon Chemicals Inc., USA	
INCOZOL	Curing agents (latent type, based on oxazolidine), PUR systems; moisture scavengers	Industrial Copolymers Ltd., UK	
INDOFAST	Organic pigments	Bayer	
INTERCIDE	Microbiocides	Akzo Nobel	

TRADE NAME	NATURE AND/OR APPLICATION	PRODUCER OR SUPPLIER	REMARKS
INTERLITE	Stabilisers for PVC (including grades for recycling)	Akcros Chemicals	
INTEROX	Range of peroxide catalysts and crosslinking agents	Peroxid Chemie, Germany	
INTERSTAB	Ranges of heat stabilisers for PVC, and lubricants for PVC and other plastics	Akzo Nobel; Akcros Chemicals	Formerly produced by Interstab Chemicals Inc.
INTERSTAT	Antistatic agents	Akzo Nobel	
INTERWAX	Lubricants	Akzo Nobel	
IONOL	Antioxidants	Shell Chemicals	E.g. IONOL CP; IONOX 330 - cf. BS 2782: Method 434A: 1975
IONOX	Antioxidants	Shell Chemicals	
IRGACOLOR	Pigments (inorganic type)	Ciba-Geigy	
IRGALITH	Pigments (organic type)	Ciba-Geigy	
IRGANOX	Range of antioxidants, including metal deactivators (e.g. IGRANOX MD 1024)	Ciba-Geigy	Cf. ASTM D 5524-94 and D 1996-92
IRGASAN	Biocides	Ciba-Geigy	E.g. IRGASAN DP 300 (2,4,4' - trichloro - 2'- hydroxydiphenyl ether)
IRGASTAB	Heat stabilisers (various types) for PVC, and UV stabilisers (nickel compounds)	Ciba-Geigy	
IRGASTAT	Antistatic agents	Ciba-Geigy	
IRGAWAX	Range of lubricants (various types)	Ciba-Geigy	
IRGAZIN	Pigments (organic type)	Ciba-Geigy	
IRIODIN	Pearlescent pigments (coated mica)	EM Industries Inc., Pigments Div., USA	
ISONATE	Isocyanate (modified MDI), reactant for PUR systems	Dow	
ISONOX	Range of antioxidants (phenolic type)	Schenectady Chemicals Inc., USA	Cf. ASTM D 5524-94 and D 1996-92
ISOTRON	Range of physical blowing agents for foaming PS, PE, and PUR	Elf Atochem	Offered as alternatives to ozone-depleting agents
JEFFAMINE	Range of curing agents (mostly based on polyoxypropylene amines) for epoxy resins	Huntsman Chemical Corp.	
JET STREAM	Ground talc fillers	The Montana Talc Co., USA	
KADOX	UV stabilisers (zinc oxide type)	Zinc Corporation of America, USA	
KANE ACE	Impact modifiers (ABS, MBS, and acrylic types) and processing aids (acrylic types) mainly for PVC	Kanegafuchi Chemical Industry Co. Ltd., Japan; Kaneka Texas Corp., USA	

TRADE NAME	NATURE AND/OR APPLICATION	PRODUCER OR SUPPLIER	REMARKS
KANTSTIK	Range of lubricants and mould-release agents for thermoplastics and thermosets	Specialty Products Co., USA	
KAOPAQUE	Aluminium silicate filler for thermoplastics	Dry Branch Kaolin Co., USA	
KAOPOLITE	Anhydrous aluminium silicate antiblocking agents	Kaopolite Inc., USA	
KELTAN	Elastomeric impact modifiers for polyolefins	DSM	
KEMAMIDE	Range of slip, antiblocking and mould-release agents (fatty-acid amide type)	Witco Corp., Polymer Additives Group, USA	E.g. KEMAMIDE E (eruca-mide); cf. also ASTM D 1996-92
KEMESTER	Slip and antiblocking agents for polyolefins	Witco Corp., Polymer Additives Group, USA	
KEMINOX	Antioxidants (phenolic type)	Chemipro Kasei Ltd. Japan	
KEMIRA	Titanium dioxide pigments	Kemira Pigments OY, Finland	
KEMISTAB	Heat stabilisers (various types) for PVC	Kemichrom SA, Spain	
KEMPORE	Chemical blowing agents (exothermic type)	Uniroyal Chemical Co. Inc., USA	Mainly ADC-based; principal applications in polyolefins
KEMTEC	Chemical blowing agents (some based on isatoic anhydride)	PMC Inc., Specialties Group Div., USA	E.g. KEMTEC 500, for PA, PET, PBT, PC, and ABS
KENPLAST ESB	Processing aid for PVC, PA, and PUR (RIM systems)	Kenrich Petrochemicals Inc., USA	
KEN-REACT	Coupling agents (titanate, pyrophosphate-derived); some flame-retardancy claimed)	Kenrich Petrochemicals Inc., USA	
KERAFAST	Pigments	Cookson Pigments Inc., USA	
KERAFORM	Colorant system for PVC	Cookson Pigments Inc., USA	
KERAPLAST	Colorant masterbatches	Cookson Pigments Inc., USA	
KETJENBLACK	Carbon powder for antistatic and conductive plastics compounds	Akzo Nobel	
KLAKUP	Surface-treated aluminium silicate fillers	Malvern Minerals Co., USA	
KOSMOS	Tin-based catalysts for PUR foam systems	Th. Goldschmidt AG, Germany	
KOTAMITE	Ground calcium carbonate fillers	ECC International	
KRASNIA	Various additives and masterbatches for plastics	Colores y Compuestos Plasticos SA, Spain	
KRATON	Styrenic block copolymers, used as impact modifiers ("D" series in PS and polyolefins; "G" series in PC and olefinic elastomers). Functionalised versions available as compatabilisers for polymer blends.	Shell Chemicals	See also Section 1 (Polymer Trade Names)
KROLOR	Silica-encapsulated pigments	Cookson Pigments Inc., USA	Resistant to heat and chemicals
KRONITEX	Flame-retardant phosphate plasticisers for PVC, cellulosics and some other thermoplastics	FMC Corp., Process Additives Div., UK and USA	

TRADE NAME	NATURE AND/OR APPLICATION	PRODUCER OR SUPPLIER	REMARKS
KRONOS	Titanium dioxide pigments	Kronos	
KSK THERMAL COLOUR	Range of thermochromic pigments and pigment concentrates	Siber Hegner Ltd., UK	
KXV	Colorant chips for PVC compositions	Permuthane Coatings (ICI Group), USA	
LANKROMARK	Range of stabilisers for PVC	Akcros Chemicals	
LBA	Chemical blowing agents (exothermic type) and concentrates	Boehringer Ingelheim KG, Germany; BI Chemicals, Henley Div., USA	E.g. LBA-55 - a liquid azo concentrate for use in ABS foam-core pipe production
LEECURE B	Range of liquid epoxy resin curing agents (BTF-based)	Leepoxy Plastics Inc., USA	
LEUCOPHOR	Optical brightening agents	Clariant Corp.	Formerly Sandoz Chemicals
LEUCOPURE	Optical brightening agents	Clariant Corp.	Formerly Sandoz Chemicals
LEVAPREN	See BAYMOD		
LIPINOL	Viscosity modifier for PVC	Huels	
LIPOXIOL	Lubricants and antistatic agents	Huels	
LISTAB	Lubricants (metal stearate type)	Chemson GmbH, Germany	
LITHOL	Pigments (organic type)	BASF	
LOOBWAX	Lubricants (various wax types)	Astor Wax Corp., USA	
LOWILITE	Range of UV stabilisers	Chemische Werke Lowi GmbH, Germany	
LOWINOX	Range of antioxidants (various types)	Chemische Werke Lowi GmbH, Germany	
LOXAMIDE	Lubricants (fatty-acid amide type)	Henkel	
LOXIOL	Range of antioxidants (various types)	Henkel	
LUBOL	Internal lubricants	L & L Industrial Chemicals, USA	
LUBRAZINC	Lubricants (zinc stearate type)	Witco Corp., Polymer Additives Group, USA	
LUBREX	Mould-release agents	Harwick Chemical Corp., USA	
LUBRIOL	Lubricant for PVC	Comiel SpA, Italy	
LUBRISTAB	Stabiliser/lubricant combinations for PVC	Comiel SpA, Italy	
LUBRIZOL	Intermediates for some PVC heat stabilisers	Lubrizol Corp., USA	
LUCIDOL	Peroxide curing and crosslinking agents	Akzo Nobel	
LUPRANATE	MDI (simple and polymeric)	BASF	
LUVOSLIP	Lubricant masterbatches (those with PE and PP carriers based on erucic and oleic acid amides)	Lehmann and Voss & Co., Germany	
LUWAX	Lubricants and mould-release agents (montan wax and PE types)	BASF	

TRADE NAME	NATURE AND/OR APPLICATION	PRODUCER OR SUPPLIER	REMARKS
LYNCUR	Curing agents (PVP-based) for epoxy resins	Maruzen America Inc., USA	
MACROLEX	Soluble colorants (dyes)	Bayer; Miles Inc., USA	Good heat stability
MAGNIFIN	Magnesium hydroxide flame-retardant, smoke-suppressant additives	Martinswerk GmbH, Germany; Lonza Inc., USA	
MAGNUM-WHITE	Smoke-suppressant fillers (based on magnesium hydroxide)	RMc Minerals Inc., USA	
MAGOTEX	Fused magnesium oxide filler	Kaopolite Inc., USA	Calcined, to promote thermal conductivity
MAPICO	Range of iron-oxide pigments	Columbian Chemicals Co., Mapico Div., USA	
MARK	Range of heat stabilisers for PVC; antioxidants and UV stabilisers for thermoplastics	Witco Corp., Polymer Additives Group, USA	Former producers: Argus Chemical Corp., then Argus Div. of Witco Corp.
MARKSTAT	Antistatic agents		
MARTINAL	Alumina trihydrate flame retardants	Martinswerk GmbH, Germany; Lonza Inc., USA	
MASTERFLAM	Range of flame-retardant and smoke-suppressant masterbatches	VAMP Srl, Italy; KMZ Chemicals, UK	
MAXITHEN	Multi-additive masterbatches (variously incorporating antioxidants, UV stabilisers, colorants, flame retardants, etc.)	Gabriel-Chemie GesmbH, Austria	
MEARLIN	Lustre colorants, including nacreous and metallic effects	The Mearl Corp., Pigments Div., USA	
MEARLITE			
MELAX	Mould-release agents	Axel Plastics Research Laboratories Inc., USA	
METABLEN	Impact modifiers (MBS type) for PVC	CECA (Elf Atochem)	
METASOL	Catalysts for PUR elastomer and foam systems	E. Merck, Germany; Calgon Corp., USA	
METASPHERES	Hollow aluminosilicate microspheres for use as fillers in liquid thermosetting resin systems	IMI Ltd., USA	
METASTAB	Heat stabiliser (organotin type) for PVC	Acima AG, Switzerland	
METATIN	Biocides (based on tributyltin)		
METRE-CHROME 1500	Colorant (liquid) dispersions for epoxy resins	Metachem Resins Corp., Mereco Products Div., USA	
METRE-CURE	Catalytic curing agents for epoxy resins		

TRADE NAME	NATURE AND/OR APPLICATION	PRODUCER OR SUPPLIER	REMARKS
MICAFLEX	Mica fillers	Eagle Quality Products Inc., USA	Offered particularly for epoxy and UP resins
MICA-LYTE	Special-effect colorants	Mica-Tek Inc., USA	
MICRAL	Alumina trihydrate flame-retardant fillers	Solem Div., J.M. Huber Corp., USA	
MICROBALOONS	Hollow glass microspheres	Emerson & Cuming Inc., USA	
MICROBLOC	Talc antiblocking agents for polyolefins	Specialty Minerals Inc., USA	
MICROCARB	Calcium carbonate fillers	Microfine Minerals Ltd., UK	
MICRO-CEL	Synthetic hydrous calcium silicate (flatting, antiblocking and processing aids)	Schuller International Inc., USA	
MICRO-CHEK	Biocides	Ferro Corp., Bedford Chemical Div., USA	
MICRODOL	Calcium carbonate fillers	Norwegian Talc, Norway and UK	
MICROFINE	Range of chemical blowing agents	High Polymer Labs, India	
MICROLITH	Dispersions of CROMOPHTAL and IRGAZIN pigments	Ciba-Geigy	
MICRO-MIX	An "engineered" filler system: mixtures of shaped particles (rods, flakes, spheres and irregular) for high loadings in various resins	America's Best Building Products, USA	
MICROMYA	Calcium carbonate fillers	Pluess-Staufer (and Omya sales offices)	
MICRO - P	A silicate filler/extender	D.J. Enterprises Inc., USA	
MICROSPIN	Resin-coated pigments for thermoplastics	Hilton Davis Chemical Co., USA	
MICROTUFF	Surface-coated talc fillers	Specialty Minerals Inc., USA	
MICRO-WHITE	Calcium carbonate fillers	ECC International	
MIDA	Colorants	Boehringer Ingelheim AG, Germany	
MILLAD	Clarifying agents for PP	Milliken Chemical Co., USA	
MILLICARB	Calcium carbonate fillers	Pluess-Staufer (and Omya sales offices)	
MINEX	Feldspar (nepheline syenite) fillers	Pluess-Staufer (and Omya sales offices)	
MINSPAR	Feldspar (nepheline syenite) fillers	K-T Feldspar Corp., USA	
MIN-U-SIL	Finely ground silica (filler for epoxy and UP resins)	US Silica Co., USA	
MIRAFLEX	Glass fibres produced from two kinds of glass (interfused for high strength and resilience)	Owens-Corning Fiberglas Corp., USA	
MISTRON	Talc fillers	Chemag GmbH, Germany	
MIXXIM	Range of antioxidants (phenolic type), metal deactivators (especially for polyolefins), UV stabilisers, and some heat stabilisers for PVC (e.g. MIXXIM DBM)	Fairmont Chemical Co.	MIXXIM BB: series for plastics processed at high temperatures
MOGUL	Surface-treated furnace carbon blacks	Cabot Corp., Special Blacks Div., USA	
MOLD-EASE	Mould-release agents	Merix Chemical Co., USA	

TRADE NAME	NATURE AND/OR APPLICATION	PRODUCER OR SUPPLIER	REMARKS
MOLDPRO	Processing aids and internal mould-release agents for various plastics	Witco Corp., Polymer Additives Group, USA	
MOLD WIZ	Internal lubricants (polymer-blend type for thermoplastics and thermosets)	Axel Plastics Research Laboratories Inc., USA	
MOLYSULFIDE	Molybdenum disulfide (lubricant additive)	Climax Performance Materials Corp., USA	
MONARCH	Furnace carbon blacks	Cabot Corp., Special Blacks Div., USA	
MONASTRAL	Range of pigments (quinacridone type)	Ciba-Geigy	
MONDUR	Isocyanates (TDI, MDI, and mixtures)	Bayer; Miles Inc., USA	
MONO-COAT	Mould-release agents	Chem-Trend Inc., Plastics and Rubber Div., USA	
MONO-PAK	Multi-additive packages for reinforced plastics	Catalyst Systems Inc., USA	
MONYKUP	Surface-treated antimony oxide flame retardants	Malvern Minerals Co., USA	
MULTRANOL	Polyols for PUR systems	Bayer; Miles Inc., USA	
MYVAPLEX **MYVEROL**	Lubricants (for PVC) and antistatic agents (for polyolefins): monoglyceride type	Eastman Chemical	
NAFTOLUBE	Range of lubricants	Chemson GmbH, Germany	
NAFTOMIX	Stabiliser/lubricant systems for PVC		
NAFTOVIN	Heat stabilisers (various types) for PVC		
NAFTOZIN	Lubricants (fatty-acid type)		
NAPHTHANIL	Colorants (naphthol-red type)	Cookson Pigments Inc., USA	
NAUGARD	Range of antioxidants (various types)	Uniroyal Chemicals Co. Inc., USA	
NAUGAWHITE	Antioxidant (alkylidene-bisphenol type)		
NIAX	Range of amine catalysts for PUR foam systems	Union Carbide Chemicals and Plastics Co. Inc., USA	
NICRON	Ground talc fillers	The Montana Talc Co., USA	
NIPOL	Nitrile rubbers, used as impact modifiers	Zeon Chemicals Inc., USA	
NITROFAST	Polymer-soluble dyes	Clariant Corp.	Formerly Sandoz Chemicals
NITROSAN	Chemical blowing agent (exothermic type)	Originally marketed by Du Pont	See also DMDNT
NL	Accelerators (cobalt-complex type) for UP resins	Akzo Nobel	
NOFACAL	Precipitated calcium carbonate fillers	Nordbayerische Farben und Mineralwerke, Germany	
NONOX	Range of antioxidants (various types)	Zeneca	Formerly ICI
NORBLOC	UV stabilisers (benzotriazole type)	Noramco Inc., USA	NORBLOC 7966 copolymerise-able with vinyl monomers
NOVACITE	Fine-particle quartz fillers	Malvern Mineral Co., USA	
NOVAKUP	Surface-treated silica fillers		

TRADE NAME	NATURE AND/OR APPLICATION	PRODUCER OR SUPPLIER	REMARKS
NOVALAR	Impact modifiers (ABS type) for PVC	Nova Polymers Inc., USA	
NOVALENE	Impact modifiers (EPDM type) for polyolefins		
NOVAMIN	Range of crosslinking agents	Condea Chemie GmbH, Germany	
NOVOPERM	Pigments (organic type)	Hoechst	
NOVOTEK	Liquid dispersions of colorants and additives	Colortek Vertriebs GmbH, Germany	
NUODEX COBALT	Promoters (cobalt-based) for UP resins	Huels	
NUXTRA COBALT			
NYACOL	Powders and liquid dispersions of collodal antimony pentoxide and zinc oxide; polymer-based concentrates of antimony pentoxide with a halogen donor (e.g. NYACOL EAM)	The PQ Corp., USA	Fine state of dispersion claimed to enhance activity
NYAD	Wollastonite fillers	Nyco Minerals Inc., USA	
NYSYNBLAK	NBR-based masterbatches: combination black concentrates and impact modifiers for ABS and PA	DSM	
NYTAL	Range of talc fillers	R. T. Vanderbilt Co. Inc., USA	
OLDOPAL	Pigments, and resins (particularly for gel coats)	Buesing & Fasch GmbH, Germany	
OMACIDE	Antimicrobial additives for PVC	Olin Corp., USA	
OMADINE	Range of biocidal additives for plastics		
OMYA	Calcium carbonate fillers for various plastics	Pluess-Staufer (and Omya sales offices)	
OMYACARB			
OMYALENE			
OMYALITE			
ONCOR	Antimony oxide flame retardants (pellets with silicate core)	Anzon Ltd., U K Anzon Inc., USA	
ONGARD	Flame-retardant smoke suppressants		E.g. ONGARD 2 - a magnesium/zinc complex
ONIFINE	Range of chemical blowing agents (modified-ADC type) for PVC (in particular crosslinked PVC foam)	Otsuka Chemical Co., Japan	US distributors: Biddel Sawyer Corp.
OPTIBLANC	Optical brightening agents	3V Sigma, Italy	
OPTIWHITE	Clay fillers	Burgess Pigment Co., USA	White-pigment effect claimed
ORACET	Soluble colorants (dyes).	Ciba-Geigy	
ORASOL	Soluble colorants (dyes) applicable in acrylics and styrenics		
P 1000	Range of luminescent colorants	Hanovia, USA	

TRADE NAME	NATURE AND/OR APPLICATION	PRODUCER OR SUPPLIER	REMARKS
PALIOGEN PALIOTOL	Pigments (organic type)	BASF	
PALOMAR	Pigments (organic type)	Bayer	
PARALOID	Range of impact modifiers for PVC ("BTA", "HT" and "KM" series) and for engineering plastics ("EXL" series)	Rohm & Haas	Initially also known as ACRYLOID in the UK
PAREL	Impact modifiers (polypropylene oxide type)	Zeon Chemicals Inc., USA	
PARICIN	Internal/external lubricants (amide wax type)	Caschem Inc., USA	
PASTE WIZ	Mould-release paste	Axel Plastics Research Laboratories Inc., USA	
PATIONIC	Range of glycerol esters (PVC lubricants; antistatic and antiblocking agents: antioxidant aids for polyolefins)	Patco Polymer Additives Div., American Ingredients Co., USA	
PERKADOX	Peroxide curing and crosslinking agents	Akzo Nobel	
PERMA-MOLD	Mould-release agents for PUR foam	Chem-Trend Inc., Plastics and Rubber Div., USA	
PEROXIMON	Range of peroxide initiators, curing and crosslinking agents	Montell Italia, Italy Himont, USA	
PERRINDO	Pigments (organic type)	Bayer	
PETROPLAS	Petroleum-based compatibilisers for various polymer blends	Isorca Inc., USA	
PHILBLACK	Carbon powders for antistatic and conductive compounds	Phillips Chemical Co., USA	
PHOS-CHEK	Range of flame retardants (based on ammonium polyphosphate), and low-smoke additives for polyolefins	Monsanto	PHOS-CHEK P/30 and P/40 also used in intumescent surface coatings
PHOSCLERE	Organophosphite heat stabilisers for PVC	Akcros Chemicals	
PLASADD PLASBLAK PLASGREY	Range of additive masterbatches Range of black masterbatches Range of grey masterbatches	Cabot Plastics International, Belgium; Cabot Corp., USA	
PLASTI-GRIT	Thermoset powders (amino and UP resins) offered as fillers for plastics	Composition Materials Co. Inc., USA	
PLASTISAN	Range of halogen-free flame retardants	3V Sigma, Italy	
PLASTISPERSE	Free-flowing colorants	Plasticolors Inc., USA	
PLASTORIT	Mica fillers	Naintsch Mineralwerke GmbH, Austria	
PLASWHITE	Range of white masterbatches	Cabot Plastics International, Belgium Cabot Corp., USA	
PLENLIZER	Heat co-stabiliser (polyol ester type) for PVC	Ajinomoto Inc., USA	

TRADE NAME	NATURE AND/OR APPLICATION	PRODUCER OR SUPPLIER	REMARKS
PLEXIGUM	Processing aids (acrylic type)	Röhm GmbH, Germany	
PLURACOL	Polyols for PUR systems	BASF	
PLUSWHITE	Precipitated white alumina trihydrate	Pluess-Staufer (and Omya sales offices)	
POLARITE	Surface-treated clay fillers	ECC International	
POLCARB	Calcium carbonate fillers	ECC International	
POLE STAR	Calcined clay fillers	ECC International	
POLYADD	Range of masterbatches	Plastichemix Industries, India	
POLYALDO	Dispersion agent for use in polyolefins	Lonza	
POLYBATCH	Multicomponent concentrates	A. Schulman Inc., USA	
POLYBLAK	Carbon-black concentrates	A. Schulman Inc., USA	
POLYBOUND	Chemical blowing agents (ADC-based), and blowing-agent concentrates	Harwick Chemical Corp., USA	
POLYCAT	Range of catalysts for PUR systems	Air Products and Chemicals Inc., USA	
POLYCIN	Polyol for PUR systems	CasChem Inc., USA	
POLYCOR	Chemical blowing agents (including exothermic/endothermic blends)	EPI Environmental Products Inc., USA	
POLY-CURE	Promoters (cobalt-based) for UP resins	OM Group Inc., USA	
POLYDIS TR	Lubricants, compatibilisers, and antiblocking agents	Structal Co., USA	
POLYDOL	Calcium/magnesium carbonate (dolomite) fillers	Sopromine SA, France	
POLYFILL	Range of plastics compounds	Plastichemix Industries, India	
POLYFLO	Range of colorants (including concentrates in a PE carrier)	Magruder Radiant Color Co., USA	
POLY-G	Polyether polyols for PUR systems	Olin Corp., USA	
POLYGEL	Thixotropic agents	3V Sigma, Italy	
POLYMICA	Muscovite mica fillers	Franklin Industrial Minerals, USA	
POLYSAR EPM	Impact modifiers (ethylene-propylene elastomers) for polyolefins	Bayer; Miles Inc., USA	
POLYSORBATE	Generic name for polyoxyethylene fatty acid esters (non-ionic surfactants), some equivalent to products in ICI TWEEN range: e.g. POLYSORBATE 40 = TWEEN 40		US equivalent generic term: USAN
POLY-SOY	Soya-bean-based reactive filler extender for UP and epoxy resins	Protein Technologies International, USA	
POLYTAL	Talc fillers	Whittaker, Clark & Daniels Inc., USA	
POLYTONE	Colorants (particles coated with polyester resin)	Hilton Davis Chemical Co., USA	
POLYTREND EST	Colorant dispersions (pastes) for UP resins	Huls America Inc., USA	
POLYWAX	Lubricants (PE wax type)	Petrolite Corp., Polymers Div., USA	

TRADE NAME	NATURE AND/OR APPLICATION	PRODUCER OR SUPPLIER	REMARKS
POROFOR	Chemical blowing agents (ADC and IA types)	Bayer; Miles Inc., USA	
PRETIOX	Titanium dioxide pigments	Precheza AS, Czech Republic	
PREVENTOL	Biocide: N-(difluoromethyl-thio)-phthalimide	Bayer	
PRIFRAC	Lubricants (fatty-acid type)	Unichema International, The Netherlands	
PRIOLUBE	Lubricants (fatty-acid ester type)	Unichema International, The Netherlands	
PRODOX	Antioxidants (various types)	PMC Inc., Specialties Group Div., USA	
PYRO-CHEK	Flame-retardant, brominated additives for styrenics and engineering plastics	Ferro Corp., Keil Chemical Div., USA	E.g. PYRO-CHEK LM, based on brominated PS
PYRONIL	Flame-retardant, aromatic bromine compounds	Elf Atochem	
Q-CEL	Hollow inorganic microspheres (main use as fillers in thermosetting resins)	PQ Corp., USA	
QUADROL	Polyols for PUR systems	BASF	
QUEENSFIL	Calcium carbonate fillers	ECC International	
QUICKSET	Ketone peroxide initiators for UP resins	Witco Corp., Polymer Additives Group, USA	
QUINCAT	Catalysts (amine type) for PUR foam systems (including RIM)	Enterprise Chemical Co., USA	
QUINDO	Pigments (organic type)	Bayer; Miles Inc., USA	
RAINBOW	Range of colorant masterbatches	Cabot Plastics International, Belgium	
RAVEN	Carbon blacks (pigments and conductive fillers)	Columbian Chemicals Co., USA	
REACTINT	Liquid, reactive polymeric colorants for PUR systems	Milliken Chemical Co., USA	
REEDLITE	Colour concentrates (heavy-metal-free) for thermoplastics	Clariant Corp., Reed Spectrum Div., USA	
REGAL	Furnace carbon blacks	Cabot Corp., Special Blacks Div., USA	
RENOL AN/CN	Range of polyamide-based colorant masterbatches	Hoechst	
REOFLAM	Range of phosphorus-containing flame retardants	FMC Corp., Process Additive Div., UK and USA	E.g. REOFLAM PB - 460 for PC and thermoplastic polyesters
REOFOS	Range of aryl phosphate flame retardants for various plastics, including flame-retardant phosphate plasticisers for PVC	FMC Corp., Process Additive Div., UK and USA	
REX-COMP	Flame-retardant, smoke-suppressant additive (magnesium-oxide based)	Polyrex Additives, Switzerland	
RHODIANOX	Antioxidants (phenolic type)	Great Lakes Chemical Corp., USA; Great Lakes Chemicals Europe Ltd., Switzerland	

TRADE NAME	NATURE AND/OR APPLICATION	PRODUCER OR SUPPLIER	REMARKS
RICON	Impact modifiers (polybutadiene and styrene/butadiene copolymer types) for polyolefins and styrenics	Ricon Resins Inc., USA	
RICOPOXY	Epoxidised polybutadiene (stabiliser for PVC pastes)	Ricon Resins Inc., USA	
RO-20	Antistatic agent concentrate	Rogers Anti-Static Chemicals, USA	
ROCKFIL	Mineral fibre reinforcement for thermoplastics	Lapinus Fibres BV, The Netherlands	
ROMHOGLAS	Aluminium-coated fibres, rovings and mats for use as thermally and electrically conductive reinforcement	Lundy Technical Center, Div. of Transtechnology Corp., USA	
RONFALOY	Impact modifiers (ABS type) for PVC	DSM	
RONOTEC	Antioxidants (some based on vitamin E)	Hoffmann-Laroche Inc., USA; Roche Products Ltd., Vitamins and Fine Chemicals Div., UK	
RUBINATE	Isocyanate (modified MDI) for PUR systems	ICI Polyurethanes, USA	
RYLEX NBC	UV stabilisers (benzophenone and nickel-compound types)	Du Pont	
S-2 GLASS	High-performance glass fibres	Owens-Corning Fiberglas SA (European), Belgium	
S-SERIES COATINGS	Hard silicone coatings for plastics	Exxene Corp., USA	
SACHTOLITH	Zinc sulfide	Sachtleben Chemie GmbH, Germany	
SAFOAM	Range of chemical blowing agents (endothermic and mixed types) with nucleating action	Reedy International Corp., USA	
SALOL	UV stabilisers (benzophenone and salicylate types)	Dow	
SAMILIZER	Antioxidants (various types)	Sumimoto Chemical Co., Japan	
SANDIN	Antistatic agents	Clariant Corp.	
SANDOFIL	Pigments and pigment concentrates	Clariant Corp.	
SANDOFLAM	Flame-retardant additive (phosphorus-based) masterbatches	Clariant Corp.	
SANDORIN	Pigments (organic type)	Clariant Corp.	Formerly Sandoz Chemicals.
SANDOSTAB	Stabilisers (various), individual and master-batches	Clariant Corp.	E.g. SANDOSTAB P - EPQ - a phosphonate-type antioxidant
SANDUVOR	UV stabiliser masterbatches (benzophenone, oxanilide, amine, and nickel-compound types)	Clariant Corp.	

TRADE NAME	NATURE AND/OR APPLICATION	PRODUCER OR SUPPLIER	REMARKS
SANTICIZER	Range of plasticisers and resin modifiers for PVC, cellulose acetate and other plastics	Monsanto	
SANTONOX	Antioxidants (phenolic type)	Monsanto	
SANTOWHITE	Antioxidants (phenolic type)	Monsanto	
SANYLEN	Colorant masterbatches	Clariant Corp.	Formerly Sandoz Chemicals
SARMAFLAM	Flame-retardant additive masterbatches	Clariant Corp.	Formerly Sandoz Chemicals
SARMATENE	Colorant masterbatches	Clariant Corp.	Formerly Sandoz Chemicals
SATINGLASS	Rovings for SMC	Vetrotex Certain Teed Corp., USA	
SATINTONE	Calcined kaolin fillers	Engelhard Corp., Specialty Minerals & Colors Group, USA	
SAYTEX	Range of flame retardants (brominated organic compounds)	Albemarle Corp.	
SCOTCHLITE GLASS BUBBLES	Hollow glass microspheres for use in thermoset composites, PUR foam systems and some PVC pastes	3M Co., Industrial Tape and Specialties Div., USA	
SEA HORSE	Calcium carbonate fillers	Swedish Whiting Co.	
SEESORB	UV stabilisers (various types)	Sun Chemical Corp., Japan	
SICO	Pigments (organic type)	BASF	
SICOLUB	Lubricants (various types)	BASF	
SICOMIN	Pigments (inorganic type)	BASF	
SICOPAL	Pigments (inorganic type)	BASF	
SICOSTAB	Heat stabilisers for PVC (various types) and UV stabilisers	BASF	
SICOTAN	Pigments (inorganic type)	BASF	
SICOTHERM	Pigments (inorganic type) - high heat resistance	BASF	
SICOTRANS	Pigments (iron oxide, transparent types)	BASF	
SIKRON	Micronised silica and cristoballite fillers	Quarzwerke GmbH, Germany	
SILBOND	Silane treated silica fillers	Quarzwerke GmbH, Germany	
SILCAT	Range of reagents for use in chemical crosslinking of PE by the silane route	Union Carbide	
SIL-CELL	Low-density (cellular) mineral filler for UP and epoxy resins and composites	Silbrico Corp., USA	
SIL-CO-SIL	Ground silica fillers	US Silica Co., USA	
SILCRON	Range of grades of ground amorphous silica gel (for use as flatting, antiblocking, anticaking, and anti-plate-out agents, and as thixotropic additive)	SCM Chemicals Inc., USA	

TRADE NAME	NATURE AND/OR APPLICATION	PRODUCER OR SUPPLIER	REMARKS
SILFIN	Special grade of SILLITIN	Franz Hoffmann & Soehne KG, Germany	
SILLIKOLOID	Colloidal silica	Franz Hoffmann & Soehne KG, Germany	
SILLTIN	Silica fillers	Franz Hoffmann & Soehne KG, Germany	
SILLUM	Flame-retardant filler blend; also alumina silicate filler	D. J. Enterprises Inc., USA	
SILTEX	Fused silica fillers	Kaopolite Inc., USA	
SILTON	Antiblocking agents (containing amorphous aluminosilicate and Na/Ca aluminosilicate powders)	International Resources Inc., USA	Produced in Japan
SILVERLINE	Range of ground talc fillers	The Montana Talc Co., USA	
SILVET	Aluminium-based pigment concentrates	Silverline Manufacturing Co., USA	
SILVEX	Range of metallic pigments (aluminium flake type)	Silverline Manufacturing Co., USA	
SIPERNAT	Range of precipitated silicas for use as antiblocking, anticaking and anti-plate-out agents	Degussa	
SMOKEBLOC	Smoke-suppressant flame retardants (based on antimony/molybdenum compound combinations)	Anzon Ltd., UK; Anzon Inc., USA	
SNOWFLAKE	Calcium carbonate fillers	ECC International	SNOWFLAKE WHITE - general-purpose filler; SNOWFLAKE PE - medium ground filler for SMC and BMC
SNOW-TEX	Kaolin filler (claimed effective as extender for TiO_2 pigment in PVC compositions)	US Silica Co., USA	
SNOW-WHITE	Anhydrous calcium sulfate fillers	US Gypsum Co., USA	
SOCAL	Precipitated calcium carbonate fillers	Deutsche Solvay-Werke GmbH, Germany	
SOFTENOL	Processing aid/surface modifier	Huels	
SOLCA-FLOC	Cellulose filler for UF, MF and PF resins	Fiber Sales & Development Corp., USA	Also distributed by Grefco Inc., USA
SONNEBORN	Range of lubricants (microcrystalline wax, petrolatum and mineral oil types)	Whitco Corp., Petroleum Specialties Group, USA	
SPECTRAFLO	Range of colorant concentrates, chemical blowing agents and mould releases, in liquid dispersions	Ferro Corp., Plastic Colorant and Dispersions Div., USA	
SPECTRAMIX	Range of colorant concentrates	Plastichemix Industries, India	
SPECTRATECH	Ranges of flame-retardant and colorant concentrates	Quantum Chemical Corp., USA	
SPESWHITE	Clay fillers	ECC International	
SPHERICEL	Hollow glass microspheres	PQ Corp., USA; Potters Ballotini Ltd., UK	
SPHERIGLASS	Solid glass microspheres	PQ Corp., USA; Potters Ballotini Ltd., UK	
SPINCOL	Colorants for imitation texture effects	FM Group Inc., USA	

TRADE NAME	NATURE AND/OR APPLICATION	PRODUCER OR SUPPLIER	REMARKS
SPINFLAM	Intumescent flame-retardant additives	Montell Italia, Italy; Himont USA	
STABIOL	Heat stabilisers (calcium/zinc type) for PVC	Henkel	
STANCLERE	Range of organotin heat stabilisers for PVC and curing agents for PUR systems	Akcros Chemicals	E.g. STANCLERE TL is DBTDL
STAN-TONE	Colorant dispersions	Harwick Chemical Corp., USA	
STAPA	Aluminium paste for metallic colour effects in thermoplastics	Obron Atlantic Corp., USA	
STATEXAN	Antistatic agents	Bayer; Miles Inc., USA	
STAVINOR	Heat stabiliser (metal-soap type) for PVC	CECA (Elf Atochem)	
STEREON	Impact modifiers (styrene/butadiene block copolymers) for ABS, PS, PE and PP	Firestone Synthetic Rubber and Latex Co., USA	
STERLINE	Ground talc fillers	The Montana Talc Co., USA	Food-contact grades
STERLING	Carbon blacks (furnace type)	Cabot Corp., Special Blacks Div., USA	
STRANDEX	Heat stabilisers (lead type) for PVC	Anzon Ltd., UK; Anzon Inc., USA	Special granular, non-dusting form
STYRID	Styrene-suppressant additives	Specialty Products Co., USA	
SULFOCEL	Chemical blowing agent (OBBSH)	Fairmont Chemical Co., USA	
SUMISORB	UV stabilisers (various types)	Sumimoto Chemical Co., Japan	
SUPERCOAT	Fine-ground, coated calcium carbonate	ECC International	
SUPER FLOSS	Diatomaceous silica offered as antiblocking agent for polyethylene films	Celite Corp., USA	
SUPERMITE	Fine-ground calcium carbonate	ECC International	
SUPER-PFLEX	Surface-treated calcium carbonate fillers	Celite Corp., USA	
SUZOREX **SUZORITE** **SUZORITE E**	Mica fillers (surface-treated grades) Mica fillers Electroconductive (nickel-plated) mica flakes for EMI shielding compositions	Suzorite Mica Products Inc., USA	
SYLOBLOC **SYLOID**	Synthetic silica fillers and additives for use in CAB, PVC, polyolefins, EVA, and acrylics	W.R. Grace & Co., Grace Div., USA	Combinations of SYLOBLOC with amides used as slip agents
SYNCROLUBE	Range of lubricants	Croda Universal, USA and UK	
SYNLUBE	Mould-release agents (various types)	Synair Corp., USA	
SYNOX	Antioxidants (phenolic type)	Great Lakes Chemicals Corp., USA	
SYNPRO **SYNPRO-FLO** **SYNPRON** **SYNPRO-WARE**	Metal-stearate lubricants and mould release agents Heat stabilisers (various non-lead types) for PVC Dispersions of mineral flame retardants in phosphate esters and halogenated compounds	Synthetic Products Co., USA	

TRADE NAME	NATURE AND/OR APPLICATION	PRODUCER OR SUPPLIER	REMARKS
SYNTASE	UV stabilisers (benzophenone type)	Great Lakes Chemicals Corp., USA	
SYNTHAMID	Cell-stabilising agent for PVC foams produced by mechanical frothing	Dr. Th. Boehme, Germany	
SYNTHEWAX	Lubricants (fatty-acid amide type)	Comiel SpA, Italy	
TEBESTAT	Antistatic agents	Dr. Th. Boehme, Germany	
TEDIMON	Isocyanates	EniChem	
TEFLON MP	Fluoroadditives for lubricity and wear resistance	Du Pont	
TEGO AB-M	Chemical blowing agent for PUR foam systems		
TEGO ANTIFLAMM	Flame retardant for PUR foam systems		
TEGOAMIN	Amine catalyst for PUR foam systems		
TEGOMER	Reactive siloxanes	Th. Goldschmidt AG, Germany	
TEGOMER AC	Dispersing agent for fillers and pigments		
TEGOPREN	Modified siloxanes		
TEGOSIL	Mould release (external, silicone-based)		
TEGOSTAB	Stabilisers for PUR foam systems		
TEGOTENS	Lubricants (glycerol/fatty-acid ester type)		
TENOX	Antioxidants	Eastman Chemical	
TERCAROL	Polyols for PUR systems	EniChem	
TETRATHAL	Tetrachlorophthalic anhydride: flame retardant for PU resins	Monsanto	
TEXACAT	Range of various catalysts (amine types) mainly for PUR foam systems	Huntsman Chemical Corp	Formerly Texaco Chemical Co.,
TEXACURE	Reactive amines		
TEXADD	Additives (including some catalysts) for PUR systems		
THANCAT	Amine catalysts	Condea Chemie GmbH, Germany	
THANOL	Polyols for PUR systems	Arco Chemical Co., USA	
THERMACURE	Range of peroxide initiators for PU resins	Cook Composites and Polymers Co., USA	
THERMAL COLOUR	Thermochromic pigments	Siber Hegner Ltd., UK	Developed by Kirokusozai Sogo Kenkyusho Co., Japan
THERMAX	Carbon blacks (various grades)	Cancarb Ltd., Canada	
THERM-CHEK	Heat stabilisers for PVC	Ferro Corp., Bedford Chemical Div., USA	
THERMOFOAM	Range of miscellaneous foaming-agent concentrates for thermoplastics	Thermofil Inc., USA	

TRADE NAME	NATURE AND/OR APPLICATION	PRODUCER OR SUPPLIER	REMARKS
THERMOGUARD	Range of flame retardants (various types including antimony oxides and brominated epoxy compounds)	Elf Atochem	
THERMOLITE	Heat stabilisers (various types) for PVC		
THERMOLUBE	Processing-aid concentrates for thermoplastics	Thermofil Inc., USA	
THERMOPLAST	Soluble colorants (dyes) of good heat stability	BASF	
THORNEL	Electrically conductive carbon fibres	Amoco Chemicals	Transparent to x-rays
TILCOM	Lubricants and processing aids (organic titanate and zirconate types)	Tioxide, UK and USA	Also offered as printing-ink adhesion promoters
TIMONOX	Antimony trioxide	Cookson Mathey, UK; Cookson Pigments Inc., USA	
TINUVIN	Range of antioxidants and UV stabilisers (various types)	Ciba-Geigy	Cf. ASTM D 1996 - 92 and D 5524 - 94
TIOLON	Mould-release agent (PTFE type)	Tiodize Co. Inc., USA	
TIONA	Titanium dioxide pigments	SCM Chemicals Ltd., UK	
TIOXIDE	Titanium dioxide pigments	Tioxide	
TIPAQUE	Titanium dioxide pigments	Ishiharo Sangyo Kaisha Ltd., Japan	
TITA-FRANCE	Titanium dioxide pigments	Rhône-Poulenc Chemicals, France	
TITANKUP	Titanium dioxide pigments (surface-treated)	Malvern Minerals Co., USA	
TITANOX	Titanium dioxide pigments	Kronos	
TI-PURE	Titanium dioxide pigments	Du Pont	
TONOX	Curing agents (aromatic diamine type) for epoxy resins	Uniroyal Chemical Co. Inc., USA	
TOPANOL	Antioxidants (amine and phenolic types)	Zeneca	Formerly ICI
TORAYCA	Carbon fibres (produced from PAN precursors)	Toray Industries (America) Inc., USA	
TOSPEARL	Silicone-based, particulate antiblocking agents	GE Silicones Europe; Toshiba Silicone, Japan	
TRANSLINK	Surface-coated kaolin fillers	Engelhard Corp., Specialty Minerals and Colors Group, USA	
TREFIL	Silicone-treated mineral fillers		
TREMIN	Silicone-treated wollastonite fillers	Quarzwerke GmbH, Germany	
TREMINEX	Silicone-treated feldspar (nepheline syenite) fillers		
TRIGONOX	Range of peroxide catalysts and crosslinking agents	Akzo Nobel	
TRIKUP	Surface-treated alumina trihydrate flame retardants	Malvern Minerals Co., USA	

TRADE NAME	NATURE AND/OR APPLICATION	PRODUCER OR SUPPLIER	REMARKS
TRONOX	Titanium dioxide pigments	Kerr-McGee Chemical Corp., USA	
TUF-LUBE	Mould-release agents (fluorocarbon type)	Specialty Products Co., USA	
TV-2	Broad-spectrum biostat	Sanitized Inc., USA	
TWEEN	See POLYSORBATE		
TYTANPOL	Titanium dioxide pigments	Zaklady Chemiczne Police, Poland	
ULTRACARB	Flame-retardant, smoke-suppressant fillers based on a mixture of huntite ($Mg_3Ca(CO_3)_4$) and hydromagnesite ($Mg_4(CO_3)_3(OH)_2 \cdot 3H_2O$)	Microfine Minerals Ltd., UK	
ULTRACEL	Polyols for PUR systems	Arco Chemical Co., USA and UK	
ULTRAFLEX	Surface-treated calcium carbonate fillers	Specialty Minerals Inc., USA	
ULTRANOX	Range of antioxidants (including phosphite and phenolic types) and antioxidant blends	GE Specialty Chemicals	
UNICELL	Chemical blowing agents (various types) for most thermoplastics	Stanchem Inc., Canada	Manufactured by Don Jin Chemical Industry Co., South Korea
UNIFLEX	Lubricants (fatty-acid ester type)	Union Camp Corp., USA	
UNIGARD	Flame-retardant compounds	Union Carbide Corp., Specialty Chemicals Div., USA	
UNIMAX	Colorant concentrates for thermoplastics	Gabriel-Chemie GmbH, Austria	
UNI-REZ	Range of curing agents (polyamide and amido-amine types) for epoxy resins	Union Camp Corp., USA	
UNISLIP **UNIWAX**	Lubricants (fatty-acid amide type)	Unichema International, The Netherlands; Unichema North America, USA	
UNITANE	Titanium dioxide pigments	Kemira Inc., USA	
UNITE MP	Chemically modified polyolefins for use as compatibilisers in polymer blends, tie layers in coextrusion, and as coupling promoters for fillers and reinforcements in polyolefins	Aristech Chemical Corp., Specialty Polymers Unit, USA	E.g. grafts of maleic anhydride on PP
USAN	See POLYSORBATE		
UVASIL	UV stabilisers (HALS type)	EniChem	
UVASORB	UV stabilisers (HALS type)	3V Sigma SpA, Italy	
UV-CHEK	UV stabilisers (various types)	Ferro Corp., Bedford Chemicals Div., USA	
UVINUL	UV stabilisers (benzophenone and cyanocinnamate types)	BASF	
UVITEX	Optical brightening agents	Ciba-Geigy	

TRADE NAME	NATURE AND/OR APPLICATION	PRODUCER OR SUPPLIER	REMARKS
VALU-FIL	Particulate gypsum filler with flame-retardant action (through release of water of hydration)	Allied Custom Gypsum, USA	Release of water of hydration starts at 120°C. Thus gassing may occur in melt processing above this temperature
VANCIDE	Biocide (N-trichloromethyl-mercapto-4- cyclohexene-1,2-dicarboximide)	R.T. Vanderbilt Co. Inc., USA	
VANOX	Antioxidants (various types) for polyolefins, PA and PUR	R.T. Vanderbilt Co. Inc., USA	
VANSIL	Wollastonite fillers	R.T. Vanderbilt Co. Inc., USA	
VANTALC	Talc fillers	R.T. Vanderbilt Co. Inc., USA	
VAZO	Polymerisation initiators and UP resin curing agents (nitrile-based)	Du Pont	
VECTOR	Styrenic-block copolymers for use as compatibilisers in the production of polymer blends	Dexco Polymers, USA	
VEOVA	Chemical Intermediates	Shell Chemicals	
VERSAMAG	Magnesium hydroxide (flame retardant and smoke-suppressant additive)	Morton International Inc., USA	
VESTENAMER	Processing aid	Huels	
VESTIFORM	Lubricant/processing aids	Huels	
VESTOWAX	Lubricants (various wax types)	Huels; Astor Wax Corp., USA	
VINSTAB	Chemical blowing agents (mainly ADC type); heat stabilisers for PVC	Hebron SA, Spain	
VINURAN	Acrylic processing aids for PVC	BASF	
VINYFOR	Acrylic processing aids for PVC	Eiwa Chemical Idustry Co. Ltd., Japan	
VINYLUBE	Lubricants for clear PVC compounds	Lonza	
VINYZENE	Biocides (including isothiazoline-based)	Morton International Inc., Industrial Chemicals and Additives, USA	
VIOSORB	UV stabilisers (various types)	Kyodo Chemical Co., Japan	
VIPLEX	Modifiers (hydrocarbon type) for PUR foam systems	Crowley Chemical Co., USA	Also used as plasticiser extenders
VIRCOL	Range of reactive flame retardants for PUR	Albright & Wilson	
VISTALON	EPM rubbers, used as impact modifiers inter alia	Exxon Chemical	
VORANOL	Polyols for PUR	Dow	
VULCABOND	Bonding-agent additive for polymers	Akzo Nobel	
VULCAN	Carbon powders for antistatic and conductive compounds	Cabot Corp., Special Blacks Div., USA	

TRADE NAME	NATURE AND/OR APPLICATION	PRODUCER OR SUPPLIER	REMARKS
VULCANOX	Range of antioxidants (various types)	Bayer; Miles Inc., USA	
VUL-CUP	Peroxide catalyst	Hercules	Bis (tert-butylperoxyisopropyl-benzene)
VULKOLLAN	Polyols for PUR	Bayer; Miles Inc., USA	
VYBAR	Mould-release agents and lubricants (low-melting hydrocarbon type)	Petrolite Corp., Polymers Div., USA	
VYDAX	Mould-release agents (fluoro-oligomer type)	Du Pont	
WATCHUNG	Red azo pigments	Cookson Pigments Inc., USA	
WAXOLINE	Range of dyes soluble in solvents and polymer melts	ICI Polymer Additives	Recommended for acrylics and styrenics
WESTON	Antioxidants (mainly organic phosphites) synergistic with UV stabilisers	GE Specialty Chemicals	
WHITEFLUOR	Optical brightening agents	Sumimoto Chemical Co., Japan	
WIL-LUBE	Lubricants (stearate concentrates and other types) for thermoplastics	Wilson Color Inc., USA	
WINGSTAY	Antioxidants (phenolic and amine types)	Goodyear	Range includes antiozonants
WINNOFIL S	Ultrafine precipitated, coated calcium carbonate fillers for rigid PVC and some rubbers	ICI Polymer Additives	Some impact-modifier action in PVC
WOLLASTOCOAT **WOLLASTOKUP**	} Surface-treated wollastonite fillers	Nyco Minerals Inc., USA; Malvern Minerals Co., USA	
WYTOX	Antioxidants (various types)	Uniroyal Chemical Co. Inc., USA	
XANTRIX ADS	Additive-delivery system concentrates (including antioxidants and UV stabilisers for polyolefins)	Montell Italia, Italy; Himont, USA	Range includes "visbreaking" (q.v.) peroxide additive masterbatches for PP
YSE-CURE	Range of heterocyclic diamine (spiroacetal) curing agents for epoxy resins	Ajinomoto USA Inc., USA	

TRADE NAME	NATURE AND/OR APPLICATION	PRODUCER OR SUPPLIER	REMARKS
ZEEOSPHERES	Hollow ceramic microspheres (thick-walled)	Zeelan Industries Inc., USA	Subsidiary of 3M Co.
ZELEC ECP	Conductive-particle additives (tin-oxide based) for antistatic and conductive plastics	Du Pont Specialty Chemicals., USA	Suitable for some translucent compositions
ZEOTHIX	Hydrated silica: thickener for PVC pastes, UP resins and epoxy resins	J.M. Huber Corp., Chemicals Div., USA	
ZEROGEN	Flame-retardant/smoke-suppressant additive based on magnesium hydroxide	J.M. Huber Corp., Solem Div., USA	
ZETPOL	Hydrogenated nitrile rubber (HNBR) for use as impact modifier in PVC and other plastics	Zeon Chemicals Inc., USA	
ZINSTABE	Zinc-based heat stabilisers and blowing-agent activators for PVC	Zinc Corporation of America, USA	
Z- LIGHT SPHERES	Low-density, white ceramic microspheres for use in composites (especially SMC and BMC)	Zeelan Industries Inc., USA	Subsidiary of 3M Co.
ZYLAC	Lubricants for PVC extrusions	Petrolite Corp., Polymers Div., USA	
ZYTOCAL	Calcium carbonate concentrates for polyolefins	EEC International	

REFERENCES

REFERENCES

SECTION 1

TECHNICAL TERMINOLOGY STANDARDS CITED IN PART 1

1.1 BS 1755; Part 1: 1982. *Glossary of terms used in the plastics industry - Polymer and plastics technology*. (Superseded in 1995 by BS ISO 472 : 1988. The BS version incorporates three amendments - Nos. 1,3, and 4 - made in 1993 to ISO 472 : 1988)

1.2 ISO 1382: 1982 (with Addenda 1 to 20 in the period 1982-1990). *Rubber - Vocabulary*.

1.3 ASTM D 1566-95. *Standard terminology relating to rubber*.

1.4 *Compilation of ASTM standard definitions*. 8th Edition, 1994.

1.5 ISO 472:1988 (with amendments up to 1993). *Plastics - Vocabulary* (under revision in 1996/1997)

1.6 ANSI/IEEE standard 100-1992. *IEEE standard dictionary of electrical and electronic terms*. The Institute of Electrical and Electronic Engineers Inc., USA

1.7 ASTM D 883-95a. *Standard terminology relating to plastics*.

1.8 ISO Guide 52: 1990. *Glossary of fire terms and definitions*.

1.9 CEI/IEC Standard 695-4: 1993. *Fire hazard testing - Part 4: Terminology concerning fire tests*.

1.10 ASTM E6-89(1994). *Standard terminology relating to methods of mechanical testing*.

1.11 ANSI Standard Z 94.0 - 1989. *Industrial engineering terminology. Part Z 94.7 - Material processing*.

1.12 ASTM E 176-93a. *Standard terminology of fire standards*.

1.13 ASTM D 4391-93a. *Standard terminology relating to burning behaviour of textiles*.

1.14 ASTM E 284-95a. *Standard terminology of appearance*.

1.15 ASTM E 473-85 (1988). *Standard definitions of terms relating to thermal analysis*.

1.16 ASTM D 907-93c. *Standard terminology of adhesives*.

REFERENCES

SECTION 2

STANDARDS FOR ABBRVIATED TERMS (LETTER SYMBOLS) CITED IN PART 2

2.1 ISO1043-1: 1987(E). *Plastics - Symbols - Part 1: Basic polymers and their special characteristics.* (Revised in 1997: see Ref. 2.11)

2.2 ASTM D 1600-95c. *Standard terminology for abbreviated terms relating to plastics.*

2.3 BS 3502: Part 1: 1991. *Symbols for plastics and rubber materials - Part 1: Schedule for symbols for plastics.* (A dual BS/ISO standard - text virtually identical with that of ISO 1043-1: 1987)

2.4 ASTM D 1418-95. *Standard practice for rubber and rubber latices - Nomenclature.*

2.5 ISO 1629: 1995. *Rubbers and latices - Nomenclature.*

2.6 BS 3502: Part 2: 1991. *Symbols for plastics and rubber materials - Part 2: Schedule for symbols for rubbers.* (Superseded by ISO 1629:1995 - adapted as an ISO BS standard)

2.7 ISO 1043-3: 1996. *Plastics - Symbols and abbreviated terms - Part 3: Plasticisers.*

2.8 BS 3502: Part 2: 1991. *Symbols for plastics and rubber materials - Part 3: Schedule for symbols for compounding ingredients.*

2.9 ISO 1043-2: 1988(E). *Plastics - Symbols - Part 3: Fillers and reinforcing materials.*

2.10 ISO 6472: 1994. *Rubber compounding ingredients - Abbreviations.*

2.11 ISO 1043-1: 1997 (Second Edition). *Plastics - Symbols and abbreviated terms - Part 1: Basic polymers and their special characteristics.*

2.12 ASTM D 3853-96. *Standard terminology relating to rubber and rubber latices - Abbreviations for chemicals used in compounding.*

2.13 ASTM D 5538-94 *Standard practice for thermoplastic elastomers - Terminology and abbreviations.*

REFERENCES

SECTION 3

LITERATURE REFERENCES

Anagnostopoulos, C. E., Coran, A. Y. and Gamrath H. R. (1960). *J. Appl. Polym. Sci.*, Vol. 4, No. 11. pp. 181-190.

Antelman, M. S. (1974). *The Analytical Encyclopaedia of Thermoplastic Materials.* Sadtler Research Laboratories Inc., Philadelphia, PA, USA.

Arnold, P. and Willis, H. (1972). *Identification and Analysis of Plastic Materials.* Chapter 24 in *Polymer Science* (Ed. Jenkins, A. D.) North - Holland Publishing Co., Amsterdam and London, 1972.

Ash, C. E. (1994). *J. Mater. Educ.*, Vol. 16, pp. 1-20.

Attard, G. S. and Williams, G. (1986). *Chemistry in Britain*, October, pp. 919-924.

Bardsley, A. (1995). *Nylon: The Name. Textile Horizons.* March 1995. pp. 34-35.

Bekkedahl, N. (1949). *J. Res. Nat. Bureau of Standards.* Vol. 42, pp. 145-156.

Bigg, D. C. H. (1975). *J. Appl. Polym. Sci.*, Vol. 19 pp. 3119-3125.

Billmeyer, F. W. and Saltzman, M. (1966). *Principles of Color Technology.* Interscience Publishers, New York, London, Sydney.

Braden, M. (1963). *Plastics Institute Transactions*, Vol. 31, No. 94, pp. 83-88.

Braden, M. (1968). *J Polym. Sci., Part A1*, Vol. 6, pp. 1227-1231.

Brown, R. P. (Ed.) (1988). *Handbook of Plastics Test Methods*, 3rd Edn., Longmans Scientific, and Plastics and Rubber Institute, London.

Brydson, J. A. (1972). *The Glass Transition, Melting Point and Structure.* Chapter 4 in *Polymer Science* (Ed. Jenkins A.D.), North Holland Publishing Co., Amsterdam and London, 1972.

Brydson, J. A. (1995). *Plastics Materials*, 6th Edn., Butterworths-Heinemann, London.

Ceferri, A. (1991). *Liquid Crystallinity In Polymers*, VCH, New York.

Christopher, W. F. and Fox, D. W. (1962). *Polycarbonates.* Reinhold Publishing Corporation. .

Changfoot, J., Dickson, A. G., Noel, F. and Stark, W. M. (1976). *34th ANTEC SPE Proceedings*, pp. 42-44. .

Collins, E. A., Metzger, A. P. and Furgason, R. R. (1976). *Polym. Engng. Sci.*, Vol. 16, No. 4, pp. 240-245.

Comyn, J. (Ed) (1985). *Polymer Permeability.* Elsevier Applied Science Publishers, London and New York.

Crank, J. (1967). *The Mathematics of Diffusion.* Oxford University Press, Oxford.

Crank J. and Park G. S. (Eds.) (1968). *Diffusion in Polymers.* Academic Press.

Crawford, R. J., Keating, T. G., Chee, W. Y. and Tan, Y. L. (1991). *Plast. Rubb. Comp. Proces.Appl.*, Vol. 16, No.4, pp. 263-270.

Crompton, T. R. (1977). *Chemical Analysis of Additives in Plastics*, 2nd Edn., Pergamon Press, Oxford.

Crompton, T. R. (1984). *The Analysis of Plastics.* Pergamon Press, Oxford.

Deanin, R. D. and Hauser, D. I. (1981). *Polym. - Plast. Techn. Engng.*, Vol. 17, No. 2, pp. 123-137. .

Dexter, F. D. (1953). *Modern Plastics*, Vol. 30, No. 8, pp. 125-130.

Di Raddo, R. W., Lato, M. and Garcia-Rejon, A. (1995). *Plast. Rubb. Comp. Proces. Appl.*, Vol. 23, No. 2, pp. 127-130.

Donald, A. M. and Windle, A. H. (1992). *Liquid Crystalline Polymers*. Cambridge University Press.

Fikentscher, H. (1932). *Cellulosechemie*. Vol. 13 pp. 58-69.

Fujita, H. (1961). *Fortschr. Hochpolym. - Forsch.*, Vol. 3, pp. 1-47.

Hammond, H. K. (1987). *ASTM Standardization News*, February, pp. 36-40. .

Haslam, J., Willis, H. A. and Squirrel, D. C. M. (1973). *Identification and Analysis of Plastics*, 2nd Edn., Iliffe Books, London.

Haward, R. N. (Ed.) (1973). *The Physics of Glassy Polymers*. Applied Science Publishers, London, pp. 424-451

Hazen, A. (1892). *American Chemical Journal*, Vol. 14, pp. 300-310.

Hill, H. E. and Nick, D. P. (1966). *J. Paint Technol.*, Vol. 38, No. 494, pp. 123-131.

Hollaway, L., O'Neill, M. and Thorne, A. T. (1992). *Plast. Rubb. Comp. Proc. Appl.*, Vol. 18, No. 4., pp. 269-275.

Hopfenberg, H. B. (Ed.) (1974). *Permeability of Plastics Films and Coatings*. Plenum Press, New York.

Howick. C. (1994). *Materials World*. November. pp. 580-581. .

Howick, C. (1995). *Plast. Rubb. Comp. Proces. Appl.*, Vol. 24, pp. 27-36.

The Hummel/Scholl *Atlas of Polymer and Plastics Analysis*. Vols. 1 to 3. Carl Hanser Verlag, Munich, and VCH Publishers, New York, 1988. (New editions appear at intervals).

Jenkins, F. A. and White H. E. (1953). *Fundamentals of Optics*, 2nd Edition, 2nd Impression, Section 23.8 (pp. 477-480). McGraw-Hill, New York, London, Toronto.

Kambour, R. P. (1963). *Applied Polymer Symposia*, No. 7 pp. 215-235.

Kambour, R. P. (1973). *J. Polym. Sci., Macromolecular Reviews*, Vol.7, Interscience/John Wiley, New York - London, pp. 1-154. .

Kathirgamanathan, P., Adams, P. N., Quill, K. and Underhill, A. E. (1991). *J. Mater. Chem.*, Vol. 1, No. 1, pp. 141 and 142.

Latham, J. R. and Mendham, W. E. (1973). *31st ANTEC SPE, Proceedings*, pp. 458-60.

Lawson, K. E. (1961). *Infrared Absorption of Inorganic Substances*. Reinhold, New York.

Lee, Y-D. And Chen C-M. (1987). *J. Appl. Polym. Sci.*, Vol. 33, March, pp. 1231- 1240.

Leevers, P. S., Freeman, P. N. and Arthur, M. M. (1995). *Plast. Rubb. Comp. Process*. Appl., Vol. 24, No. 3, pp. 113-121.

McArdle, C. (Ed.) (1989). *Side Chain Liquid Crystalline Polymers*, Blackie, Glasgow.

Meares, P. (1964). *Polymers: Structure and Bulk Properties*. D. Van Nostrand Co. Ltd.

Nichols, D. (1994). *Advanced Materials and Processes*, October 1994, pp. 29-32.

Rhodes, M. B. and Khaykin, B. (1991). *J. Therm. Insul.*, Vol, 14, January, pp. 175-183.

Ricklin, S. (1983). *Plast. Engng.*, Vol. 39, No. 5, pp. 29-33.

Schnell, H. (1964). Chemistry and *Physics of Polycarbonates.* Interscience Publishers.

Scott, G. (1985) *Chemistry in Britain*, Vol. 21, No. 7, pp. 648-653.

Sharples, A. *Crystallinity*, Chapter 4 in *Polymer Science* (Ed. Jenkins, A.D.), North-Holland Publishing Co., Amsterdam and London, 1972.

Stafford, G. D. and Braden, M. (1968). *J. Dent. Res.*, Vol.47, No. 2, p. 341.

Titow, W. V., Loneragan, R. J., Currell, B. R. and Johns, J. H. T. (1973). *Plastics and Polymers*, June 1973, pp. 149-156.

Titow, W. V. and Lanham, B. J. (1975). *Reinforced Thermoplastics.* Applied Science Publishers Ltd., London.

Titow, W. V. (1978). *Solvent Welding of Plastics.* Chapter 12 (pp. 181-196) in *Adhesion 2* (Ed. K.W. Allen), Applied Science Publishers Ltd., London.

Titow, W. V. (1990). PVC *Plastics: Properties, Processing and Applications.* Elsevier Science Publishers, London and New York.

Tobolsky, A.V. (1965). *Some viewpoints on Polymer Physics.* In *Structure and Properties of Polymers.* (Ed. Tobolsky, A.V.). *J. Polym. Sci.*, Part C, No. 9, pp. 170-197.

Vincent, P. I. (1974). *Short-term Strength and Impact Behaviour.* Chapter 5 in *Thermoplastics Properties and Design* (Ed. Ogorkiewicz, R. M.). John Wiley and Sons, London, New York, Sydney, Toronto.

Warren, P. (1984). 33rd *Wire and Cable Symposium, US Army Communications, Research and Development Command* (CORADCOM), *Proceedings* p. 33.

Weast, R. C. and Astle, M. J. (Eds.). *CRC Handbook of Chemistry and Physics*, 60th Edn.(1979-1980), CRC Press Inc., pp. D-195 and D-196.

Wessling, B. (1995). *Spektrum der Wissenschaft* (German edition of *Scientific American*), October, 110-116.

Wickson, E. J. (Ed.) (1993). *Handbook of PVC Formulating.* John Wiley and Sons, New York.

Williams, M. L., Landel, R. F. and Ferry, J. D. (1955). *J. Am. Chem. Soc.*, Vol. 77, 3701.

Wilson, A. S., Bowley, H. J. and Gerrard, D. L. (1988). *Plast. Rubb. Int.*, Vol. 13, No. 1, pp. 23-25.

Wolters, M. (1983). *Plast. Rubb. Proces. Appl.*, Vol. 3, pp. 323-330.

Yarsley, V. E. and Flavell, W. (1956). *Cellulosic Plastics:* Part 1 - *Cellulose Acetate, Cellulose Ethers, and Regenerated Cellulose.* Plastics Monograph No. 6C., The Plastics Institute.

Yasuda, H. (1966), in *Polymer Handbook* (Eds. J. Brandrup and E.H. Immergut), Interscience, New York.

Zimmerman, R. L. and Austin, T. H. (1988). *J. Cell. Plast.*, Vol. 24, May, pp. 256-265.